CLASSICAL DYNAMICS

DONALD T. GREENWOOD
Professor of Aerospace Engineering
University of Michigan

DOVER PUBLICATIONS, INC.
Mineola, New York

Bibliographical Note

This Dover edition, first published in 1997, is an unabridged and corrected republication of the work first published by Prentice-Hall, Inc., Englewood Cliffs, New Jersey, in 1997 as part of the *Prentice-Hall International Series In Dynamics.*

Library of Congress Cataloging-in-Publication Data

Greenwood, Donald T.
    Classical dynamics / Donald T. Greenwood.
       p.   cm.
    Originally published: Englewood Cliffs, N. J. : Prentice-Hall, c1977, in series : Prentice-Hall international series in dynamics.
    Includes bibliographical references and index.
    ISBN 0-486-69690-1 (pbk.)
    1. Dynamics.   2. Relativistic mechanics.   I. Title
QA845.G827   1997   97–15387
531'.11'01515—dc21   CIP

Manufactured in the United States of America
Dover Publications, Inc., 31 East 2nd Street, Mineola, N.Y. 11501

CONTENTS

PREFACE

Nearly two hundred years have elapsed since Lagrange published his *Mécanique Analytique* (1788) which laid the foundations of analytical dynamics. Later discoveries, most notably those of Hamilton and Jacobi, contributed to a theory of dynamics having unusual elegance and beauty. The next great advance came early in this century when Einstein presented a new view of the physical world with the publication of the first of his papers on relativity in 1905. In the intervening years, these foundations have been studied and improved upon, particularly with respect to the mathematical methods and notation, as well as in their logical and experimental basis. Furthermore, a wider use has been made of the advanced theories of classical mechanics as our society has become more sophisticated in its technology. It is appropriate, then, that graduate students in science and engineering should have a strong background in these more abstract and intellectually satisfying areas of dynamical theory. This is the subject of the present textbook.

The topics covered represent a somewhat expanded version of the material in an advanced dynamics course at the University of Michigan. It is assumed that the incoming students are familiar with the principles of vectorial mechanics and have some facility in the use of this theory for the analysis of systems of particles and for rigid body rotation in two and three dimensions. Furthermore, they should be familiar with Lagrange's equation and preferably have had some experience in its application to relatively difficult problems. At present, the typical incoming student in this course at Michigan has had some exposure to dynamics in a freshman physics course, as well as an introductory dynamics course of four semester hours and an intermediate course, also of four hours. The intermediate and advanced courses form a two-term sequence.

Because the student is assumed to be familiar with the more elementary topics in dynamics, these are not covered in any detail in this book. The first chapter, *Introductory Concepts*, is included, however, in order to establish some of the notation and the more important definitions to be used later. For example, the ideas of virtual work and d'Alembert's principle are introduced here. If the student has already attained a proficiency in these areas, then little time needs to be spent on this chapter.

Chapter 2 uses d'Alembert's principle as a starting point for the derivation of Lagrange's equations of motion. The explicit form and nature of these equations are discussed in detail for holonomic and nonholonomic systems. The discussion of particular applications of Lagrange's equations is continued in Chapter 3 where the idea of impulsive constraints is introduced. The study of impulsive motion also provides the opportunity for a brief discussion of quasi-coordinates. Further topics include discussions and comparisons of gyroscopic systems and dissipative systems, as well as velocity-dependent potentials. Some of these applications may be omitted without a loss of continuity in the remaining chapters.

In Chapter 4 the calculus of variations is introduced in the study of dynamics. The most emphasis is given to Hamilton's principle, but other results such as the principle of least action are studied. Noncontemporaneous variations, as well as the usual contemporaneous variations, are considered in a general evaluation of the canonical integral. The Hamiltonian viewpoint of dynamics is given careful consideration in the discussions of the canonical equations of motion and phase space. Much of the groundwork for the theory of the next two chapters is presented here.

Chapters 5 and 6 are concerned primarily with the theory of canonical transformations. It was decided to consider the Hamilton-Jacobi theory before a more general discussion of canonical transformations, rather than attempting the reverse order. This allows the student to obtain some rather concrete procedures for solving problems, and it is hoped that there will be a further motivation for the rather extensive theory of canonical transformations which follows.

The final chapter is an introductory presentation of special relativity with the Lagrangian and Hamiltonian formulations included. Enough examples and problems are presented to encourage a good working familiarity with the subject at this level.

The author would recommend, in general, that the problems at the end of each chapter be used as much as time will allow. The problems are of varying difficulty and should help greatly in solidifying the principal concepts of classical dynamics. Occasionally, theoretical results which were not included in the text because of limited space are presented instead as problems.

A major portion of this book was written during a sabbatical leave at the University of California at San Diego. I wish to thank Professors J. W. Miles and R. E. Roberson of the Department of Aerospace and Mechanical Engineering Sciences for helping to make my stay there enjoyable and fruitful. I am particularly appreciative of the help of my wife who did all the typing and helped with the proofreading.

DONALD T. GREENWOOD

Ann Arbor, Michigan

1

INTRODUCTORY CONCEPTS

Dynamics is the study of the motions of interacting bodies. It describes these motions in terms of postulated laws. *Classical* dynamics is restricted to those systems of interacting bodies for which quantum mechanical effects are negligible; that is, it applies primarily to macroscopic phenomena. The non-relativistic theories and methods of men such as Newton, Euler, Lagrange, and Hamilton are included, as well as the more recent relativistic dynamics of Einstein.

In this chapter we introduce a few of the basic concepts of nonrelativistic classical dynamics and shall begin to form the notational framework to be used throughout the book. Some of the material should be familiar to the reader and is not presented in detail. Other topics are explained more carefully in order to clarify the more important definitions and assumptions.

1-1. THE MECHANICAL SYSTEM

Let us consider a mechanical system consisting of N particles, where a *particle* is an idealized material body having its mass concentrated at a point. The motion of a particle is therefore the motion of a point in space. Since a point has no geometrical dimensions we cannot specify the orientation of a particle, nor can we associate any particular rotational motion with it. In this nonrelativistic treatment, that is, for all but the final chapter, we shall assume that the mass of each particle remains constant.

Equations of Motion. The differential equations of motion for a system of N particles can be obtained by applying Newton's laws of motion to the particles individually. For a single particle of mass m which is subject to a force $\mathbf{F}$ we obtain from Newton's second law the vector equation

$$\mathbf{F} = m\mathbf{a} \tag{1-1}$$

or

$$\mathbf{F} = \frac{d\mathbf{p}}{dt} = \dot{\mathbf{p}} \tag{1-2}$$

where the *linear momentum* $\mathbf{p}$ is given by

$$\mathbf{p} = m\mathbf{v} \tag{1-3}$$

and where the acceleration **a** (or $\dot{\mathbf{v}}$) is measured relative to an *inertial frame of reference.*

The existence of an *inertial*, or *Newtonian*, reference frame is a fundamental postulate of Newtonian dynamics. As an example of an inertial reference frame, consider a frame with its origin at the sun and assume that it is non-rotating with respect to the so-called "fixed" stars. It can be shown that any other reference frame that is not rotating, but that is translating with a uniform velocity relative to a given inertial frame, is itself an inertial frame. Hence, the existence of a single inertial reference frame implies the existence of an infinity of other inertial frames which are equally valid (but not necessarily equally convenient) for the description of the motion of a particle, using the principles of Newtonian dynamics.

Let us assume, then, that we have found such a suitable inertial frame and that the vector $\mathbf{r}_i$ specifies the position of the *i*th particle relative to that frame. The equations of motion for the system of N particles can be written with the aid of Eq. (1-1).

$$m_i\ddot{\mathbf{r}}_i = \mathbf{F}_i + \mathbf{R}_i \qquad (i = 1, 2, \ldots, N) \qquad (1\text{-}4)$$

where m_i is the mass of the *i*th particle and where we have broken the total force acting on this particle into two vector components, $\mathbf{F}_i$ and $\mathbf{R}_i$. $\mathbf{F}_i$ is called the *applied force* and $\mathbf{R}_i$ is the *constraint force*. Briefly, $\mathbf{R}_i$ is that force which ensures that the geometrical constraints are followed in the motion of the *i*th particle. The applied force $\mathbf{F}_i$ represents the sum of all other forces acting on the *i*th particle. A more detailed discussion of constraints and the associated forces is given in Secs. 1-3 and 1-4.

In general, the forces that act on a body may be classified according to their mode of application as follows: (1) *contact* forces and (2) *body*, or *field*, forces. Contact forces are transmitted to the body by a direct mechanical push or pull. Body forces, on the other hand, are associated with action at a distance and are represented by gravitational, electrical, or other fields. It frequently occurs that body forces are applied throughout a body, but contact forces are applied only at its boundary surface. The forces $\mathbf{R}_i$ associated with the geometrical constraints are always contact forces. However, the applied forces $\mathbf{F}_i$ may be of either the body or contact type, or a combination of the two.

Instead of writing a single vector equation such as (1-4) for each particle, it is sometimes more convenient to write three scalar equations. Using the Cartesian coordinates (x_i, y_i, z_i) to represent the position of the *i*th particle, we obtain

$$\begin{aligned} m_i\ddot{x}_i &= F_{ix} + R_{ix} \\ m_i\ddot{y}_i &= F_{iy} + R_{iy} \qquad (i = 1, 2, \ldots, N) \\ m_i\ddot{z}_i &= F_{iz} + R_{iz} \end{aligned} \qquad (1\text{-}5)$$

where F_{ix} and R_{ix} are the x components of $\mathbf{F}_i$ and $\mathbf{R}_i$, respectively, and where F_{iy}, R_{iy}, F_{iz}, R_{iz} are defined similarly.

The notation of Eq. (1-5) is somewhat unwieldy, however. In order to simplify the writing of the equations, let us denote the Cartesian coordinates of the first particle by (x_1, x_2, x_3), of the second particle by (x_4, x_5, x_6), and so on. Then, noting that the mass of the kth particle is

$$m_{3k-2} = m_{3k-1} = m_{3k} \tag{1-6}$$

we can write the equations of motion in the form

$$m_i \ddot{x}_i = F_i + R_i \qquad (i = 1, 2, \ldots, 3N) \tag{1-7}$$

where F_i and R_i are the x_i components of the applied forces and the constraint forces, respectively. As particular examples of this notation, we see that F_5 is the y component of the applied force acting on the second particle, and R_{3N} is the z component of the constraint force acting on the Nth particle.

For the case in which there are no constraints, the force components F_i are expressed as functions of position, velocity, and time; the R's are zero. Hence the system of N particles is described by $3N$ second-order differential equations which are, in general, nonlinear. Although these equations may be solvable, in theory, for the position of each particle as a function of time, the equations are not often completely integrable in closed form. Frequently the practical solution involves the use of a computer.

If there are m constraints acting on the system, then the forces may be functions, not only of position, velocity, and time, but also of m additional variables known as Lagrange multipliers. In this case there are a total of $(3N + m)$ variables to be obtained as functions of time, using the $3N$ differential equations of motion and the m equations of constraint.

Thus we see that the writing of the equations of motion for a system of N particles, using Cartesian coordinates, may result in a formidable set of nonlinear ordinary differential equations. In certain cases, however, the analysis can be simplified considerably by the use of a different set of coordinates, involving fewer constraints, or perhaps eliminating the constraints altogether. The proper choice of coordinates and the use of further transformations of variables for the purpose of simplifying the analysis are topics which are discussed extensively in the remainder of the book.

Units. The equations of motion, whether they are written in the vector form of Eq. (1-4) or in the scalar form of Eq. (1-7), require that the variables be expressed in a consistent set of units. By *consistent* we mean that the quantities on both sides of each equation must be expressed in the same, or equivalent, units. If we consider the *dimensions* of the units used in the equations of motion, we find that mass, length, time, and force are present. Because the equations of motion must exhibit dimensional homogeneity, however, these four dimensions are not independent. In fact, any one dimension can be expressed in terms of the other three.

Certain systems of units known as *absolute* systems use mass, length, and time as the fundamental dimensions. For example, the mks system uses the meter as the fundamental unit of length, the kilogram as the fundamental unit of mass, and the second as the fundamental unit of time. The unit of force, the newton, is a *derived* unit and is equivalent to 1 kg m/sec². In general, we shall use this system whenever explicit units are mentioned.

Another common system of units is the *English gravitational system* in which units having the dimensions of force, length, and time are considered to be fundamental. Here the foot is the fundamental unit of length, the second is the fundamental unit of time, and the pound is the fundamental unit of force. The fundamental unit of mass is the slug and is a derived unit. It is equal to 1 lb sec²/ft.

1-2. GENERALIZED COORDINATES

Degrees of Freedom. An important characteristic of a given mechanical system is its number of degrees of freedom. The number of degrees of freedom is equal to the number of coordinates minus the number of independent equations of constraint. For example, if the configuration of a system of N particles is described using $3N$ Cartesian coordinates, and if there are l independent equations of constraint relating these coordinates, then there are $(3N - l)$ degrees of freedom.

To illustrate the idea of degrees of freedom, suppose that three particles are connected by rigid rods to form a triangular body with the particles at its corners. The configuration of the system is specified by giving the locations of the three particles, that is, by 9 Cartesian coordinates. But each rigid rod is represented mathematically by an independent equation of constraint. So $3N - l = 9 - 3 = 6$, and the system has six degrees of freedom.

The triangular body is an example of a rigid body, and has the same number of degrees of freedom as a general rigid body. One can see this by noting that the triangle can be imagined to be embedded in any given rigid body. In this case, each possible configuration of the triangle determines the configuration of the rigid body, and vice versa.

It is important to realize that the number of degrees of freedom is a characteristic of the system itself, and does not depend upon the particular set of coordinates used in its description. For example, the configuration of the previous triangular body might be specified by giving the three Cartesian coordinates of an arbitrary point in the body and also a set of three Eulerian angles describing its orientation. In this case, there are six coordinates and no constraints, again yielding six degrees of freedom.

Frequently it is advantageous to search for such a set of *independent* coordinates with which to describe the configuration of a system. In this case, there are as many coordinates as degrees of freedom, and the analysis contains a minimum number of variables.

Generalized Coordinates. We have seen that various sets of coordinates can be used to express the configuration of a given system. Furthermore, these sets do not necessarily have the same number of coordinates nor the same number of constraints. Nevertheless, the number of coordinates minus the number of independent equations of constraint is always equal to the number of degrees of freedom.

Now consider two sets of coordinates which describe the same system. At any given time, the values of each set of coordinates are simply a group of numbers. The process of obtaining one set of numbers from the other is known as a *coordinate transformation.*

As we think of the wide variety of possible coordinate transformations, any set of parameters which gives an unambiguous representation of the configuration of the system will serve as a system of coordinates in a more general sense. These parameters are known as *generalized coordinates.* All the common types of coordinates can serve as generalized coordinates, but many other parameters can also be used. For example, motion of a certain generalized coordinate might involve translation of one portion of the system and rotation of another portion.

Generalized coordinates usually have a readily visualized geometrical significance, and are often chosen on this basis. Furthermore, it is helpful in most analyses to choose a set of generalized coordinates which are *independent.* If the generalized coordinates specify the configuration of the system and can be varied independently without violating the constraints, then the number of generalized coordinates is equal to the number of degrees of freedom.

Straightforward procedures, such as the use of Lagrange's equations, exist for obtaining the differential equations of motion in terms of generalized coordinates. As we shall discover in the discussion of Chap. 2, the use of independent generalized coordinates allows the analysis of the motion of most systems to be made without solving for the forces of constraint.

Returning now to a consideration of the transformation equations relating the Cartesian coordinates $x_1, x_2, \ldots, x_{3N}$ to the generalized coordinates $q_1, q_2, \ldots, q_n$, we will assume that these equations are of the form

$$x_1 = x_1(q_1, q_2, \ldots, q_n, t)$$
$$x_2 = x_2(q_1, q_2, \ldots, q_n, t)$$

$$\begin{matrix} \cdot & \cdot & \cdot \\ \cdot & \cdot & \cdot \\ \cdot & \cdot & \cdot \end{matrix} \qquad (1\text{-}8)$$

$$x_{3N} = x_{3N}(q_1, q_2, \ldots, q_n, t)$$

It is possible that each system of coordinates has equations of constraint associated with it. If the x's have l equations of constraint and the q's have m equations of constraint, then, equating the number of degrees of freedom in each case, we find that

$$3N - l = n - m \qquad (1\text{-}9)$$

It is desirable that one and only one set of q's corresponds to each possible configuration of the system. In other words, there should be a one-to-one correspondence between points in the allowable domain of the x's and points in the allowable domain of the q's for each value of time. The necessary and sufficient condition that one can solve for the q's as functions of the x's and t is that the Jacobian determinant of the transformation be nonzero.

As an example, suppose that the $3N$ x's have l equations of constraint of the form

$$f_j(x_1, x_2, \ldots, x_{3N}, t) = \alpha_j \qquad (j = 3N - l + 1, \ldots, 3N) \qquad (1\text{-}10)$$

Let the n generalized coordinates be chosen so that they are independent, that is, the number of degrees of freedom is

$$n = 3N - l$$

Now define an additional set of l q's and identify them with the l constant functions f_j of Eq. (1–10).

$$q_j = f_j(x_1, x_2, \ldots, x_{3N}, t) = \alpha_j \qquad (j = n + 1, \ldots, 3N) \qquad (1\text{-}11)$$

Then the transformation equations of (1-8) can be considered to be of the form

$$\begin{aligned}
x_1 &= x_1(q_1, q_2, \ldots, q_{3N}, t) \\
x_2 &= x_2(q_1, q_2, \ldots, q_{3N}, t) \\
&\quad \vdots \\
x_{3N} &= x_{3N}(q_1, q_2, \ldots, q_{3N}, t)
\end{aligned} \qquad (1\text{-}12)$$

If the Jacobian determinant is nonzero, that is, if

$$\frac{\partial(x_1, x_2, \ldots, x_{3N})}{\partial(q_1, q_2, \ldots, q_{3N})} \neq 0 \qquad (1\text{-}13)$$

then Eq. (1-8) or Eq. (1-12) can be solved for the q's as functions of the x's and time.

$$q_j = f_j(x_1, x_2, \ldots, x_{3N}, t) \qquad (j = 1, 2, \ldots, n) \qquad (1\text{-}14)$$

The remaining constant q's for $j = n + 1, \ldots, 3N$ were given by Eqs. (1-10) and (1-11).

Example 1-1. As a simple example of a transformation from Cartesian to generalized coordinates, consider a particle which is constrained to move on a fixed circular path of radius a, as shown in Fig. 1-1. The equation of constraint is

$$(x_1^2 + x_2^2)^{1/2} = a$$

Let a single generalized coordinate q_1 represent the one degree of freedom. This polar angle can vary freely without violating the constraint. In accordance with Eq. (1-11), let us define a second generalized coordinate q_2 which is constant.

$$q_2 = a$$

The transformation equations are

$$x_1 = q_2 \cos q_1$$

$$x_2 = q_2 \sin q_1$$

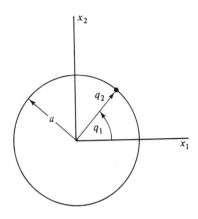

Fig. 1-1. A particle on a fixed circular path.

The Jacobian for this transformation is

$$\frac{\partial(x_1, x_2)}{\partial(q_1, q_2)} = \begin{vmatrix} \dfrac{\partial x_1}{\partial q_1} & \dfrac{\partial x_1}{\partial q_2} \\[2mm] \dfrac{\partial x_2}{\partial q_1} & \dfrac{\partial x_2}{\partial q_2} \end{vmatrix} = -q_2$$

Hence, the q's may be expressed as functions of the x's except when the Jacobian is zero at $q_2 = 0$. In this case the radius of the circle is zero and the angle q_1 is undefined. These transformation equations are

$$q_1 = \tan^{-1} \frac{x_2}{x_1}$$

$$q_2 = (x_1^2 + x_2^2)^{1/2}$$

where we arbitrarily take $0 \leq q_1 < 2\pi$ and $0 < q_2 < \infty$ in order that the q's will be single-valued functions of the x's. These transformation equations apply at all points on the finite $x_1 x_2$ plane except at the origin.

Configuration Space. We have seen that the configuration of a system of N particles is specified by giving the values of its $3N$ Cartesian coordinates. If the system has l independent equations of constraint of the form of Eq. (1-10), then it is possible to find n independent generalized coordinates $q_1, q_2, \ldots, q_n$, where $n = 3N - l$. Hence a set of n numbers, namely, the values of the n q's, completely specifies the configuration of the system. It is convenient to think of these n numbers as the coordinates of a *single point* in an n-dimensional space known as *configuration space*. In other words, the configuration of any mechanical system having a finite number of degrees of freedom is represented as a single point in an n-dimensional q-space. We may also consider a vector $\mathbf{q}$ to be drawn from the origin to the given configuration point. This $\mathbf{q}$ vector has the corresponding n q's as its components in a Euclidean (rectangular) space of n dimensions.

As a given mechanical system changes its configuration with time, the configuration point traces out a curve in q-space. For the usual case of independent q's, the curve will be continuous but otherwise unconstrained. But if there are constraints which are expressed as functions of the q's, the configuration point moves on a hypersurface having fewer than n dimensions.

The concept of a configuration space or q-space is used frequently in our further discussions of analytical dynamics.

1-3. CONSTRAINTS

We have seen that a system of N particles may have less than $3N$ degrees of freedom because of the presence of constraints. These constraints put geometrical restrictions upon the possible motions of the system and result in corresponding forces of constraint. Now let us consider the classification and mathematical description of constraints in greater detail.

Holonomic Constraints. Suppose the configuration of a system is specified by the n generalized coordinates $q_1, q_2, \ldots, q_n$ and assume that there are k independent equations of constraint of the form

$$\phi_j(q_1, q_2, \ldots, q_n, t) = 0 \qquad (j = 1, 2, \ldots, k) \qquad (1\text{-}15)$$

A constraint which can be expressed in this fashion is known as a *holonomic constraint*.

A system whose constraint equations, if any, are all of the holonomic form given in Eq. (1-15) is called a *holonomic system*. As an example of a holonomic system, consider the motion in the xy plane of the two particles shown in Fig. 1-2. These particles are connected by a rigid rod of length l; hence the corresponding equation of constraint is

$$(x_2 - x_1)^2 + (y_2 - y_1)^2 - l^2 = 0$$

In this case there are four coordinates and one equation of constraint, yielding three degrees of freedom. One could use this equation to eliminate one of the variables from the equations of motion. This procedure often entails algebraic difficulties, however, and is rarely used. Instead, let us search for a set of independent generalized coordinates, since it is known that these coordinates exist for all holonomic systems. For example, we can choose the Cartesian coordinates (x, y) of the center of the rod and the angle θ between the rod and the x axis as the generalized coordinates.

We have assumed that the length l of the rod is constant, and therefore the holonomic constraint equation does not contain time. Constraints of this sort in which the time t does not appear explicitly are known as *scleronomic*

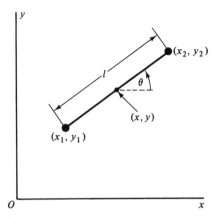

Fig. 1-2. Two particles connected by a rod of length l.

constraints. On the other hand, if the length l had been given as an explicit function of time, the constraint would have been classed as *rheonomic*. In the usual case, a rheonomic constraint is a moving constraint.

The terms *scleronomic* and *rheonomic* can also be applied to a mechanical *system*. A system is *scleronomic* if (1) none of the constraint equations contain t explicitly and (2) the transformation equations (1-8) give the x's as functions of the q's only. If any of the constraint equations or the transformation equations contain time explicitly, the system is *rheonomic*.

To explain this point further, it sometimes occurs that the generalized coordinates can be chosen in such a manner that there are no equations of constraint, or perhaps only scleronomic constraints, and yet the transformation equations contain time explicitly. An example is a particle constrained to move on a rigid wire which is rotating uniformly about a fixed axis, the single generalized coordinate being the position of the particle relative to the wire. Here there are no equations of constraint, but the system is rheonomic.

So far we have discussed two methods that can be used in the analysis of systems with holonomic constraints, namely, the elimination of variables using the constraint equations and the use of independent generalized coordinates. A third approach, which can be applied to either holonomic or nonholonomic systems, is the *Lagrange multiplier method*. This method represents the constraints by introducing the corresponding constraint forces which are expressed in terms of k variable parameters λ_j known as *Lagrange multipliers*. This method will be explained further in Sec. 2-1.

Nonholonomic Constraints. Now let us consider a system of m constraints which are written as *nonintegrable* differential expressions of the form

$$\sum_{i=1}^{n} a_{ji}\, dq_i + a_{jt}\, dt = 0 \qquad (j = 1, 2, \ldots, m) \qquad (1\text{-}16)$$

where the a's are, in general, functions of the q's and t. Constraints of this type are known as *nonholonomic constraints*.

As a result of the nonintegrable nature of these differential equations, one cannot obtain functions of the form given in Eq. (1-15) and use these to eliminate some of the variables. Nor is it possible to find a set of independent generalized coordinates. Hence, nonholonomic systems always require more coordinates for their description than there are degrees of freedom.

As an example of a nonholonomic system, consider again the two particles and rigid rod of Fig. 1-2. We assume that the particles can slide on the horizontal xy plane without friction. The system is changed, however, by the addition of a nonholonomic constraint in the form of knife-edge supports at the two particles, as in Fig. 2-6. These supports move with the system and are oriented perpendicular to the direction of the rod in such a manner that they allow no velocity component along the rod at either particle. Hence, the velocity of the center of the rod must be perpendicular to the rod, resulting in the constraint equation

$$\dot{x} = -\dot{y} \tan \theta$$

or

$$\cos \theta \, dx + \sin \theta \, dy = 0 \tag{1-17}$$

This expression is not an exact differential, that is, no function $\Phi(x, y, \theta)$ exists such that Eq. (1-17) is of the form

$$d\Phi = \frac{\partial \Phi}{\partial x} dx + \frac{\partial \Phi}{\partial y} dy + \frac{\partial \Phi}{\partial \theta} d\theta = 0 \tag{1-18}$$

Furthermore, Eq. (1-17) cannot be multiplied by any integrating factor to yield an exact differential. Hence, it is not integrable.

More generally, it can be shown† that the necessary and sufficient condition for the integrability of the differential equation

$$a_x \, dx + a_y \, dy + a_\theta \, d\theta = 0 \tag{1-19}$$

is that

$$a_x \left(\frac{\partial a_y}{\partial \theta} - \frac{\partial a_\theta}{\partial y} \right) + a_y \left(\frac{\partial a_\theta}{\partial x} - \frac{\partial a_x}{\partial \theta} \right) + a_\theta \left(\frac{\partial a_x}{\partial y} - \frac{\partial a_y}{\partial x} \right) = 0 \tag{1-20}$$

where the a's are functions of x, y, and θ. Applying this criterion to Eq. (1-17), we confirm that the expression is not integrable.

The system consisting of two particles and a rigid rod illustrates an important kinematic difference between holonomic and nonholonomic constraints. This difference occurs with respect to *accessibility*. If we first consider two unconstrained particles moving on the xy plane, we note that there is a four-dimensional configuration space, corresponding to the four independent coor-

---

†See E. L. Ince, *Ordinary Differential Equations* (New York: Dover Publications, Inc., 1956), p. 54.

dinates used to describe the configuration of the system. The addition of a holonomic constraint in the form of a rigid rod connecting the particles results in a reduction of the number of degrees of freedom from four to three. Since there are now three independent generalized coordinates, the configuration space is also reduced to three dimensions. But any point in configuration space is accessible from any other point; that is, any possible configuration can be reached from any other configuration.

Now consider the effect of the addition of a nonholonomic constraint represented by knife edges at the particles. This constraint restricts the particle velocities to the direction of the perpendicular to the rod. The number of degrees of freedom is reduced to two, but the number of required generalized coordinates remains at three. Furthermore, any point in the three-dimensional configuration space is accessible from any other point. In general, the kinematic effect of a nonholonomic constraint is to constrain the *direction* of the allowable motions at any given point of configuration space. But this does not reduce the number of dimensions in the configuration space, nor does it limit the variety of configurations available to the system. This last result is a direct consequence of the nonintegrability of the differential form; for if a function of the form

$$\Phi(q_1, q_2, \ldots, q_n, t) = c \qquad (1\text{-}21)$$

could be found, it would represent a hypersurface (embedded in q-space) on which the configuration point would be constrained to move, thereby limiting its region of accessibility.

Example 1-2. A classical example of a nonholonomic constraint occurs when there is rolling contact without slipping. For instance, consider a vertical disk of radius r which rolls without slipping on the horizontal xy plane, as shown in Fig. 1-3. Let us choose as generalized coordinates the point of contact (x, y), the angle of rotation ϕ of the disk about a perpendicular axis through its center, and the angle α between the plane of the disk and the yz plane. The requirement of rolling without slipping implies that

$$\begin{aligned} dx - r \sin \alpha \, d\phi &= 0 \\ dy - r \cos \alpha \, d\phi &= 0 \end{aligned} \qquad (1\text{-}22)$$

since $r \, d\phi$ is a differential element of displacement along the path traced by the point of contact, and α is the angle between the tangent to this path and the y axis.

In this example, there are four coordinates and two independent equations of constraint, resulting in two degrees of freedom. Notice, however, that the entire four-dimensional configuration space is accessible; that is, it is possible to arrive at any configuration (x, y, ϕ, α) from any other configuration by properly choosing the path.

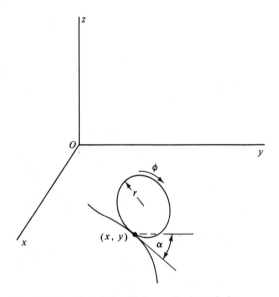

Fig. 1-3. A vertical disk rolling on a horizontal plane.

Previously we found that the integrability of a differential form requires it to be an exact differential, or be capable of becoming an exact differential through multiplication by an integrating factor which is some function of the variables. The necessary and sufficient conditions for the exactness of the differential form of Eq. (1-16) are that

$$\frac{\partial a_{ji}}{\partial q_k} = \frac{\partial a_{jk}}{\partial q_i}$$
$$\frac{\partial a_{ji}}{\partial t} = \frac{\partial a_{jt}}{\partial q_i} \qquad (i, k = 1, 2, \ldots, n) \qquad (1\text{-}23)$$

Since the coefficients of $d\phi$ in Eq. (1-22) are functions of α, but the coefficients of $d\alpha$ are zero, it is clear that no integrating factor can be found such that the resulting expression meets the conditions of Eq. (1-23). Hence both constraints are nonholonomic.

Unilateral Constraints. The constraints which have been discussed so far have all been *bilateral*; that is, if one imagines a small allowable displacement from any configuration of the system, the negative of this displacement is also allowable, assuming any fixed value of time. Such bilateral constraints are always expressed as an equality.

Now consider a constraint which is written in the form of an inequality involving a function of the q's and possibly time, such as

$$f(q_1, q_2, \ldots, q_n, t) \leq 0 \qquad (1\text{-}24)$$

This implies that the configuration point is restricted to a certain region of an n-dimensional configuration space which may vary with time. If the configuration point lies on the boundary of the allowable region, the unilateral nature of the constraint is apparent because the negative of a permitted small displacement generally will lie outside the region and therefore will not be allowed.

As a specific example of a unilateral constraint, suppose that a free particle is contained within a fixed hollow sphere of radius r which is centered at the origin of a Cartesian coordinate system. Then, using (x, y, z) as the coordinates of the particle, the unilateral constraint is given by

$$x^2 + y^2 + z^2 - r^2 \leq 0 \qquad (1\text{-}25)$$

If the particle is within the sphere and not touching its surface, the inequality holds and the particle moves freely with three degrees of freedom. On the other hand, if the particle moves along the sphere during some interval, the equality applies and the particle moves freely on a two-dimensional surface. Finally, if the particle hits the sphere and rebounds in accordance with some coefficient of restitution, the instant of impact provides the boundary point between two periods of free motion, the initial conditions of a given period being calculated from the final conditions of the previous period.

In any event, the motion of the particle is obtained by considering a sequence of *holonomic* systems, where switching occurs whenever the particle hits the spherical surface or leaves it. Thus the unilateral or inequality form of constraint is holonomic in nature.

1-4. VIRTUAL WORK

The concept of virtual work is fundamental in the study of analytical mechanics. It is directly associated with the application of energy methods in the derivation of the equations of motion, as well as being an important concept in the study of stability. Since virtual work is associated with a virtual displacement, let us consider first the nature of a virtual displacement.

Virtual Displacement. Suppose the configuration of a system of N particles is given by the $3N$ Cartesian coordinates $x_1, x_2, \ldots, x_{3N}$ which are measured relative to an inertial frame and may be subject to constraints. At any given time, let us assume that the coordinates move through infinitesimal displacements $\delta x_1, \delta x_2, \ldots, \delta x_{3N}$ which are virtual or imaginary in the sense that they are assumed to occur without the passage of time, and do not necessarily conform to the constraints. This small change $\delta \mathbf{x}$ in the configuration of the system is known as a *virtual displacement*.

In the usual case, a virtual displacement conforms to the *instantaneous constraints*, that is, any moving constraints are assumed to be stopped during the virtual displacement. For example, suppose the system is subject to k holonomic constraints

$$\phi_j(x_1, x_2, \ldots, x_{3N}, t) = 0 \qquad (j = 1, 2, \ldots, k) \qquad (1\text{-}26)$$

Let us take the total differential of ϕ_j and obtain

$$d\phi_j = \sum_{i=1}^{3N} \frac{\partial \phi_j}{\partial x_i} \, dx_i + \frac{\partial \phi_j}{\partial t} \, dt = 0 \qquad (j = 1, 2, \ldots, k) \qquad (1\text{-}27)$$

A virtual displacement which conforms to these constraints has the δx's related by the k equations

$$\sum_{i=1}^{3N} \frac{\partial \phi_j}{\partial x_i} \, \delta x_i = 0 \qquad (j = 1, 2, \ldots, k) \qquad (1\text{-}28)$$

Here we have replaced the dx's in Eq. (1-27) by δx's and have omitted the dt term because the time is held fixed during a virtual displacement.

In a similar fashion, let us now assume that the system has m nonholonomic constraints of the form

$$\sum_{i=1}^{3N} a_{ji} \, dx_i + a_{jt} \, dt = 0 \qquad (j = 1, 2, \ldots, m) \qquad (1\text{-}29)$$

Any virtual displacement which conforms to these constraints must have the δx's related by the m equations

$$\sum_{i=1}^{3N} a_{ji} \, \delta x_i = 0 \qquad (j = 1, 2, \ldots, m) \qquad (1\text{-}30)$$

The question arises whether a virtual displacement can also be a possible real displacement, described by a set of dx's, and assumed to occur during the time increment dt. In other words, under what conditions can $\delta \mathbf{x}$ be replaced by $d\mathbf{x}$? A comparison of Eqs. (1-27) and (1-28) shows that any holonomic constraints must also be scleronomic, that is, the condition

$$\frac{\partial \phi_j}{\partial t} = 0 \qquad (j = 1, 2, \ldots, k) \qquad (1\text{-}31)$$

must apply. Similarly, any nonholonomic constraints must meet the condition

$$a_{jt} = 0 \qquad (j = 1, 2, \ldots, m) \qquad (1\text{-}32)$$

Since these conditions are not met in the general case, it is clear that a virtual displacement is not, in general, a possible real displacement.

It is sometimes convenient to assume that a set of δx's conforming to the instantaneous constraints occurs during an interval δt. The corresponding ratios of the form $\delta x / \delta t$ have the dimensions of velocity and are known as *virtual velocities*. In general, virtual velocities are not possible velocities for the actual system. It is only when Eqs. (1-31) and (1-32) apply that a virtual velocity consistent with the constraints is also a possible velocity.

Thus far in the discussion of virtual displacements we have used Cartesian coordinates. Now let us consider a system whose configuration is given by the minimum number of generalized coordinates. Thus, any constraints will be nonholonomic and can be expressed in the form

$$\sum_{i=1}^{n} a_{ji} \, dq_i + a_{jt} \, dt = 0 \qquad (j = 1, 2, \ldots, m) \tag{1-33}$$

or, alternatively,

$$\sum_{i=1}^{n} a_{ji} \dot{q}_i + a_{jt} = 0 \qquad (j = 1, 2, \ldots, m) \tag{1-34}$$

where the a's are functions of the q's and t.

Any virtual displacement consistent with the constraints must meet the conditions

$$\sum_{i=1}^{n} a_{ji} \, \delta q_i = 0 \qquad (j = 1, 2, \ldots, m) \tag{1-35}$$

The corresponding *generalized virtual velocity* **u** has components u_i which satisfy

$$\sum_{i=1}^{n} a_{ji} u_i = 0 \qquad (j = 1, 2, \ldots, m) \tag{1-36}$$

Now let us consider once again the necessary conditions for the virtual velocity of any point of the system to be a possible velocity. Comparing Eqs. (1-34) and (1-36) we see that

$$a_{jt} = 0 \qquad (j = 1, 2, \ldots, m) \tag{1-37}$$

In addition, however, the transformation equations relating the x's and the q's must not contain t explicitly; that is, there can be no moving constraints since moving constraints induce actual velocity components other than the allowable virtual velocities u_i.

Virtual Work. Let us return now to a system of N particles whose configuration is given by the Cartesian coordinates $x_1, x_2, \ldots, x_{3N}$. Suppose that force components $F_1, F_2, \ldots, F_{3N}$ are applied at the corresponding coordinates in a positive sense. The *virtual work* δW of these forces in a virtual displacement $\delta \mathbf{x}$ is given by

$$\delta W = \sum_{j=1}^{3N} F_j \, \delta x_j \tag{1-38}$$

An alternate form of the expression for the virtual work is

$$\delta W = \sum_{i=1}^{N} \mathbf{F}_i \cdot \delta \mathbf{r}_i \tag{1-39}$$

where $\mathbf{F}_i$ is the force applied at the ith particle, and where $\mathbf{r}_i$ is the position vector of this particle. We see from the vector formulation that the virtual work does not depend upon the use of any particular coordinate system, assuming, of course, that the motion is measured relative to an inertial reference frame.

In the expressions for virtual work, it is important to realize that the forces are assumed to remain *constant* throughout the virtual displacement. This is true even if the actual force changes drastically as the result of an infinitesimal displacement. A sudden change of force with position could occur, for example, in certain nonlinear systems.

Another point to notice is that the virtual work expressions are defined to be *linear* in the virtual displacements. In other words, the virtual work is similar to a first variation.

Now consider a system which is subject to constraints. Let the total force acting on the ith particle be separated into an *applied force* $\mathbf{F}_i$ and a *constraint force* $\mathbf{R}_i$. The virtual work of the constraint forces is

$$\delta W_c = \sum_{i=1}^{N} \mathbf{R}_i \cdot \delta \mathbf{r}_i \tag{1-40}$$

Many of the constraints that commonly occur are of a class known as *workless constraints*. We can define a workless constraint as follows: *A workless constraint is any bilateral constraint such that the virtual work of the corresponding constraint forces is zero for any virtual displacement which is consistent with the constraints.* It can be seen that, for a system having only workless constraints, the virtual work δW_c is equal to zero, or

$$\sum_{i=1}^{N} \mathbf{R}_i \cdot \delta \mathbf{r}_i = 0 \tag{1-41}$$

where the virtual displacements $\delta \mathbf{r}_i$ are consistent with the instantaneous constraints.

Examples of workless constraints are (1) rigid interconnections between particles, (2) sliding motion on a frictionless surface, and (3) rolling contact without slipping. Let us consider these examples in more detail.

First assume that two particles are connected by a rigid massless rod, as in Fig. 1-4(a). Because of Newton's third law, the forces exerted by the rod on the particles m_1 and m_2 are equal, opposite, and collinear. Hence

$$\mathbf{R}_2 = R_2 \mathbf{e}_r = -\mathbf{R}_1 \tag{1-42}$$

where $\mathbf{e}_r$ is a unit vector directed along the rod, as shown. Furthermore, since the rod is rigid, the displacement components of the particles in the direction of the rod must be equal, or

$$\mathbf{e}_r \cdot \delta \mathbf{r}_1 = \mathbf{e}_r \cdot \delta \mathbf{r}_2 \tag{1-43}$$

Therefore, the virtual work of the constraint forces is zero.

$$\delta W_c = \mathbf{R}_1 \cdot \delta \mathbf{r}_1 + \mathbf{R}_2 \cdot \delta \mathbf{r}_2 = 0 \tag{1-44}$$

Now consider a body B which slides without friction on a fixed surface S, as shown in Fig. 1-4(b). The constraint force $\mathbf{R}$ is normal to the surface

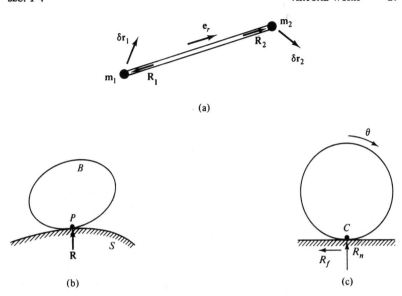

(a)

(b)   (c)

Fig. 1-4. Examples of workless constraints.

at the contact point P, but any virtual displacement of P involves sliding in
the tangent plane at that point. Hence no work is done by the constraint
force **R** in a virtual displacement.

As a third example, consider a vertical circular disk which rolls without
slipping along a straight horizontal path, Fig. 1-4(c). The total force of the
surface acting on the disk can be separated into a normal component R_n and
a frictional component R_f, the latter being directed tangent to the surface.
These force components pass through the instantaneous center C. The instantaneous center, however, does not move as a result of a virtual displacement
$\delta\theta$ if we make the usual assumption that infinitesimals of higher order are
neglected. In other words, the virtual velocity of C is zero. Hence the virtual
work of the constraint forces is zero.

Although this example involved the particular case of a disk rolling on a
plane, a similar argument would apply to the rolling contact of any body on
a fixed surface.

The examples we have given include the most common types of workless
constraints, but many others are possible. A rigid rod whose length is given
as an explicit function of time is a workless constraint because time is considered to remain fixed during a virtual displacement. Similarly, a frictionless
surface moving as a given function of time would also be workless. In these
cases, the constraint forces do work on the system in an actual displacement,
but not in the assumed virtual displacement.

It should be emphasized that workless constraints do no work on the system *as a whole* in an arbitrary virtual displacement. Quite possibly, however, the workless constraint forces will do work on individual particles of the system. For example, constraint forces are responsible for the transfer of energy from one particle of a rigid body to another as the particles move with changing speed in some general motion of the body.

Unless a statement is made to the contrary, we shall assume in any further discussions of constraints that the term *constraint force* should be interpreted as a *workless* constraint force. In cases such as sliding constraints with friction, the tangential force of friction is lumped with the applied force F_i, and the normal component is treated as a workless constraint force in the usual manner.

Unilateral constraints are not classed as workless constraints because allowable virtual displacements can be found in which the virtual work of the constraint forces is not zero. In the discussion of the principle of virtual work we shall consider this point further.

Principle of Virtual Work. One of the important applications of the idea of virtual work arises in the study of the static equilibrium of mechanical systems. Suppose we consider a *scleronomic* system of N particles. If this system is in static equilibrium, then, for each particle,

$$\mathbf{F}_i + \mathbf{R}_i = 0 \tag{1-45}$$

Therefore the virtual work done by all the forces in moving through an arbitrary virtual displacement consistent with the constraints is zero.

$$\sum_{i=1}^{N} (\mathbf{F}_i + \mathbf{R}_i) \cdot \delta \mathbf{r}_i = \sum_{i=1}^{N} \mathbf{F}_i \cdot \delta \mathbf{r}_i + \sum_{i=1}^{N} \mathbf{R}_i \cdot \delta \mathbf{r}_i = 0 \tag{1-46}$$

If we now assume that all the constraints are workless, and if the $\delta \mathbf{r}_i$ are reversible virtual displacements consistent with the constraints, then

$$\sum_{i=1}^{N} \mathbf{R}_i \cdot \delta \mathbf{r}_i = 0 \tag{1-47}$$

From Eqs. (1-46) and (1-47), we conclude that

$$\delta W = \sum_{i=1}^{N} \mathbf{F}_i \cdot \delta \mathbf{r}_i = 0 \tag{1-48}$$

We have shown that if a system of particles with workless constraints is in static equilibrium, then it follows that the virtual work of the *applied forces* is zero for any virtual displacement consistent with the constraints.

Now assume that the same system of particles is initially motionless, but is *not in equilibrium*. Then one or more of the particles must have a net force applied to it, and in accordance with Newton's law of motion, it will start to move in the direction of that force. Since any motion must be compatible with the constraints (which are assumed to be fixed), we can always choose a virtual displacement in the direction of the actual motion at each point. In this case the virtual work is positive, that is,

$$\sum_{i=1}^{N} \mathbf{F}_i \cdot \delta \mathbf{r}_i + \sum_{i=1}^{N} \mathbf{R}_i \cdot \delta \mathbf{r}_i > 0 \qquad (1\text{-}49)$$

But again the constraints are workless and Eq. (1-47) applies. Hence

$$\delta W = \sum_{i=1}^{N} \mathbf{F}_i \cdot \delta \mathbf{r}_i > 0 \qquad (1\text{-}50)$$

A reversal of the $\delta\mathbf{r}$'s would yield a negative virtual work for this system. But in any event, if the system is not in equilibrium, it is always possible to find a set of virtual displacements consistent with the constraints which will result in the virtual work of the applied forces being nonzero.

These results can be summarized in the *principle of virtual work: The necessary and sufficient condition for the static equilibrium of an initially motionless scleronomic system which is subject to workless constraints is that zero virtual work be done by the applied forces in moving through an arbitrary virtual displacement satisfying the constraints.*

As an example of the application of the principle of virtual work, consider the system shown in Fig. 1-5. Two frictionless blocks of equal mass m are connected by a massless rigid rod. Using x_1 and x_2 as coordinates, solve for the force F_2 if the system is in static equilibrium.

This example shows a scleronomic system with workless constraints. The external constraint forces are the wall and floor reactions R_1 and R_2; the internal constraint forces are the equal and opposite compressive forces in the rod. We note that the total virtual work of these constraint forces is zero.

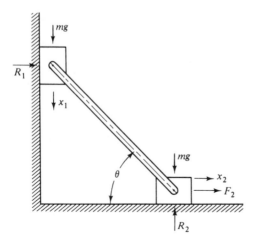

Fig. 1-5. A frictionless system which is constrained to move in the vertical plane.

The *applied* forces are the gravitational forces acting on the blocks and the external force F_2. Hence, using the principle of virtual work, we see that the required condition for static equilibrium is that

$$mg\, \delta x_1 + F_2\, \delta x_2 = 0 \tag{1-51}$$

But δx_1 and δx_2 are related by an equation of constraint. Since the displacement components along the rod must be equal at the two ends, we have

$$\sin \theta\, \delta x_1 - \cos \theta\, \delta x_2 = 0 \tag{1-52}$$

Solving Eqs. (1-51) and (1-52), we obtain

$$F_2 = -mg \cot \theta$$

This is the force required to keep the initially motionless system in static equilibrium.

The principle of virtual work was derived for systems with bilateral constraints. But the idea of a virtual displacement and the associated virtual work is quite general and can be applied to unilateral systems. So let us consider the system of Fig. 1-6, consisting of a cube of mass m which is resting in

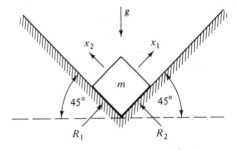

Fig. 1-6. A system with unilateral constraints.

static equilibrium at a corner formed by two frictionless, mutually perpendicular planes. Assume that any motion is restricted to the vertical plane.

The unilateral constraint equations are

$$x_1 \ge 0, \qquad x_2 \ge 0 \tag{1-53}$$

where we assume that $x_1 = x_2 = 0$ when the cube is in its equilibrium position. In this case, the only applied forces are due to gravity and consist of the components in the directions of x_1 and x_2, respectively.

$$F_1 = F_2 = -\frac{1}{\sqrt{2}}\, mg \tag{1-54}$$

Therefore the virtual work of the applied forces is

$$\delta W = F_1\, \delta x_1 + F_2\, \delta x_2 = -\frac{mg}{\sqrt{2}}(\delta x_1 + \delta x_2) \tag{1-55}$$

Thus, the virtual work $\delta W \leq 0$ for any virtual displacement consistent with the unilateral constraints.

In general, for an initially motionless system containing frictionless fixed constraints, which may be unilateral, the necessary and sufficient condition for static equilibrium is that the virtual work of the applied forces be equal to or less than zero, that is,

$$\delta W \leq 0 \qquad (1\text{-}56)$$

for all virtual displacements consistent with the constraints.

Now let us calculate the virtual work of the constraint forces. Noting that

$$R_1 = R_2 = \frac{1}{\sqrt{2}} mg \qquad (1\text{-}57)$$

we obtain

$$\delta W_c = R_1 \, \delta x_1 + R_2 \, \delta x_2 \geq 0 \qquad (1\text{-}58)$$

where we recall that R_1 and R_2 are assumed to remain constant during the virtual displacement. In this example of a system with unilateral constraints, we see that there can be nonzero virtual work by the constraint forces in an allowable virtual displacement. Hence, unilateral constraints cannot be classed as workless, even though they may be frictionless.

It should be pointed out that the constraint forces R_1 and R_2 can be found by using the principle of virtual work. The procedure followed is to set the total virtual work of *all* forces equal to zero, that is,

$$\left(R_1 - \frac{mg}{\sqrt{2}} \right) \delta x_1 + \left(R_2 - \frac{mg}{\sqrt{2}} \right) \delta x_2 = 0 \qquad (1\text{-}59)$$

Here we assume that the δx's are *not constrained*, and therefore are reversible and independent. Hence each coefficient of Eq. (1-59) must be zero, and we obtain

$$R_1 = R_2 = \frac{1}{\sqrt{2}} mg$$

In the study of unilateral constraints, one finds that the constraint forces can change suddenly as the constraint function reaches or leaves its limiting value. A similar sudden change can occur in Coulomb friction forces, but in this case the force is considered as a discontinuous function of the sliding *velocity*. Let us examine now the virtual work of a system containing elements having Coulomb friction.

Our approach is to separate the total reaction force at a sliding surface into a normal workless component and a tangential frictional component which is considered as an *applied* force. Let us assume that each Coulomb friction force opposes relative sliding motion and has a magnitude equal to the coefficient of friction μ multiplied by the normal force at the given surface. With these assumptions, an initially motionless system is in static equilibrium if and only if the virtual work δW of the applied forces is given by

$$\delta W \leq 0 \tag{1-60}$$

for all virtual displacements consistent with the constraints.

Note that the actual friction forces of a system in static equilibrium are *not equal*, in general, to the forces used in the virtual work expression, but are smaller in magnitude. This implies that a finite additional force is required to cause a system with Coulomb friction to break free of the equilibrium condition and start to move. In contrast, a frictionless system with bilateral constraints requires only an infinitesimal change in an applied force in order to cause it to move from its position of static equilibrium.

D'Alembert's Principle. Let us consider again a system of N particles and write the equation of motion for each particle in the form

$$\mathbf{F}_i + \mathbf{R}_i - m_i\ddot{\mathbf{r}}_i = 0 \qquad (i = 1, 2, \ldots, N) \tag{1-61}$$

where, as before, $\mathbf{F}_i$ is the applied force and $\mathbf{R}_i$ is the constraint force acting on the ith particle. The term $-m_i\ddot{\mathbf{r}}_i$ has the dimensions of force and is known as the *inertial force* acting on the ith particle, where m_i is the constant mass and $\ddot{\mathbf{r}}_i$ is its acceleration relative to an inertial frame.

It is customary to call $\mathbf{F}_i$ and $\mathbf{R}_i$ *real* or *actual forces* in contrast to the inertial forces. Hence Eq. (1-61) states that the sum of *all* the forces, real and inertial, acting on *each* particle of a system is zero. This result is sometimes known as *d'Alembert's principle*.[†]

The requirement that the sum of all the forces at each particle be zero is similar to the necessary condition for static equilibrium. Since the principle of *virtual work* applies to systems in static equilibrium, let us use the principle on this force system, including the inertial forces. The total work done by all the forces in an arbitrary virtual displacement is

$$\delta W = \sum_{i=1}^{N} (\mathbf{F}_i + \mathbf{R}_i - m_i\ddot{\mathbf{r}}_i) \cdot \delta \mathbf{r}_i = 0 \tag{1-62}$$

If we now assume that the $\mathbf{R}_i$ are workless constraint forces, and if we choose the $\delta \mathbf{r}_i$ to be reversible virtual displacements consistent with the constraints, then we obtain from Eqs. (1-47) and (1-62) that

$$\sum_{i=1}^{N} (\mathbf{F}_i - m_i\ddot{\mathbf{r}}_i) \cdot \delta \mathbf{r}_i = 0 \tag{1-63}$$

[†]A somewhat different statement was made by J. d'Alembert in his *Traité de Dynamique* (1743). Although he referred to velocities rather than forces, he stated in essence that the constraint forces, meaning interaction forces, form a system in static equilibrium. As a *consequence* of this principle, the applied and inertial forces together form a system that also is in equilibrium in the sense of the virtual work expression of Eq. (1-63). See Lindsay and Margenau, *Foundations of Physics* (New York: Dover Publications, Inc., 1957), pp. 102–112.

This equation is the *Lagrangian form of d'Alembert's principle* and is one of the most important equations of classical dynamics.

As a result of including the forces of inertia in this application of the principle of virtual work, the validity of the principle is extended to dynamic as well as static systems. Notice that Eq. (1-63) does not contain the forces of constraint, which are often unknown, but requires only the applied forces F_i. Also, the equation applies to rheonomic as well as scleronomic systems, provided that the δr's are consistent with the instantaneous constraints.

Example 1-3. A particle of mass m is suspended by a massless wire of length

$$r = a + b \cos \omega t \qquad (a > b > 0) \tag{1-64}$$

to form a spherical pendulum. Find the equations of motion.

Let us use the spherical coordinates θ and ϕ, where θ is measured from the upward vertical, as shown in Fig. 1-7. The angle ϕ is measured between a vertical reference plane passing through the support point O and the vertical plane containing the pendulum.

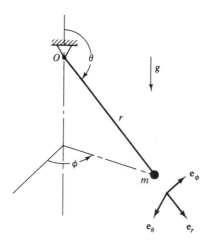

Fig. 1-7. A spherical pendulum of variable length.

The general expression for the acceleration of a particle whose spherical coordinates are (r, θ, ϕ) is as follows:

$$\begin{aligned}
\ddot{\mathbf{r}} = {} & (\ddot{r} - r\dot{\theta}^2 - r\dot{\phi}^2 \sin^2 \theta)\mathbf{e}_r \\
& + (r\ddot{\theta} + 2\dot{r}\dot{\theta} - r\dot{\phi}^2 \sin \theta \cos \theta)\mathbf{e}_\theta \\
& + (r\ddot{\phi} \sin \theta + 2\dot{r}\dot{\phi} \sin \theta + 2r\dot{\theta}\dot{\phi} \cos \theta)\mathbf{e}_\phi
\end{aligned} \tag{1-65}$$

where $\mathbf{e}_r$, $\mathbf{e}_\theta$, and $\mathbf{e}_\phi$ are unit vectors forming an orthogonal triad.

A virtual displacement consistent with the instantaneous constraint is

$$\delta\mathbf{r} = r\,\delta\theta\,\mathbf{e}_\theta + r\sin\theta\,\delta\phi\,\mathbf{e}_\phi \qquad (1\text{-}66)$$

Furthermore, the applied gravitational force is

$$\mathbf{F} = -mg\cos\theta\,\mathbf{e}_r + mg\sin\theta\,\mathbf{e}_\theta \qquad (1\text{-}67)$$

Substituting from Eqs. (1-65), (1-66), and (1-67) into Eq. (1-63), we obtain

$$mr\,[g\sin\theta - (r\ddot{\theta} + 2\dot{r}\dot{\theta} - r\dot{\phi}^2\sin\theta\cos\theta)]\,\delta\theta$$
$$- mr\sin\theta[r\ddot{\phi}\sin\theta + 2\dot{r}\dot{\phi}\sin\theta + 2r\dot{\theta}\dot{\phi}\cos\theta]\,\delta\phi = 0 \qquad (1\text{-}68)$$

Since $\delta\theta$ and $\delta\phi$ are independent virtual displacements, their coefficients must each be zero. Dividing out the common nonzero factors, and substituting for r and its derivatives from Eq. (1-64), we obtain

$$
\begin{aligned}
(a + b\cos\omega t)\ddot{\theta} - 2b\omega\dot{\theta}\sin\omega t & \\
- (a + b\cos\omega t)\dot{\phi}^2\sin\theta\cos\theta &= g\sin\theta \\
(a + b\cos\omega t)\ddot{\phi}\sin\theta - 2b\omega\dot{\phi}\sin\omega t\sin\theta & \\
+ 2(a + b\cos\omega t)\dot{\theta}\dot{\phi}\cos\theta &= 0
\end{aligned}
\qquad (1\text{-}69)
$$

These are the differential equations of motion for this system.

Generalized Force. In our previous discussions of virtual work, we have been concerned with the work done by applied forces (or their equivalent orthogonal components) in moving through a certain virtual displacement. For example, if a given set of forces $F_1, F_2, \ldots, F_{3N}$ is applied to a system of N particles, the virtual work of these forces is

$$\delta W = \sum_{j=1}^{3N} F_j\,\delta x_j \qquad (1\text{-}70)$$

Now let us suppose that the $3N$ ordinary Cartesian coordinates $x_1, x_2, \ldots, x_{3N}$ are related to the n generalized coordinates $q_1, q_2, \ldots, q_n$ by transformation equations of the form of Eq. (1-8). If we differentiate this equation and set $\delta t = 0$ (since we are considering a virtual displacement), we obtain the following:

$$\delta x_j = \sum_{i=1}^{n} \frac{\partial x_j}{\partial q_i}\,\delta q_i \qquad (j = 1, 2, \ldots, 3N) \qquad (1\text{-}71)$$

where the coefficients $\partial x_j/\partial q_i$ are, in general, functions of the q's and t. Substituting this expression for δx_j into Eq. (1-70), we obtain

$$\delta W = \sum_{j=1}^{3N}\sum_{i=1}^{n} F_j \frac{\partial x_j}{\partial q_i}\,\delta q_i \qquad (1\text{-}72)$$

Let us define the *generalized force* Q_i by the equation

$$Q_i = \sum_{j=1}^{3N} F_j \frac{\partial x_j}{\partial q_i} \qquad (i = 1, 2, \ldots, n) \qquad (1\text{-}73)$$

Then, substituting from Eq. (1-73) into Eq. (1-72) and changing the order of summation, we obtain

$$\delta W = \sum_{i=1}^{n} Q_i \, \delta q_i \qquad (1\text{-}74)$$

Comparing the virtual work expressions of Eqs. (1-70) and (1-74), we see that they are of the same mathematical form. Previously we defined the F's as the ordinary force components applied at the corresponding x's in the positive sense. From Eq. (1-70) it can be seen that F_j is also equal to the virtual work per unit displacement of δx_j for the case in which all the other δx's are zero. In a similar fashion, we can consider the generalized force Q_i to be the virtual work done by all the F's acting on the system per unit displacement of δq_i, assuming the other δq's are zero. Here we make the usual assumptions that the virtual displacement is small enough to have an insignificant effect on the geometry of the system, and the forces remain constant during this virtual displacement.

The dimensions of a generalized force depend upon the dimensions of the corresponding generalized coordinate. But in all cases $Q_i \, \delta q_i$ must have the dimensions of work or energy. So if q_i represents a linear displacement, the corresponding Q_i is an ordinary force. On the other hand, if q_i is an angle, then the corresponding Q_i is a moment. In some cases, a generalized coordinate may represent a deflection form in which both translations and rotations occur at various parts of the system. If we take the q_i in this case to be a dimensionless ratio, then the corresponding Q_i has the dimensions of energy.

Usually the generalized coordinates are chosen in such a manner that they are independent. If there are constraints, however, these constraints are ignored in making the required virtual displacement in which only one of the δq's is nonzero. This does not mean that the *constraint forces* can be ignored, for the R's as well as the F's will contribute to the generalized force Q_i under these conditions. For example, generalized constraint forces occur in nonholonomic systems because it is impossible to choose independent generalized coordinates. These generalized constraint forces are usually expressed in terms of Lagrange multipliers and will be discussed in Sec. 2-1.

The concept of a generalized force is very useful in the statement of the principle of virtual work. Suppose we consider an initially motionless holonomic system having workless, fixed constraints. If its configuration is expressed in terms of *independent* generalized coordinates, then *the necessary and sufficient condition for static equilibrium is that all the Q's due to the applied forces be zero.*

One is tempted at this point to find expressions for the generalized *inertial* forces and to use these with the principle of virtual work to arrive at the general dynamical equations, that is, the differential equations of motion for the system in terms of generalized coordinates and forces. This is a valid approach which can be used to derive Lagrange's equation, but we will post-pone this derivation until the next chapter.

Example 1-4. Three particles are connected by two rigid rods having a joint between them to form the system shown in Fig. 1-8. A vertical force F

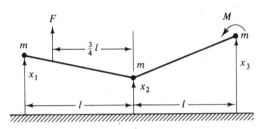

Fig. 1-8. A system with an applied force and moment.

and a moment M are applied as shown. The configuration of the system is given by the ordinary coordinates (x_1, x_2, x_3) or by the generalized coordinates (q_1, q_2, q_3), where

$$x_1 = q_1 + q_2 + \tfrac{1}{2}q_3$$
$$x_2 = q_1 - q_3 \qquad\qquad (1\text{-}75)$$
$$x_3 = q_1 - q_2 + \tfrac{1}{2}q_3$$

Find the generalized forces Q_1, Q_2, and Q_3. Assume small motions.

First, let us check whether the transformation equations of Eq. (1-75) yield independent q's. Evaluating the Jacobian of the transformation, we obtain

$$\frac{\partial(x_1, x_2, x_3)}{\partial(q_1, q_2, q_3)} = \begin{vmatrix} 1 & 1 & \tfrac{1}{2} \\ 1 & 0 & -1 \\ 1 & -1 & \tfrac{1}{2} \end{vmatrix} = -3$$

Since this determinant is nonzero, we see that the q's are independent. Eq. (1-75) can be solved for the q's as functions of the x's, yielding

$$q_1 = \tfrac{1}{3}(x_1 + x_2 + x_3)$$
$$q_2 = \tfrac{1}{2}(x_1 - x_3) \qquad\qquad (1\text{-}76)$$
$$q_3 = \tfrac{1}{3}(x_1 - 2x_2 + x_3)$$

Thus, for any set of x's, we obtain a corresponding unique set of q's.

The generalized forces are obtained by considering small virtual displace-ments of each of the generalized coordinates, whose deflection forms are shown in Fig. 1-9. Notice that an increase in q_1 represents a pure translation, while q_2 is associated with a rotation about the center, and q_3 denotes a defor-mation or bending of the system. If we consider the translational displacement

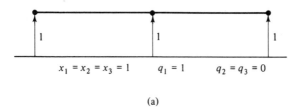

(a)

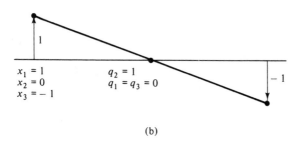

(b)

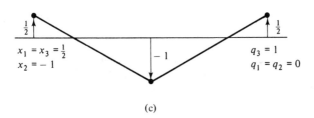

(c)

Fig. 1-9. Deflection forms corresponding to the generalized coordinates.

of the point of application of the applied force F, and also the rotation of the applied moment M, we obtain the following expression for the virtual work:

$$\delta W = F \delta q_1 + \left(\frac{3}{4}F - \frac{M}{l}\right)\delta q_2 + \left(\frac{1}{8}F + \frac{3M}{2l}\right)\delta q_3 \qquad (1\text{-}77)$$

Comparing Eqs. (1-74) and (1-77) we find that the generalized forces are

$$Q_1 = F$$
$$Q_2 = \frac{3}{4}F - \frac{M}{l} \qquad (1\text{-}78)$$
$$Q_3 = \frac{1}{8}F + \frac{3M}{2l}$$

Another approach is to obtain the Q's by using Eq. (1-73) directly. First note that the force F can be replaced by a force $3F/4$ at x_1 and a force $F/4$

at x_2. Also, the moment M can be replaced by equal and opposite forces of magnitude M/l acting in the directions of $-x_2$ and x_3. With these substitutions of equipollent† force systems, we find that

$$F_1 = \frac{3}{4}F$$

$$F_2 = \frac{1}{4}F - \frac{M}{l} \tag{1-79}$$

$$F_3 = \frac{M}{l}$$

Then, using Eq. (1-73) and remembering that the partial derivatives have been calculated previously in evaluating the Jacobian, we obtain expressions for the Q's which are identical with those given in Eq. (1-78).

In this example, we have chosen a particular set of generalized coordinates which have the geometrical significance of representing translation, rotation, and deformation of the system. But this is not necessary. Any set of independent parameters which specifies the configuration of the system may serve as generalized coordinates. The corresponding generalized forces are then obtained by finding the virtual work of the applied forces per unit δq.

1-5. ENERGY AND MOMENTUM

Potential Energy. Let us consider a single particle whose position is given by the Cartesian coordinates (x, y, z). Suppose that the total force **F** acting on the particle has the components

$$F_x = -\frac{\partial V}{\partial x}$$

$$F_y = -\frac{\partial V}{\partial y} \tag{1-80}$$

$$F_z = -\frac{\partial V}{\partial z}$$

where the *potential energy function* $V(x, y, z)$ is a single-valued function of *position only*; that is, it is not a function of velocity or time. A force **F** meeting these conditions is known as a *conservative force*.

Now let us consider the work dW done by the force **F** as it moves through an infinitesimal displacement $d\mathbf{r}$. We have

$$dW = \mathbf{F} \cdot d\mathbf{r} = F_x\,dx + F_y\,dy + F_z\,dz \tag{1-81}$$

---

†Two force systems acting on a given rigid body are *equipollent* if they have the same total force and the same total moment with respect to an arbitrary point.

and, substituting from Eq. (1-80), we obtain

$$dW = -\frac{\partial V}{\partial x}\, dx - \frac{\partial V}{\partial y}\, dy - \frac{\partial V}{\partial z}\, dz = -dV(x, y, z) \qquad (1\text{-}82)$$

Thus we see that dW is an exact differential. If we consider next the work W done by the force $\mathbf{F}$ as the particle moves over a certain path between points A and B, we find that

$$W = \int_A^B \mathbf{F} \cdot d\mathbf{r} = -\int_A^B dV = V_A - V_B \qquad (1\text{-}83)$$

Since the potential energy is a function of position only, we conclude that the work done on the particle depends upon the initial and final positions, but is *independent of the specific path* connecting these points. A further conclusion arises if A and B coincide, namely, that the work done in moving around any closed path is zero. Thus we have

$$\oint \mathbf{F} \cdot d\mathbf{r} = 0 \qquad (1\text{-}84)$$

for any conservative force $\mathbf{F}$.

Work and Kinetic Energy. Suppose we define the *kinetic energy* T of a particle of mass m by

$$T = \tfrac{1}{2}mv^2 \qquad (1\text{-}85)$$

where v is the velocity of the particle relative to an inertial reference frame. Let us consider the line integral of Eq. (1-83) which gives the work done on the particle by the total force $\mathbf{F}$ as the particle moves over a certain path from A to B. In accordance with Newton's law of motion, we can replace $\mathbf{F}$ by $m\ddot{\mathbf{r}}$ and obtain

$$W = m \int_A^B \ddot{\mathbf{r}} \cdot d\mathbf{r} = \frac{1}{2} m \int_A^B \frac{d}{dt}(\dot{\mathbf{r}} \cdot \dot{\mathbf{r}})\, dt = \frac{1}{2} m \int_A^B d(v^2)$$

where each integral is evaluated over the same path. Then, using the definition of kinetic energy, we obtain

$$W = \tfrac{1}{2}m(v_B^2 - v_A^2) = T_B - T_A \qquad (1\text{-}86)$$

Equation (1-86) is a statement of an important general principle of dynamics, namely, the *principle of work and kinetic energy: The increase in the kinetic energy of a particle as it moves from one arbitrary point to another is equal to the work done by the forces acting on the particle during the given interval.*

Note that the force $\mathbf{F}$ may arise from any source; it need not be conservative. Furthermore, force components which remain normal to the particle velocity $\mathbf{v}$ do no work and can be neglected in applying the principle.

Conservation of Energy. If the only forces acting on a given particle are *conservative*, then Eq. (1-83) applies and, with the aid of Eq. (1-86), we obtain

$$V_A - V_B = T_B - T_A$$

or

$$V_A + T_A = V_B + T_B = E \qquad (1\text{-}87)$$

Since the points A and B are arbitrary, we conclude that the total mechanical energy E remains constant during the motion of the particle. This is the *principle of conservation of energy.*

Now let us consider the more general case of a *system* of N particles whose configuration is specified by the Cartesian coordinates $x_1, x_2, \ldots, x_{3N}$. If the only forces which do work on the system during its motion are given by

$$F_j = -\frac{\partial V}{\partial x_j} \qquad (1\text{-}88)$$

where the potential energy $V(x_1, x_2, \ldots, x_{3N})$ is a single-valued function of position only, then the total energy E is again conserved.

Frequently it is convenient to specify the configuration of a system of particles by using generalized coordinates. Suppose, for example, that the x's and q's are related by

$$x_j = x_j(q_1, q_2, \ldots, q_n) \qquad (j = 1, 2, \ldots, 3N) \qquad (1\text{-}89)$$

Then we can use Eqs. (1-73) and (1-88) to obtain an expression for the generalized force Q_i associated with the conservative force field.

$$Q_i = -\sum_{j=1}^{3N} \frac{\partial V}{\partial x_j}\frac{\partial x_j}{\partial q_i} = -\frac{\partial V}{\partial q_i} \qquad (1\text{-}90)$$

where the potential energy V is now expressed as a function of the q's. Each Q_i may be considered to be a component of a *generalized force vector* $\mathbf{Q}$ in an n-dimensional configuration space. If no other generalized forces do work on the system, then, as in Eq. (1-83), we can write

$$W = \int_A^B \mathbf{Q} \cdot d\mathbf{q} = -\int_A^B dV = V_A - V_B$$

where the points A and B are now considered as end-points of the path in q-space. Once again W is independent of the path between the given end-points, and the total energy is conserved.

Now let us consider briefly the case where Eq. (1-89) applies, but the potential energy function V is an explicit function of *time* as well as the q's. Typical situations in which V can assume this form include (1) a system with moving constraints, (2) a system in which a parameter such as a certain stiffness is an explicit function of time, and (3) a system with a time-varying field such as an electric field. In this case the work done by the force $\mathbf{Q}$ in an infinitesimal displacement $d\mathbf{q}$ is no longer equal to an exact differential, but differs by a term $(\partial V/\partial t)\, dt$. Hence the work done on the system in going from configuration A to configuration B in an actual motion depends upon the path as well as the time, and the force $\mathbf{Q}$ is *not conservative*.

Equilibrium and Stability. Consider a system of N particles whose applied forces are conservative and are obtained from a potential energy function of the form $V(x_1, x_2, \ldots, x_{3N})$. From Eqs. (1-38) and (1-88) we see that the virtual work of these applied forces is

$$\delta W = -\sum_{j=1}^{3N} \frac{\partial V}{\partial x_j} \delta x_j = -\delta V$$

which we note is linear in the δx's and is, therefore, the first variation of the potential energy. Then, using the principle of virtual work, we find that the necessary and sufficient condition for the static equilibrium of this system is that

$$\delta V = 0 \tag{1-91}$$

for every virtual displacement consistent with the constraints.

If the potential energy is expressed in terms of the generalized coordinates $q_1, q_2, \ldots, q_n$, then

$$\delta V = \sum_{i=1}^{n} \frac{\partial V}{\partial q_i} \delta q_i \tag{1-92}$$

For a *holonomic* system having *independent* q's, the condition that $\delta V = 0$ for an arbitrary virtual displacement requires that the coefficients be zero at the equilibrium configuration; that is,

$$\frac{\partial V}{\partial q_i} = 0 \qquad (i = 1, 2, \ldots, n) \tag{1-93}$$

But these conditions imply that the potential energy is at a stationary value. Therefore, we conclude that an equilibrium configuration of a conservative holonomic system with workless fixed constraints must occur at a position where the potential energy has a stationary value.

Next let us consider the question of the *stability* of this system at a position of static equilibrium. If we expand the potential energy in a Taylor series about a reference value V_0, we obtain

$$V = V_0 + \left(\frac{\partial V}{\partial q_1}\right)_0 \delta q_1 + \left(\frac{\partial V}{\partial q_2}\right)_0 \delta q_2 + \cdots + \frac{1}{2}\left(\frac{\partial^2 V}{\partial q_1^2}\right)_0 (\delta q_1)^2 + \cdots$$
$$+ \left(\frac{\partial^2 V}{\partial q_1 \partial q_2}\right)_0 \delta q_1\, \delta q_2 + \cdots \tag{1-94}$$

where a zero subscript on a function implies that it is to be evaluated at the reference values of the q's. The δq's represent infinitesimal changes from this reference configuration.

Now assume that we choose an equilibrium configuration as the reference position. From Eq. (1-93) we see that all the coefficients $(\partial V/\partial q_i)_0$ are zero. Therefore, the potential energy expression contains no terms of first order in the δq's. Assuming small δq's, we can write

$$\Delta V = V - V_0 = \frac{1}{2}\left(\frac{\partial^2 V}{\partial q_1^2}\right)_0 (\delta q_1)^2 + \left(\frac{\partial^2 V}{\partial q_1 \, \partial q_2}\right)_0 \partial q_1 \, \partial q_2$$
$$+ \frac{1}{2}\left(\frac{\partial^2 V}{\partial q_2^2}\right)_0 (\partial q_2)^2 + \cdots \qquad (1\text{-}95)$$

where ΔV is the change in the potential energy from its value at equilibrium. Here we use ΔV rather than δV to indicate that terms of higher order than δq are included.

If $\Delta V > 0$ for every possible virtual displacement having at least one of the δq's nonzero, then the reference position is one of minimum potential energy corresponding to *stable* equilibrium. On the other hand, if a virtual displacement can be found such that $\Delta V < 0$, the equilibrium position is *unstable*. A third possibility is that $\Delta V \geq 0$ for all possible virtual displacements, but $\Delta V = 0$ for some virtual displacements in which the δq's are not all zero. This is the case of *neutral stability*.[†]

If we consider only the *quadratic terms* in the δq's, the static stability of the system can be determined in certain cases. For example, if ΔV is a positive definite quadratic form,[‡] the system is stable. If ΔV is negative definite, negative semidefinite, or indefinite, the system is unstable. The remaining possibilities, namely, that the quadratic form is positive semidefinite or identically zero, do not determine the stability; higher-order terms must also be considered.

From a more practical viewpont, suppose we consider the system to be initially in equilibrium but subject to small random disturbances. In this case, a stable system will remain near the reference position; a system with neutral stability will drift slowly in such a manner that V remains essentially zero; and an unstable system will move away from the reference position with an increasing velocity.

Kinetic Energy of a System. Consider a system of N particles, and let $\mathbf{r}_i$ be the position vector of the ith particle relative to a point O fixed in an inertial frame (Fig. 1-10). With respect to this inertial frame, which we shall consider to be fixed, the total kinetic energy of the system is the sum of the individual kinetic energies of the particles, namely,

$$T = \frac{1}{2}\sum_{i=1}^{N} m_i \, \dot{\mathbf{r}}_i^2 \qquad (1\text{-}96)$$

[†]Sometimes neutral stability is considered as a form of instability. See, for example, the discussion of gyroscopic stability in Sec. 3-3.

[‡]See Eq. (2-11) and the discussion following Eq. (2-239).

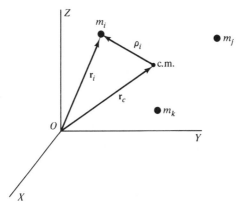

Fig. 1-10. Position vectors for a system of particles.

where we use the notation that

$$\dot{\mathbf{r}}_i^2 = \dot{\mathbf{r}}_i \cdot \dot{\mathbf{r}}_i = v_i^2$$

From Fig. 1-10, we see that

$$\mathbf{r}_i = \mathbf{r}_c + \boldsymbol{\rho}_i \qquad (1\text{-}97)$$

Hence we obtain

$$T = \tfrac{1}{2} \sum_{i=1}^{N} m_i(\dot{\mathbf{r}}_c^2 + 2\dot{\mathbf{r}}_c \cdot \dot{\boldsymbol{\rho}}_i + \dot{\boldsymbol{\rho}}_i^2) \qquad (1\text{-}98)$$

But

$$\sum_{i=1}^{N} m_i \boldsymbol{\rho}_i = 0 \qquad (1\text{-}99)$$

since $\boldsymbol{\rho}_i$ is measured from the center of mass. Also, $\dot{\mathbf{r}}_c$ does not enter into the summation and can be factored out. Therefore, using the notation that m is the total mass of the system, we see that Eq. (1-98) reduces to

$$T = \tfrac{1}{2} m \dot{\mathbf{r}}_c^2 + \tfrac{1}{2} \sum_{i=1}^{N} m_i \dot{\boldsymbol{\rho}}_i^2 \qquad (1\text{-}100)$$

The last term of Eq. (1-100) can be considered as the kinetic energy of the system relative to its mass center; that is, it is the kinetic energy of the system as viewed by an observer translating with the center of mass but not rotating. Now we can state *König's theorem: The total kinetic energy of a system is equal to the sum of (1) the kinetic energy due to a particle having a mass equal to the total mass of the system and moving with the velocity of the center of mass and (2) the kinetic energy due to the motion of the system relative to its center of mass.*

Eq. (1-100) was derived for a system of particles, but it can be adapted to the case of a rigid body in general motion. Suppose we consider a small volume element dV having a density ρ (Fig. 1-11). Each element of the body

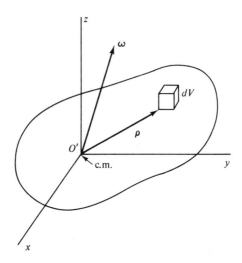

Fig. 1-11. A typical volume element in a rotating rigid body.

will, in general, be translating and rotating, the only possible exception being that an instantaneous axis of rotation might exist in the body, and the elements along this line might then have no translational velocity at the given instant. Nevertheless, the dimensions of a typical volume element can be chosen to be so small that its rotational kinetic energy is negligible compared with its translational kinetic energy. Hence, in the limit, each element of the rigid body can be considered as a *particle* of infinitesimal mass and Eq. (1-100) can be modified as follows:

$$T = \tfrac{1}{2}m\dot{\mathbf{r}}_c^2 + \tfrac{1}{2}\int_V \rho\dot{\boldsymbol{\rho}}^2\, dV \qquad (1\text{-}101)$$

where $\boldsymbol{\rho}$ is the position vector of the volume element relative to the *mass center*. The first term on the right side is called the *translational* kinetic energy of the rigid body; the second is the *rotational* kinetic energy.

Let us consider the rotational kinetic energy in more detail. If we take the reference point O' at the center of mass and assume that the body is rotating with an angular velocity $\boldsymbol{\omega}$, we see that

$$\dot{\boldsymbol{\rho}} = \boldsymbol{\omega} \times \boldsymbol{\rho} \qquad (1\text{-}102)$$

and therefore

$$\dot{\boldsymbol{\rho}}^2 = \dot{\boldsymbol{\rho}}\cdot\boldsymbol{\omega} \times \boldsymbol{\rho} = \boldsymbol{\omega}\cdot\boldsymbol{\rho} \times \dot{\boldsymbol{\rho}} \qquad (1\text{-}103)$$

Hence the rotational kinetic energy of the rigid body can be written in the form

$$T_{\text{rot}} = \tfrac{1}{2}\boldsymbol{\omega}\cdot\int_V \rho\boldsymbol{\rho} \times (\boldsymbol{\omega} \times \boldsymbol{\rho})\, dV \qquad (1\text{-}104)$$

If we substitute an equivalent expression for the triple vector product, we obtain

$$T_{rot} = \tfrac{1}{2}\boldsymbol{\omega}\cdot\int_V \rho[\boldsymbol{\rho}^2\boldsymbol{\omega} - (\boldsymbol{\rho}\cdot\boldsymbol{\omega})\boldsymbol{\rho}]\,dV \tag{1-105}$$

Now let us take a Cartesian coordinate system with its origin at the center of mass and assume that it rotates with the body. Then, in terms of the unit vectors **i**, **j**, **k**, we have

$$\boldsymbol{\rho} = x\mathbf{i} + y\mathbf{j} + z\mathbf{k} \tag{1-106}$$

and

$$\boldsymbol{\omega} = \omega_x\mathbf{i} + \omega_y\mathbf{j} + \omega_z\mathbf{k} \tag{1-107}$$

Substituting from Eqs. (1-106) and (1-107) into (1-105), we obtain

$$T_{rot} = \tfrac{1}{2}\boldsymbol{\omega}\cdot\int_V \rho\{[(y^2 + z^2)\omega_x - (xy\omega_y + xz\omega_z)]\mathbf{i} + \cdots\}\,dV \tag{1-108}$$

which reduces to

$$\begin{aligned}T_{rot} = &\tfrac{1}{2}I_{xx}\omega_x^2 + \tfrac{1}{2}I_{yy}\omega_y^2 + \tfrac{1}{2}I_{zz}\omega_z^2 \\ &+ I_{xy}\omega_x\omega_y + I_{xz}\omega_x\omega_z + I_{yz}\omega_y\omega_z\end{aligned} \tag{1-109}$$

or

$$T_{rot} = \tfrac{1}{2}\sum_i\sum_j I_{ij}\omega_i\omega_j \tag{1-110}$$

where the *moments of inertia* are

$$\begin{aligned}I_{xx} &= \int_V \rho(y^2 + z^2)\,dV \\ I_{yy} &= \int_V \rho(x^2 + z^2)\,dV \\ I_{zz} &= \int_V \rho(x^2 + y^2)\,dV\end{aligned} \tag{1-111}$$

and the *products of inertia* are

$$\begin{aligned}I_{xy} &= I_{yx} = -\int_V \rho xy\,dV \\ I_{xz} &= I_{zx} = -\int_V \rho xz\,dV \\ I_{yz} &= I_{zy} = -\int_V \rho yz\,dV\end{aligned} \tag{1-112}$$

Using matrix notation, the rotational kinetic energy can be written in the form

$$T_{rot} = \tfrac{1}{2}\boldsymbol{\omega}^T\mathbf{I}\boldsymbol{\omega} \tag{1-113}$$

or, writing the matrices in detail,

$$T_{rot} = \tfrac{1}{2}\begin{Bmatrix}\omega_x \\ \omega_y \\ \omega_z\end{Bmatrix}^T\begin{bmatrix}I_{xx} & I_{xy} & I_{xz} \\ I_{yx} & I_{yy} & I_{yz} \\ I_{zx} & I_{zy} & I_{zz}\end{bmatrix}\begin{Bmatrix}\omega_x \\ \omega_y \\ \omega_z\end{Bmatrix} \tag{1-114}$$

Another useful form is obtained by noting that if $\boldsymbol{\omega}$ has the same direction as one of the coordinate axes at a given moment, then the expression for T_{rot} reduces to a single term, namely,

$$T_{\text{rot}} = \tfrac{1}{2}I\omega^2 \qquad (1\text{-}115)$$

where I is the moment of inertia about an axis which is in the direction of $\boldsymbol{\omega}$ and passes through the mass center.

We have obtained expressions for the kinetic energy of a system in terms of its motion relative to a fixed point, and also in terms of its motion relative

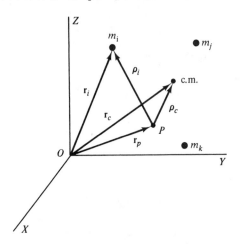

Fig. 1-12. Position vectors for a system of particles, using an arbitrary reference point.

to its center of mass. Now let us find an expression for the kinetic energy in terms of the motion with respect to an arbitrary reference point P (Fig. 1-12). In this case,

$$\mathbf{r}_i = \mathbf{r}_p + \boldsymbol{\rho}_i \qquad (1\text{-}116)$$

Then, substituting from Eq. (1-116) into the basic kinetic energy expression of Eq. (1-96), we obtain

$$T = \tfrac{1}{2} \sum_{i=1}^{N} m_i (\dot{\mathbf{r}}_p + \dot{\boldsymbol{\rho}}_i) \cdot (\dot{\mathbf{r}}_p + \dot{\boldsymbol{\rho}}_i)$$

$$= \tfrac{1}{2} m \dot{\mathbf{r}}_p^2 + \tfrac{1}{2} \sum_{i=1}^{N} m_i \dot{\boldsymbol{\rho}}_i^2 + \dot{\mathbf{r}}_p \cdot \sum_{i=1}^{N} m_i \dot{\boldsymbol{\rho}}_i \qquad (1\text{-}117)$$

But the center of mass location $\boldsymbol{\rho}_c$ relative to P is

$$\boldsymbol{\rho}_c = \frac{1}{m} \sum_{i=1}^{N} m_i \boldsymbol{\rho}_i \qquad (1\text{-}118)$$

Hence,

$$T = \tfrac{1}{2}m\dot{r}_p^2 + \tfrac{1}{2}\sum_{i=1}^{N} m_i\dot{\rho}_i^2 + \dot{r}_p \cdot m\dot{\rho}_c \tag{1-119}$$

Thus we find that the total kinetic energy is the sum of three parts: (*1*) *the kinetic energy due to a particle having a mass m and moving with the reference point P*, (2) *the kinetic energy of the system due to its motion relative to P, and* (3) *the scalar product of the velocity of the reference point and the linear momentum of the system relative to the reference point*. It can be seen that Eq. (1-119) reduces to Eq. (1-100) for the particular case where the reference point P is taken at the center of mass and $\rho_c = 0$.

Now suppose that we consider the kinetic energy of a rigid body in terms of its motion relative to an arbitrary reference point P. The kinetic energy due to its motion relative to P is

$$T_{\text{rel}} = \tfrac{1}{2}m\dot{\rho}_c^2 + \tfrac{1}{2}\sum_i \sum_j I_{ij}\omega_i\omega_j \tag{1-120}$$

where the moments and products of inertia are taken with respect to the mass center. Then the total kinetic energy is

$$T = \tfrac{1}{2}m\dot{r}_p^2 + \tfrac{1}{2}m\dot{\rho}_c^2 + \tfrac{1}{2}\sum_i \sum_j I_{ij}\omega_i\omega_j + \dot{r}_p \cdot m\dot{\rho}_c \tag{1-121}$$

If it turns out that the motion of the rigid body relative to P is a pure rotation about P, that is, P is fixed in the body, then the kinetic energy of relative motion can be written in the form

$$T_{\text{rel}} = \tfrac{1}{2}\sum_i \sum_j I_{ij}\omega_i\omega_j \tag{1-122}$$

where the moments and products of inertia are now taken with respect to P.

Angular Momentum. Let us consider once again a system of N particles, as shown in Fig. 1-10. The total angular momentum **H** with respect to a fixed point O is

$$\mathbf{H} = \sum_{i=1}^{N} \mathbf{r}_i \times m_i\dot{\mathbf{r}}_i \tag{1-123}$$

that is, it is the sum of the moments about O of the individual linear momenta of the particles, assuming that each vector $m_i\dot{\mathbf{r}}_i$ has a line of action passing through the corresponding particle.

Now let us substitute the expression for $\mathbf{r}_i$ from Eq. (1-97) into Eq. (1-123), obtaining

$$
\begin{aligned}
\mathbf{H} &= \sum_{i=1}^{N} (\mathbf{r}_c + \boldsymbol{\rho}_i) \times m_i(\dot{\mathbf{r}}_c + \dot{\boldsymbol{\rho}}_i) \\
&= \mathbf{r}_c \times m\dot{\mathbf{r}}_c + \mathbf{r}_c \times \sum_{i=1}^{N} m_i\dot{\boldsymbol{\rho}}_i + \sum_{i=1}^{N} m_i\boldsymbol{\rho}_i \times \dot{\mathbf{r}}_c + \sum_{i=1}^{N} \boldsymbol{\rho}_i \times m_i\dot{\boldsymbol{\rho}}_i
\end{aligned} \tag{1-124}
$$

Using Eq. (1-99), the two middle terms are zero and the result can be simplified to

$$\mathbf{H} = \mathbf{r}_c \times m\dot{\mathbf{r}}_c + \sum_{i=1}^{N} \boldsymbol{\rho}_i \times m_i \dot{\boldsymbol{\rho}}_i \qquad (1\text{-}125)$$

where the last term on the right is the angular momentum $\mathbf{H}_c$ about the center of mass, that is,

$$\mathbf{H}_c = \sum_{i=1}^{N} \boldsymbol{\rho}_i \times m_i \dot{\boldsymbol{\rho}}_i \qquad (1\text{-}126)$$

To summarize, *the angular momentum of a system of particles of total mass m about a fixed point O is equal to the angular momentum about O of a single particle of mass m which is moving with the center of mass plus the angular momentum of the system about the center of mass.*

If we apply this result to the case of a rigid body (Fig. 1-11) in arbitrary motion, we find that the total angular momentum with respect to a fixed point O is

$$\mathbf{H} = \mathbf{r}_c \times m\dot{\mathbf{r}}_c + \mathbf{H}_c \qquad (1\text{-}127)$$

where

$$\mathbf{H}_c = \int_V \rho \boldsymbol{\rho} \times (\boldsymbol{\omega} \times \boldsymbol{\rho})\, dV \qquad (1\text{-}128)$$

In terms of the moments and products of inertia about the center of mass, we can obtain the body-axis components of $\mathbf{H}_c$ from the matrix equation

$$\mathbf{H}_c = \mathbf{I}\boldsymbol{\omega} \qquad (1\text{-}129)$$

Also, comparing Eqs. (1-104) and (1-128), we see that the rotational kinetic energy can be written in the form

$$T_{\text{rot}} = \tfrac{1}{2}\boldsymbol{\omega} \cdot \mathbf{H}_c \qquad (1\text{-}130)$$

in agreement with Eq. (1-113).

We have obtained expressions for the angular momentum of a system of particles with respect to a fixed point and with respect to the center of mass. Now let us consider the angular momentum with respect to an arbitrary point P (Fig. 1-12). More explicitly, let us obtain the angular momentum about the point P, as viewed by a nonrotating observer moving with that point. We can define $\mathbf{H}_p$ as follows:

$$\mathbf{H}_p = \sum_{i=1}^{N} \boldsymbol{\rho}_i \times m_i \dot{\boldsymbol{\rho}}_i \qquad (1\text{-}131)$$

where we notice that $\boldsymbol{\rho}_i$ is now the position of the ith particle with respect to the reference point P. If we substitute

$$\boldsymbol{\rho}_i = \mathbf{r}_i - \mathbf{r}_c + \boldsymbol{\rho}_c \qquad (1\text{-}132)$$

and note that the center of mass location is

$$\mathbf{r}_c = \frac{1}{m} \sum_{i=1}^{N} m_i \mathbf{r}_i \qquad (1\text{-}133)$$

we obtain, after some algebraic simplification,

$$\mathbf{H}_p = \sum_{i=1}^{N} \mathbf{r}_i \times m\dot{\mathbf{r}}_i - \mathbf{r}_c \times m\dot{\mathbf{r}}_c + \boldsymbol{\rho}_c \times m\dot{\boldsymbol{\rho}}_c \tag{1-134}$$

or, using Eq. (1-123),

$$\mathbf{H}_p = \mathbf{H} - \mathbf{r}_c \times m\dot{\mathbf{r}}_c + \boldsymbol{\rho}_c \times m\dot{\boldsymbol{\rho}}_c \tag{1-135}$$

This result illustrates the general procedure for changing the reference point in an angular momentum expression, where neither point is at the center of mass. Starting with $\mathbf{H}$ taken about O, the term $-\mathbf{r}_c \times m\dot{\mathbf{r}}_c$ shifts the reference point to the center of mass. Then the addition of $\boldsymbol{\rho}_c \times m\dot{\boldsymbol{\rho}}_c$ shifts the reference point away from the center of mass to the point P. Note that even though P is not fixed in any inertial frame, the general form of the angular momentum equation is the same as would occur for a fixed reference point.

Generalized Momentum. Consider a system whose configuration is described by n generalized coordinates. Let us define the *Lagrangian function* $L(q, \dot{q}, t)$ as follows:

$$L = T - V \tag{1-136}$$

The *generalized momentum* p_i associated with the generalized coordinate q_i is defined by the equation

$$p_i = \frac{\partial L}{\partial \dot{q}_i} \tag{1-137}$$

It is, in general, a function of the q's, $\dot{q}$'s, and t. Note, however, that the Lagrangian function is, at most, quadratic in the $\dot{q}$'s. Therefore p_i is a *linear* function of the $\dot{q}$'s.

In the usual case the potential energy V is not velocity-dependent; hence $\partial V/\partial \dot{q}_i$ equals zero and

$$p_i = \frac{\partial T}{\partial \dot{q}_i} \tag{1-138}$$

if the potential energy is of the form $V(q, t)$.

As an example, consider a free particle of mass m whose position is given by the Cartesian coordinates (x, y, z). The kinetic energy is

$$T = \frac{m}{2}(\dot{x}^2 + \dot{y}^2 + \dot{z}^2)$$

and, using Eq. (1-138), we obtain

$$p_x = m\dot{x}$$

Thus, p_x is just the x component of the linear momentum.

In a similar fashion, if the position of the particle is given by the spherical coordinates (r, θ, ϕ), as in Fig. 1-7, the kinetic energy is

$$T = \frac{m}{2}(\dot{r}^2 + r^2\dot{\theta}^2 + r^2\dot{\phi}^2 \sin^2 \theta)$$

Using Eq. (1-138), we have

$$p_r = m\dot{r}$$
$$p_\theta = mr^2\dot{\theta}$$
$$p_\phi = mr^2\dot{\phi}\sin^2\theta$$

We see that p_r is the linear momentum component in the radial direction, while p_θ is the horizontal component of the *angular momentum*, that is, it is the angular momentum about the horizontal axis associated with an angular velocity $\dot{\theta}$. Similarly, since $\dot{\phi}$ is vertical, p_ϕ is the vertical component of angular momentum.

In general, p_i has the dimensions of energy divided by $\dot{q}_i$. The examples illustrate that whenever q_i denotes a distance from a certain reference point, the corresponding p_i is a linear momentum. If q_i is an angle, then p_i is an angular momentum. For more general q's, the physical significance is more obscure, but we know that p_i is linear in the $\dot{q}$'s and is therefore proportional to a weighted sum of the velocities at various points in the system.

Example 1-5. Consider once again the system of particles and rods shown in Fig. 1-8. Find expressions for the kinetic energy and the generalized momenta.

The total kinetic energy can be obtained easily in terms of $\dot{x}$'s, namely,

$$T = \frac{m}{2}(\dot{x}_1^2 + \dot{x}_2^2 + \dot{x}_3^2) \tag{1-139}$$

In order to express T as a function of the $\dot{q}$'s, we first differentiate Eq. (1-75) and obtain

$$\dot{x}_1 = \dot{q}_1 + \dot{q}_2 + \tfrac{1}{2}\dot{q}_3$$
$$\dot{x}_2 = \dot{q}_1 - \dot{q}_3 \tag{1-140}$$
$$\dot{x}_3 = \dot{q}_1 - \dot{q}_2 + \tfrac{1}{2}\dot{q}_3$$

Substituting from Eq. (1-140) into Eq. (1-139) and simplifying, we have

$$T = \frac{m}{2}\left(3\dot{q}_1^2 + 2\dot{q}_2^2 + \frac{3}{2}\dot{q}_3^2\right) \tag{1-141}$$

The generalized momenta are obtained with the aid of Eq. (1-138), the result being

$$p_1 = \frac{\partial T}{\partial \dot{q}_1} = 3m\dot{q}_1$$
$$p_2 = \frac{\partial T}{\partial \dot{q}_2} = 2m\dot{q}_2 \tag{1-142}$$
$$p_3 = \frac{\partial T}{\partial \dot{q}_3} = \frac{3}{2}m\dot{q}_3$$

In this example the kinetic energy is a quadratic function of a particularly simple form, namely, the sum of squares of $\dot{q}$'s. This results in each generalized momentum p_i being a linear function of only its corresponding generalized velocity $\dot{q}_i$; no other $\dot{q}$'s are present. The physical meaning of this mathematical simplicity is a complete lack of inertial coupling between the motions corresponding to the individual $\dot{q}$'s. In other words, the inertial forces arising from the motion of q_i do not tend to drive the system in a different coordinate q_j.

The coefficients of the $\dot{q}$'s in the expressions for the generalized momenta are known as *inertia coefficients*. By studying the deflection forms given in Fig. 1-9, it can be seen that the inertia coefficient corresponding to a certain generalized coordinate q_i is equal to the particle mass times the square of its displacement per unit q_i, summed over all particles. For example, the inertia coefficient corresponding to q_3 is

$$m\left[\left(\frac{1}{2}\right)^2 + (-1)^2 + \left(\frac{1}{2}\right)^2\right] = \frac{3}{2}m$$

Inertia coefficients will be discussed further in the next chapter.

REFERENCES

1. LANCZOS, C., *The Variational Principles of Mechanics*. Toronto: University of Toronto Press, 1949. This gives a good presentation of the basic concepts and principal viewpoints of analytical mechanics.

2. SYNGE, J. L. and B. A. GRIFFITH, *Principles of Mechanics*, 3rd ed. New York: McGraw-Hill Book Company, 1959. An excellent introduction is given to the basic concepts of dynamics, as well as the more advanced topics of analytical mechanics.

3. WHITTAKER, E. T., *A Treatise on the Analytical Dynamics of Particles and Rigid Bodies*, 4th ed. New York: Dover Publications, Inc., 1944. This classical text gives an excellent treatment of the ideas of virtual displacements, generalized coordinates, and constraints.

4. MEIROVITCH, L., *Methods of Analytical Dynamics*. New York: McGraw-Hill Book Company, 1970. A good treatment is given of the material in this chapter. See, in particular, the sections on virtual work and d'Alembert's principle.

PROBLEMS

1-1. A simple pendulum consists of a particle of mass m and a massless string of length l that is attached at a point O which moves along the horizontal x axis with a displacement $x_0 = A \sin \omega t$. Assume that the positive y axis points upward and the coordinates (x, y) represent the location of the particle. If we consider the angle θ of the pendulum (measured counterclockwise from the direction of the negative y axis) as a generalized coordinate, write the transformation equations of the form of Eq. (1-8). What is the constraint equation relating (x, y, t)?

1-2. A rigid rod of length l undergoes small motion in which the coordinates (x_1, x_2) represent the vertical displacements of the ends. The configuration is also given by the generalized coordinates (z, θ), where z is the vertical displacement of the center and θ is the rotation angle. What are the transformation equations? For the given applied forces at the ends, evaluate the generalized forces Q_z and Q_θ.

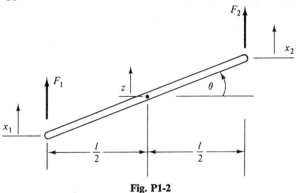

Fig. P1-2

1-3. Particles 1 and 2 are connected by a rigid rod of length l. The configuration of the system can be given by the Cartesian coordinates (x_1, y_1, x_2, y_2) or by the

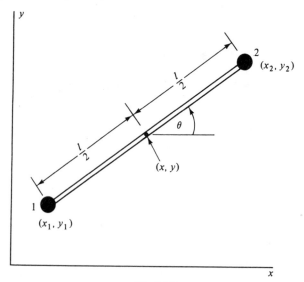

Fig. P1-3

generalized coordinates (x, y, θ). Write the transformation equations giving the Cartesian coordinates in terms of the generalized coordinates. Next define a fourth generalized coordinate $q_4 \equiv l$ and evaluate the Jacobian $\partial(x_1, y_1, x_2, y_2)/\partial(x, y, \theta, q_4)$. Solve for the generalized coordinates in terms of the Cartesian coordinates.

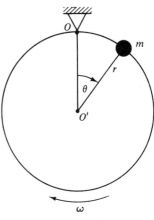

1-4. A particle can slide on a rigid wire which is bent in the form of a circle of radius r. This wire rotates about point O on its circumference with an angular velocity ω. Assuming that the position of the particle is specified by an angle θ measured from the line OO', find the kinetic energy and the generalized momentum p_θ.

Fig. P1-4

1-5. A thin uniform rod of mass m and length l is constrained to move in the xy plane with end A remaining on the x axis. Using (x, θ) as generalized coordinates, find expressions for the kinetic energy and the generalized momentum p_θ.

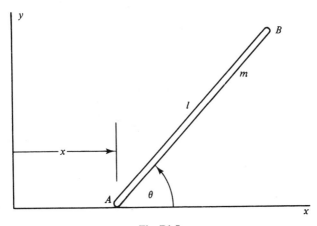

Fig. P1-5

1-6. A disk of radius r and mass m can roll without slipping on a thin rod which rotates about a fixed point O at a constant rate ω. Obtain an expression of the form $T(q, \dot{q})$ for the total kinetic energy of the disk. (See p. 44).

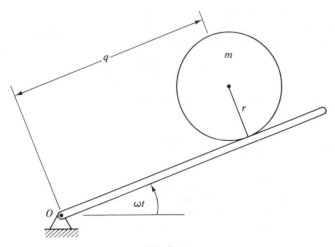

Fig. P1-6

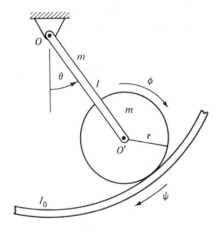

Fig. P1-7

1-7. The given system has a thin rigid bar of length l and mass m which can rotate about O. A uniform disk of radius r and mass m is attached to the bar by a pivot at O' and rolls without slipping on the inside surface of a rotating cylinder having a moment of inertia I_0 about its central axis at O. Find a constraint equation which expresses $\dot{\phi}$ as a function of $\dot{\theta}$ and $\dot{\psi}$, where these angular rates are absolute. Obtain an expression for the total kinetic energy in terms of $\dot{\theta}$ and $\dot{\psi}$, and solve for the generalized momenta p_θ and p_ψ.

1-8. Show that if the angular momentum of a system of N particles is defined by using the *absolute* velocity $\dot{\mathbf{r}}_i$ of each particle rather than the *relative* velocity $\dot{\boldsymbol{\rho}}_i$ of Eq. (1-126); that is, if the angular momentum about the center of mass is defined by the equation

$$\mathbf{H}_c = \sum_{i=1}^{N} \boldsymbol{\rho}_i \times m_i \dot{\mathbf{r}}_i$$

the value of $\mathbf{H}_c$ is not changed.

1-9. A particle of mass m can slide without friction on a fixed circular wire of radius r which lies in a vertical plane. Using d'Alembert's principle and the equation of constraint, show that $y\ddot{x} - x\ddot{y} - gx = 0$.

1-10. Particle A of mass $2m$ and particle B of mass m are connected by a massless rod of length l. Particle A is constrained to move along the horizontal x axis while particle B can move only along the vertical y axis. What is the equation of constraint relating x and y? Use d'Alembert's principle to obtain the equation of motion $2y\ddot{x} - x\ddot{y} - gx = 0$.

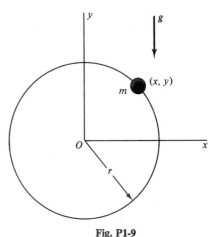

Fig. P1-9

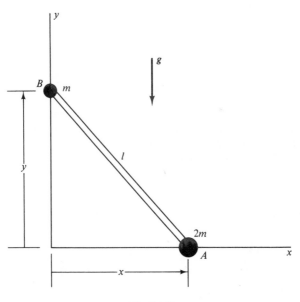

Fig. P1-10

1-11. A flywheel having a moment of inertia I can rotate freely in the horizontal plane about a vertical axis through O. A particle of mass m is attached to O by a spring of stiffness k and unstressed length x_0. The particle slides without friction in a radial groove in the flywheel. Use d'Alembert's principle to obtain the differential equations of motion in terms of the coordinates (x, θ). Show that one of these equations implies that the total angular momentum is conserved.

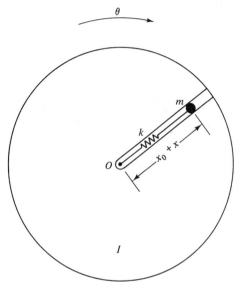

Fig. P1-11

1-12. Two particles having masses m and $2m$ are connected by a massless rod to form a dumbbell. It can slide without friction in a circular bowl of radius r. Consider a virtual displacement $\delta\theta$ and use the principle of virtual work to obtain the value of θ at the position of static equilibrium.

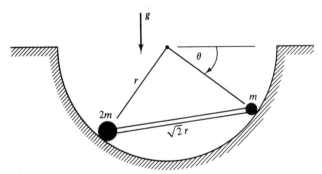

Fig. P1-12

1-13. Two thin rods, each of mass m and length l, are pinned together at their upper ends. A particle of mass m is suspended by massless strings connected to the midpoints of the rods, as shown. Assume planar motion and use the method of virtual work to find the position of static equilibrium in the interval $0 < \theta < \pi/6$. Is it stable?

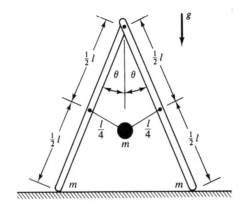

Fig. P1-13

2

LAGRANGE'S EQUATIONS

There are two general approaches to the subject of classical dynamics. These approaches are usually given the names *vectorial dynamics* and *analytical dynamics*. Vectorial dynamics is based on a direct application of Newton's law of motion. It concentrates on the forces and motions associated with the individual parts of the system, and on the interactions among these parts. On the other hand, analytical dynamics is more concerned with the system as a whole, and uses descriptive *scalar* functions such as the kinetic and potential energies. By performing certain operations on these functions, it is often possible to obtain a complete set of equations of motion without solving explicitly for the constraint forces acting on the various parts of the system. Furthermore, new insights are obtained into the variational principles which are so important to a thorough understanding of dynamics.

With the derivation and application of Lagrange's equations, we begin an emphasis on the analytical approach to dynamics. This approach will be shown to be very useful in attacking the more complex and difficult problems of classical dynamics in a systematic way.

2-1. DERIVATION OF LAGRANGE'S EQUATIONS

Kinetic Energy. Let us consider a system of N particles whose positions relative to an inertial frame are given by the Cartesian coordinates $x_1, x_2, \ldots,$ x_{3N}. The total kinetic energy of the system is found from Eq. (1-96) or, equivalently, from

$$T = \tfrac{1}{2} \sum_{k=1}^{3N} m_k \dot{x}_k^2 \qquad (2\text{-}1)$$

where $m_1 = m_2 = m_3$ is the mass of the first particle, and (x_1, x_2, x_3) specifies its position. Similarly, $m_4 = m_5 = m_6$ is the mass of the second particle, and so forth.

Now let us express the kinetic energy in terms of the generalized coordinates $q_1, q_2, \ldots, q_n$. Using transformation equations giving the x's as functions of the q's and time, we have

$$x_k = x_k(q, t) \qquad (k = 1, \ldots, 3N) \qquad (2\text{-}2)$$

where we assume that these functions are twice differentiable with respect to the q's and t. We find that

$$\dot{x}_k(q, \dot{q}, t) = \sum_{i=1}^{n} \frac{\partial x_k}{\partial q_i} \dot{q}_i + \frac{\partial x_k}{\partial t} \tag{2-3}$$

Note that $\dot{x}_k$ is *linear* in the $\dot{q}$'s, and that $\partial x_k/\partial q_i$ and $\partial x_k/\partial t$ are functions of the q's and t.

If we substitute from Eq. (2-3) into Eq. (2-1), we obtain

$$T(q, \dot{q}, t) = \frac{1}{2} \sum_{k=1}^{3N} m_k \left(\sum_{i=1}^{n} \frac{\partial x_k}{\partial q_i} \dot{q}_i + \frac{\partial x_k}{\partial t} \right)^2 \tag{2-4}$$

Let us group the terms according to their degree in the $\dot{q}$'s, using the notation

$$T = T_2 + T_1 + T_0 \tag{2-5}$$

where T_2 is a homogeneous quadratic function of the $\dot{q}$'s, T_1 is a homogeneous linear function of the $\dot{q}$'s, and T_0 includes the remaining terms which are functions of the q's and t.

More explicitly, we find that T_2 is of the form

$$T_2 = \frac{1}{2} \sum_{i=1}^{n} \sum_{j=1}^{n} m_{ij} \dot{q}_i \dot{q}_j \tag{2-6}$$

where

$$m_{ij} = m_{ji} = \sum_{k=1}^{3N} m_k \frac{\partial x_k}{\partial q_i} \frac{\partial x_k}{\partial q_j} \tag{2-7}$$

Also,

$$T_1 = \sum_{i=1}^{n} a_i \dot{q}_i \tag{2-8}$$

where

$$a_i = \sum_{k=1}^{3N} m_k \frac{\partial x_k}{\partial q_i} \frac{\partial x_k}{\partial t} \tag{2-9}$$

Finally,

$$T_0 = \frac{1}{2} \sum_{k=1}^{3N} m_k \left(\frac{\partial x_k}{\partial t} \right)^2 \tag{2-10}$$

Note that the coefficients m_{ij} and a_i, as well as T_0, are functions of the q's and t.

Assuming that $m_k > 0$ for all k, we see from Eq. (2-1) that the total kinetic energy T is a positive definite quadratic function of the $\dot{x}$'s. In other words, T is zero only if all the $\dot{x}$'s are zero; if any of the $\dot{x}$'s is nonzero, the kinetic energy is positive.

Now suppose we express T as a function of the q's, $\dot{q}$'s, and t. It is still true for any real system that the kinetic energy is zero only if the system is motionless; otherwise it is positive. Since the q's are usually chosen such that the $\dot{q}$'s are all zero if and only if the system is motionless, T is usually a positive definite function of the $\dot{q}$'s. But this is not always the case, particularly if there are moving constraints.

Let us consider T_2 in more detail. We see from Eqs. (2-4), (2-6), and (2-7) that T_2 is the *total* kinetic energy for the case in which all the partial derivatives $\partial x_k / \partial t$ are zero, that is, for a system in which any moving constraints or moving reference frames are held fixed. Then, assuming that one or more nonzero $\dot{q}$'s implies the motion of one or more particles of the system, and vice versa, we conclude that T_2 must be a *positive definite quadratic function* of the $\dot{q}$'s.

The positive definite nature of T_2 restricts the possible values of the *inertia coefficients* m_{ij}. If we consider the symmetric $n \times n$ *generalized inertia matrix* **m**, the necessary and sufficient conditions that T_2 be positive definite are that

$$m_{11} > 0, \quad \begin{vmatrix} m_{11} & m_{12} \\ m_{21} & m_{22} \end{vmatrix} > 0, \dots, \quad \begin{vmatrix} m_{11} & m_{12} & \cdots & m_{1n} \\ m_{21} & m_{22} & & \cdot \\ \cdot & & & \cdot \\ \cdot & & & \cdot \\ \cdot & & & \cdot \\ m_{n1} & \cdot & \cdot & \cdot & m_{nn} \end{vmatrix} > 0 \qquad (2\text{-}11)$$

This is equivalent to the requirement that the determinant of the matrix and all the principal minors be positive. One of the consequences is that all the inertia coefficients m_{ii} along the main diagonal must be positive, as can be seen directly from Eq. (2-7).

From Eqs. (2-9) and (2-10) we observe that T_1 and T_0 are nonzero only for the case of *rheonomic* systems. It follows, then, that the kinetic energy T of a *scleronomic* system is a homogeneous quadratic function of the $\dot{q}$'s. In this case we see from Eq. (2-7) that the inertia coefficients m_{ij} are functions of the q's, but not of time.

Since T_1 is linear in the $\dot{q}$'s, it is apparent that it can be positive or negative. On the other hand, T_0 is positive or zero.

Lagrange's Equations. As a starting point for the derivation of Lagrange's equations, let us consider a system of N particles and write d'Alembert's principle, Eq. (1-63), in the form

$$\sum_{k=1}^{3N} (F_k - m_k \ddot{x}_k)\, \delta x_k = 0 \qquad (2\text{-}12)$$

where F_k is the *applied* force component associated with x_k, that is, it includes all the real forces acting on the given particle except the workless constraint forces.

Using Eq. (2-2), the virtual displacement δx_k can be expressed in terms of the δq's as follows:

$$\delta x_k = \sum_{i=1}^{n} \frac{\partial x_k}{\partial q_i}\, \delta q_i \qquad (2\text{-}13)$$

Hence, from Eqs. (2-12) and (2-13) we obtain

$$\sum_{k=1}^{3N} \sum_{i=1}^{n} \left(F_k \frac{\partial x_k}{\partial q_i} - m_k \ddot{x}_k \frac{\partial x_k}{\partial q_i} \right) \delta q_i = 0 \qquad (2\text{-}14)$$

Now, it is evident from Eq. (2-3) that

$$\frac{\partial \dot{x}_k}{\partial \dot{q}_i} = \frac{\partial x_k}{\partial q_i} \qquad (2\text{-}15)$$

Also, noting that the order of differentiation can be changed, we obtain

$$\frac{d}{dt} \left(\frac{\partial x_k}{\partial q_i} \right) = \sum_{j=1}^{n} \frac{\partial^2 x_k}{\partial q_j \partial q_i} \dot{q}_j + \frac{\partial^2 x_k}{\partial t\, \partial q_i} = \frac{\partial \dot{x}_k}{\partial q_i} \qquad (2\text{-}16)$$

Then we can write the generalized momentum p_i in the form

$$p_i = \frac{\partial T}{\partial \dot{q}_i} = \sum_{k=1}^{3N} m_k \dot{x}_k \frac{\partial \dot{x}_k}{\partial \dot{q}_i} \qquad (2\text{-}17)$$

and, using Eqs. (2-15) and (2-16), we obtain

$$\frac{d}{dt} \left(\frac{\partial T}{\partial \dot{q}_i} \right) = \sum_{k=1}^{3N} m_k \ddot{x}_k \frac{\partial x_k}{\partial q_i} + \sum_{k=1}^{3N} m_k \dot{x}_k \frac{\partial \dot{x}_k}{\partial q_i} \qquad (2\text{-}18)$$

But

$$\frac{\partial T}{\partial q_i} = \sum_{k=1}^{3N} m_k \dot{x}_k \frac{\partial \dot{x}_k}{\partial q_i} \qquad (2\text{-}19)$$

Hence, from Eqs. (2-18) and (2-19), we have

$$\sum_{k=1}^{3N} m_k \ddot{x}_k \frac{\partial x_k}{\partial q_i} = \frac{d}{dt} \left(\frac{\partial T}{\partial \dot{q}_i} \right) - \frac{\partial T}{\partial q_i} \qquad (2\text{-}20)$$

The generalized force Q_i was previously defined to be

$$Q_i = \sum_{k=1}^{3N} F_k \frac{\partial x_k}{\partial q_i} \qquad (2\text{-}21)$$

Then, using Eqs. (2-14), (2-20), and (2-21), we obtain

$$\sum_{i=1}^{n} \left[Q_i - \frac{d}{dt} \left(\frac{\partial T}{\partial \dot{q}_i} \right) + \frac{\partial T}{\partial q_i} \right] \delta q_i = 0 \qquad (2\text{-}22)$$

which is essentially a restatement of the Lagrangian form of d'Alembert's principle in terms of generalized coordinates.

Thus far we have not made any restrictions on the δq's, except that they must conform to the instantaneous constraints. This restriction was necessary in order to neglect the virtual work of the constraint forces. Now let us make the additional assumptions that the system is *holonomic* and its configuration is described by a set of *independent* generalized coordinates. If the δq's are independent, then the coefficient of each δq_i in Eq. (2-22) must be zero. Hence,

$$\frac{d}{dt} \left(\frac{\partial T}{\partial \dot{q}_i} \right) - \frac{\partial T}{\partial q_i} = Q_i \qquad (i = 1, 2, \ldots, n) \qquad (2\text{-}23)$$

These n equations are known as *Lagrange's equations* and are written here in one of their principal forms. As we shall demonstrate, they consist of n second-order nonlinear differential equations.

In attempting to gain some physical insight into the meaning of Lagrange's equations, let us note first that, since the δq's are independent, all the generalized constraint forces are zero. Then we see that the generalized applied force Q_i is equal and opposite to a *generalized inertial force* given by

$$\frac{\partial T}{\partial q_i} - \frac{d}{dt}\left(\frac{\partial T}{\partial \dot{q}_i}\right)$$

where $-d/dt(\partial T/\partial \dot{q}_i)$ represents the negative rate of change of the generalized momentum.

We have been considering a holonomic system whose configuration is given by a set of independent generalized coordinates. Now let us make the additional assumption that all the generalized forces are derivable from a potential function $V(q, t)$ as follows:

$$Q_i = -\frac{\partial V}{\partial q_i} \qquad (2\text{-}24)$$

Then Lagrange's equations can be written in the form

$$\frac{d}{dt}\left(\frac{\partial T}{\partial \dot{q}_i}\right) - \frac{\partial T}{\partial q_i} + \frac{\partial V}{\partial q_i} = 0 \qquad (i = 1, 2, \ldots, n) \qquad (2\text{-}25)$$

Let us recall from Eq. (1-136) that the Lagrangian function $L(q, \dot{q}, t)$ is

$$L = T - V \qquad (2\text{-}26)$$

Then, since V is not a function of the $\dot{q}$'s, we find from Eqs. (2-25) and (2-26) that

$$\frac{d}{dt}\left(\frac{\partial L}{\partial \dot{q}_i}\right) - \frac{\partial L}{\partial q_i} = 0 \qquad (i = 1, 2, \ldots, n) \qquad (2\text{-}27)$$

This is the *standard form* of Lagrange's equation for a *holonomic system*.

Here we have derived a method of obtaining a complete set of differential equations of motion for the system by operating on a single scalar function $L(q, \dot{q}, t)$. Hence, it can be seen that the Lagrangian function must contain all the necessary information concerning its possible motions. Furthermore, the form of Eq. (2-27) does not depend upon which particular set of generalized coordinates is chosen to describe the system, provided that the q's are independent.

Another form of Lagrange's equations can be written for systems in which the generalized forces are not wholly derivable from a potential function. Let

$$Q_i = -\frac{\partial V}{\partial q_i} + Q_i' \qquad (2\text{-}28)$$

Then we obtain from Eqs. (2-23), (2-26), and (2-28) that

$$\frac{d}{dt}\left(\frac{\partial L}{\partial \dot{q}_i}\right) - \frac{\partial L}{\partial q_i} = Q_i' \qquad (i = 1, 2, \ldots, n) \qquad (2\text{-}29)$$

where the Q_i' are those generalized forces not derivable from a potential function. Examples of typical Q' forces are friction forces and time-varying forcing functions.

Form of the Equations of Motion. Now let us consider more explicitly the form of the equations of motion which result from the application of Lagrange's equations in the standard form given by Eq. (2-27).

We note first that the generalized momentum is *linear* in the $\dot{q}$'s. Referring to Eqs. (2-6) and (2-8), we have

$$p_i = \frac{\partial T}{\partial \dot{q}_i} = \sum_{j=1}^{n} m_{ij}\dot{q}_j + a_i \tag{2-30}$$

where m_{ij} and a_i are functions of the q's and t. Hence, it follows that the equations of motion are *linear* in the $\ddot{q}$'s, since all the terms containing $\ddot{q}$'s arise from differentiating Eq. (2-30) with respect to time. Performing this differentiation, we have

$$\frac{d}{dt}\left(\frac{\partial T}{\partial \dot{q}_i}\right) = \sum_{j=1}^{n} m_{ij}\ddot{q}_j + \sum_{j=1}^{n} \dot{m}_{ij}\dot{q}_j + \dot{a}_i \tag{2-31}$$

where

$$\dot{m}_{ij} = \sum_{l=1}^{n} \frac{\partial m_{ij}}{\partial q_l}\dot{q}_l + \frac{\partial m_{ij}}{\partial t} \tag{2-32}$$

and

$$\dot{a}_i = \sum_{j=1}^{n} \frac{\partial a_i}{\partial q_j}\dot{q}_j + \frac{\partial a_i}{\partial t} \tag{2-33}$$

Also,

$$\sum_{j=1}^{n} \dot{m}_{ij}\dot{q}_j = \sum_{j=1}^{n}\sum_{l=1}^{n} \frac{\partial m_{ij}}{\partial q_l}\dot{q}_l\dot{q}_j + \sum_{j=1}^{n} \frac{\partial m_{ij}}{\partial t}\dot{q}_j$$

$$= \frac{1}{2}\sum_{j=1}^{n}\sum_{l=1}^{n}\left(\frac{\partial m_{ij}}{\partial q_l} + \frac{\partial m_{il}}{\partial q_j}\right)\dot{q}_j\dot{q}_l + \sum_{j=1}^{n} \frac{\partial m_{ij}}{\partial t}\dot{q}_j \tag{2-34}$$

$$\frac{\partial T_2}{\partial q_i} = \frac{1}{2}\sum_{j=1}^{n}\sum_{l=1}^{n} \frac{\partial m_{jl}}{\partial q_i}\dot{q}_j\dot{q}_l \tag{2-35}$$

$$\frac{\partial T_1}{\partial q_i} = \sum_{j=1}^{n} \frac{\partial a_j}{\partial q_i}\dot{q}_j \tag{2-36}$$

where we note that several dummy indices have been changed.

Finally, substituting from Eqs. (2-31) through (2-36) into Lagrange's equations, we obtain

$$\sum_{j=1}^{n} m_{ij}\ddot{q}_j + \frac{1}{2}\sum_{j=1}^{n}\sum_{l=1}^{n}\left(\frac{\partial m_{ij}}{\partial q_l} + \frac{\partial m_{il}}{\partial q_j} - \frac{\partial m_{jl}}{\partial q_i}\right)\dot{q}_j\dot{q}_l$$

$$+ \sum_{j=1}^{n}\left(\frac{\partial m_{ij}}{\partial t} + \frac{\partial a_i}{\partial q_j} - \frac{\partial a_j}{\partial q_i}\right)\dot{q}_j + \frac{\partial a_i}{\partial t} - \frac{\partial T_0}{\partial q_i} + \frac{\partial V}{\partial q_i} = 0$$

$$(i = 1, 2, \ldots, n) \tag{2-37}$$

The notation can be shortened by using a *Christoffel symbol of the first kind* which is applied here to the quadratic form T_2. Let

$$[jl, i] = \frac{1}{2}\left(\frac{\partial m_{ij}}{\partial q_l} + \frac{\partial m_{il}}{\partial q_j} - \frac{\partial m_{jl}}{\partial q_i}\right) \qquad (2\text{-}38)$$

Then Eq. (2-37) can be written in the form

$$\sum_{j=1}^{n} m_{ij}\ddot{q}_j + \sum_{j=1}^{n}\sum_{l=1}^{n}[jl, i]\dot{q}_j\dot{q}_l + \sum_{j=1}^{n}\gamma_{ij}\dot{q}_j$$

$$+ \sum_{j=1}^{n}\frac{\partial m_{ij}}{\partial t}\dot{q}_j + \frac{\partial a_i}{\partial t} - \frac{\partial T_0}{\partial q_i} + \frac{\partial V}{\partial q_i} = 0$$

$$(i = 1, 2, \ldots, n) \qquad (2\text{-}39)$$

where γ_{ij} is an element of a skew-symmetric matrix and is given by

$$\gamma_{ij} = -\gamma_{ji} = \frac{\partial a_i}{\partial q_j} - \frac{\partial a_j}{\partial q_i} \qquad (2\text{-}40)$$

The n equations of Eq. (2-37) or (2-39) are the equations of motion. Although these equations are nonlinear in general, they are linear in the $\ddot{q}$'s. Furthermore, since T_2 has been shown to be a positive definite quadratic form in the $\dot{q}$'s, the matrix **m** is also positive definite and has an inverse. Therefore, it is always possible to solve for the $\ddot{q}$'s in terms of the q's, $\dot{q}$'s, and t. If this is done, the resulting equations of motion are of the form

$$\ddot{q}_i + f_i(q, \dot{q}, t) = 0 \qquad (i = 1, 2, \ldots, n) \qquad (2\text{-}41)$$

Nonholonomic Systems. The derivation of Lagrange's equations for a holonomic system required that the generalized coordinates be independent. For a *nonholonomic system*, however, there must be more generalized coordinates than the number of degrees of freedom. Therefore, the δq's are no longer independent if we assume a virtual displacement consistent with the constraints. For example, if there are m nonholonomic constraint equations of the form

$$\sum_{i=1}^{n} a_{ji}\,dq_i + a_{jt}\,dt = 0 \qquad (j = 1, 2, \ldots, m) \qquad (2\text{-}42)$$

the δq's must meet the following conditions:

$$\sum_{i=1}^{n} a_{ji}\,\delta q_i = 0 \qquad (j = 1, 2, \ldots, m) \qquad (2\text{-}43)$$

Now let us assume once again that each generalized *applied* force Q_i is obtained from a potential function, as in Eq. (2-24). The constraints are assumed to be workless, so the *generalized constraint forces* C_i must meet the condition

$$\sum_{i=1}^{n} C_i\,\delta q_i = 0 \qquad (2\text{-}44)$$

for any virtual displacement consistent with the constraints.

Now suppose we multiply Eq. (2-43) by a factor λ_j known as a *Lagrange multiplier* and obtain the m equations

$$\lambda_j \sum_{i=1}^{n} a_{ji}\,\delta q_i = 0 \qquad (j = 1, 2, \ldots, m) \tag{2-45}$$

Next, subtract the sum of these m equations from Eq. (2-44). Interchanging the order of summation, we obtain

$$\sum_{i=1}^{n} \left(C_i - \sum_{j=1}^{m} \lambda_j a_{ji} \right) \delta q_i = 0 \tag{2-46}$$

Up to this point, the λ's have been considered to be arbitrary, while the δq's must conform to the constraints of Eq. (2-43). But if we choose the λ's such that

$$C_i = \sum_{j=1}^{m} \lambda_j a_{ji} \qquad (i = 1, 2, \ldots, n) \tag{2-47}$$

then the coefficients of the δq's are zero, and Eq. (2-46) will apply for *any* set of δq's. In other words, the δq's can be chosen independently.

With these assumptions, we can equate the generalized force C_i with Q_i' and, using Eqs. (2-29) and (2-47), we obtain

$$\frac{d}{dt}\left(\frac{\partial L}{\partial \dot{q}_i} \right) - \frac{\partial L}{\partial q_i} = \sum_{j=1}^{m} \lambda_j a_{ji} \qquad (i = 1, 2, \ldots, n) \tag{2-48}$$

This is the *standard form* of Lagrange's equation for a *nonholonomic system*.

In addition to these n equations of motion, we have the m nonholonomic constraint equations which can be written in the form

$$\sum_{i=1}^{n} a_{ji}\dot{q}_i + a_{jt} = 0 \qquad (j = 1, 2, \ldots, m) \tag{2-49}$$

Thus we have a total of $(n + m)$ equations with which to solve for the $(n + m)$ independent variables, namely, the n q's and the m λ's.

If we consider what has been accomplished by the Lagrange multiplier method, we see from Eqs. (2-47) and (2-48) that the constraints enter the equations of motion in the form of *constraint forces* rather than in geometric terms. Also, we can understand the physical significance of the λ's by noting that they are linearly related to the constraint forces.

The standard form of Lagrange's equation for a nonholonomic system, Eq. (2-48), can also be applied to a *holonomic system* in which there are more generalized coordinates than degrees of freedom. For example, suppose there are m holonomic constraint equations of the form

$$\phi_j(q_1, q_2, \ldots, q_n, t) = 0 \qquad (j = 1, 2, \ldots, m) \tag{2-50}$$

Taking the total differential of ϕ_j we obtain

$$d\phi_j = \sum_{i=1}^{n} \frac{\partial \phi_j}{\partial q_i}\,dq_i + \frac{\partial \phi_j}{\partial t}\,dt = 0 \tag{2-51}$$

which is of the form of Eq. (2-42), where we let

$$a_{ji} = \frac{\partial \phi_j}{\partial q_i}, \qquad a_{jt} = \frac{\partial \phi_j}{\partial t} \tag{2-52}$$

If the Lagrange multiplier method is applied to this system and the resulting differential equations are completely solved, the result will be that the q's and λ's are expressed as explicit functions of time. Then, with the aid of Eq. (2-47), the generalized constraint forces C_i can also be obtained as explicit functions of time.

In general, however, holonomic systems will be described in terms of independent q's, thereby avoiding any equations of constraint. The Lagrange multiplier method is normally used with holonomic systems only if one desires to solve for the constraint forces.

2-2. EXAMPLES

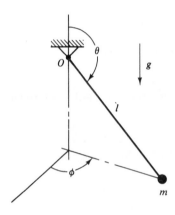

Fig. 2-1. A spherical pendulum.

Example 2-1. Find the differential equations of motion for a spherical pendulum of length l (Fig. 2-1).

Using spherical coordinates, we see that the kinetic energy of the particle is

$$T = \tfrac{1}{2}m(l^2\dot{\theta}^2 + l^2\dot{\phi}^2 \sin^2 \theta) \tag{2-53}$$

and the potential energy is

$$V = mgl \cos \theta \tag{2-54}$$

where the support point O is at the reference level corresponding to zero potential energy.

The Lagrangian function is

$$L = T - V = \tfrac{1}{2}ml^2(\dot{\theta}^2 + \dot{\phi}^2 \sin^2 \theta) - mgl \cos \theta \tag{2-55}$$

from which we obtain, for $q_i = \theta$,

$$\frac{d}{dt}\left(\frac{\partial L}{\partial \dot{\theta}}\right) = ml^2\ddot{\theta}, \qquad \frac{\partial L}{\partial \theta} = ml^2\dot{\phi}^2 \sin \theta \cos \theta + mgl \sin \theta$$

Hence, Eq. (2-27) yields the following equation of motion:

$$ml^2\ddot{\theta} - ml^2\dot{\phi}^2 \sin \theta \cos \theta - mgl \sin \theta = 0 \tag{2-56}$$

In a similar manner,

$$\frac{d}{dt}\left(\frac{\partial L}{\partial \dot{\phi}}\right) = ml^2\ddot{\phi} \sin^2 \theta + 2ml^2\dot{\theta}\dot{\phi} \sin \theta \cos \theta$$

$$\frac{\partial L}{\partial \phi} = 0$$

and the ϕ equation of motion is

$$ml^2\ddot{\phi}\sin^2\theta + 2ml^2\dot{\theta}\dot{\phi}\sin\theta\cos\theta = 0 \qquad (2\text{-}57)$$

Equations (2-56) and (2-57) are the differential equations of motion. Note that these equations are nonlinear, although $\ddot{\theta}$ and $\ddot{\phi}$ appear linearly. Note also that the equations of motion are identical with the equations obtained by setting the coefficients of the virtual displacements $\delta\theta$ and $\delta\phi$ equal to zero in Eq. (1-68) for the case $r = l$. This indicates that the Lagrangian procedure has resulted in terms proportional to the θ and ϕ components of the inertial and gravitational forces acting on the particle.

The ϕ equation of motion is immediately integrable in this example because $\partial L/\partial\phi = 0$, and therefore

$$\frac{d}{dt}\left(\frac{\partial L}{\partial\dot{\phi}}\right) = 0 \qquad (2\text{-}58)$$

Hence,

$$\frac{\partial L}{\partial\dot{\phi}} = ml^2\dot{\phi}\sin^2\theta = p_\phi \qquad (2\text{-}59)$$

where p_ϕ is a constant equal to the generalized momentum conjugate to ϕ. In this instance, p_ϕ is the angular momentum about a vertical axis through the support point O.

Example 2-2. A double pendulum consists of two particles suspended by massless rods, as shown in Fig. 2-2. Assuming that all motion takes place in a vertical plane, find the differential equations of motion. Linearize these equations, assuming small motions.

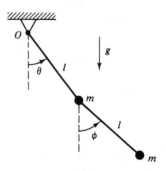

First let us obtain an expression for the kinetic energy. The absolute velocity of the lower particle is equal to the vector sum of (1) the absolute velocity of the upper particle and (2) the velocity of the lower particle relative to the upper particle. Since the two velocity vectors differ in direction by the angle $(\phi - \theta)$, we can use the cosine law to obtain the magnitude v of the vector sum. We have, then, that the velocity of the lower particle is

Fig. 2-2. A double pendulum.

$$v = l[\dot{\theta}^2 + \dot{\phi}^2 + 2\dot{\theta}\dot{\phi}\cos(\phi - \theta)]^{1/2}$$

and therefore the total kinetic energy is

$$T = \tfrac{1}{2}ml^2[2\dot{\theta}^2 + \dot{\phi}^2 + 2\dot{\theta}\dot{\phi}\cos(\phi - \theta)] \qquad (2\text{-}60)$$

Choosing the reference level for potential energy at O, we obtain

$$V = -mgl(2 \cos \theta + \cos \phi) \tag{2-61}$$

The Lagrangian function is

$$L = \tfrac{1}{2}ml^2[2\dot{\theta}^2 + \dot{\phi}^2 + 2\dot{\theta}\dot{\phi} \cos (\phi - \theta)] + mgl(2 \cos \theta + \cos \phi) \tag{2-62}$$

The θ equation is obtained from

$$\frac{\partial L}{\partial \dot{\theta}} = ml^2[2\dot{\theta} + \dot{\phi} \cos (\phi - \theta)]$$

$$\frac{d}{dt}\left(\frac{\partial L}{\partial \dot{\theta}}\right) = ml^2[2\ddot{\theta} + \ddot{\phi} \cos (\phi - \theta) - \dot{\phi}(\dot{\phi} - \dot{\theta}) \sin (\phi - \theta)]$$

$$\frac{\partial L}{\partial \theta} = ml^2\dot{\theta}\dot{\phi} \sin (\phi - \theta) - 2mgl \sin \theta$$

which, upon substitution into Lagrange's equation, yields

$$ml^2[2\ddot{\theta} + \ddot{\phi} \cos (\phi - \theta) - \dot{\phi}^2 \sin (\phi - \theta)] + 2mgl \sin \theta = 0 \quad (2\text{-}63)$$

In a similar fashion, the ϕ equation is obtained from

$$\frac{\partial L}{\partial \dot{\phi}} = ml^2[\dot{\phi} + \dot{\theta} \cos (\phi - \theta)]$$

$$\frac{d}{dt}\left(\frac{\partial L}{\partial \dot{\phi}}\right) = ml^2[\ddot{\phi} + \ddot{\theta} \cos (\phi - \theta) - \dot{\theta}(\dot{\phi} - \dot{\theta}) \sin (\phi - \theta)]$$

$$\frac{\partial L}{\partial \phi} = -ml^2\dot{\theta}\dot{\phi} \sin (\phi - \theta) - mgl \sin \phi$$

yielding

$$ml^2[\ddot{\phi} + \ddot{\theta} \cos (\phi - \theta) + \dot{\theta}^2 \sin (\phi - \theta)] + mgl \sin \phi = 0 \tag{2-64}$$

In this example, the kinetic energy is a homogeneous quadratic function of the $\dot{q}$'s; hence, T_1 and T_0 are zero. One can obtain the generalized inertia matrix $\mathbf{m}$ from the kinetic energy expression of Eq. (2-60). It is

$$\mathbf{m} = ml^2 \begin{bmatrix} 2 & \cos (\phi - \theta) \\ \cos (\phi - \theta) & 1 \end{bmatrix} \tag{2-65}$$

By using the criteria of Eq. (2-11), it is apparent that T is positive definite. Having obtained $\mathbf{m}$, one can substitute into the explicit form of the equations of motion given in Eq. (2-37) in order to check the actual equations of motion given in Eqs. (2-63) and (2-64).

Now let us linearize the differential equations of motion for the case of small motions. In other words, let us assume that θ, ϕ, and their time derivatives are much smaller than one. With these assumptions, we can use the approximations

$$\cos (\phi - \theta) \cong 1, \qquad \sin (\phi - \theta) \cong \phi - \theta$$

and the equations of motion become

$$ml^2(2\ddot{\theta} + \ddot{\phi}) + 2mgl\theta = 0$$
$$ml^2(\ddot{\theta} + \ddot{\phi}) + mgl\phi = 0 \tag{2-66}$$

where we have neglected higher-order terms in the small quantities.

Example 2-3. A block of mass m_2 can slide on another block of mass m_1 which, in turn, slides on a horizontal surface, as shown in Fig. 2-3(a). Using x_1 and x_2 as coordinates, obtain the differential equations of motion. Solve for the accelerations of the two blocks as they move under the influence of gravity, assuming that all surfaces are frictionless. Find the force of interaction between the blocks.

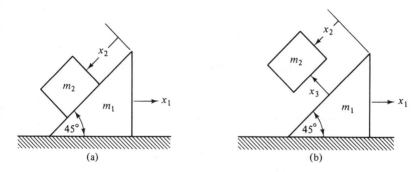

Fig. 2-3. A system of sliding blocks.

First we observe that x_1 is the absolute displacement of m_1, but x_2 is the displacement of m_2 relative to m_1. To obtain the absolute velocity v_2 of m_2, we can use the cosine law to add the velocity of m_1 and the velocity of m_2 relative to m_1. We find that

$$v_2^2 = \dot{x}_1^2 + \dot{x}_2^2 - 2\dot{x}_1\dot{x}_2 \cos 45°$$

and therefore

$$T = \tfrac{1}{2}m_1\dot{x}_1^2 + \tfrac{1}{2}m_2(\dot{x}_1^2 + \dot{x}_2^2 - \sqrt{2}\,\dot{x}_1\dot{x}_2) \tag{2-67}$$

Any changes in potential energy arise from changes in x_2, so we can take

$$V = -\frac{1}{\sqrt{2}}\,m_2 g x_2 \tag{2-68}$$

Hence

$$L = \frac{1}{2}m_1\dot{x}_1^2 + \frac{1}{2}m_2(\dot{x}_1^2 + \dot{x}_2^2 - \sqrt{2}\,\dot{x}_1\dot{x}_2) + \frac{1}{\sqrt{2}}\,m_2 g x_2 \tag{2-69}$$

The x_1 equation is obtained as follows:

$$\frac{d}{dt}\left(\frac{\partial L}{\partial \dot{x}_1}\right) = m_1\ddot{x}_1 + m_2\left(\ddot{x}_1 - \frac{1}{\sqrt{2}}\ddot{x}_2\right)$$

$$\frac{\partial L}{\partial x_1} = 0$$

and, using Lagrange's equation, we obtain

$$(m_1 + m_2)\ddot{x}_1 - \frac{1}{\sqrt{2}} m_2\ddot{x}_2 = 0 \tag{2-70}$$

Note that, since $\partial L/\partial x_1 = 0$, the generalized momentum corresponding to x_1 is conserved, that is, the horizontal linear momentum is constant. Next, we find that

$$\frac{d}{dt}\left(\frac{\partial L}{\partial \dot{x}_2}\right) = m_2\ddot{x}_2 - \frac{1}{\sqrt{2}} m_2\ddot{x}_1$$

$$\frac{\partial L}{\partial x_2} = \frac{1}{\sqrt{2}} m_2 g$$

and the x_2 equation is

$$-\frac{1}{\sqrt{2}} m_2\ddot{x}_1 + m_2\ddot{x}_2 - \frac{1}{\sqrt{2}} m_2 g = 0 \tag{2-71}$$

Equations (2-70) and (2-71) are the differential equations of motion for the system. They can be solved for the accelerations $\ddot{x}_1$ and $\ddot{x}_2$, yielding

$$\ddot{x}_1 = \frac{m_2 g}{2m_1 + m_2}$$

$$\ddot{x}_2 = \sqrt{2}\,\frac{(m_1 + m_2)g}{2m_1 + m_2} \tag{2-72}$$

Now suppose we wish to use the Lagrange multiplier method to solve for the interaction force between the blocks. This interaction force is normal to the frictionless contact surface and may be considered as the generalized constraint force corresponding to a coordinate x_3 which is shown in Fig. 2-3(b). Although we are using three generalized coordinates, there are only two degrees of freedom in the actual system because there is one equation of holonomic constraint, namely,

$$x_3 = 0 \tag{2-73}$$

This equation can be expressed in the form

$$\dot{x}_3 = 0 \tag{2-74}$$

which is similar to an equation of nonholonomic constraint. Comparing Eq. (2-74) with Eq. (2-49), we see that

$$\begin{aligned}
a_{11} &= 0 \\
a_{12} &= 0 \\
a_{13} &= 1 \\
a_{1t} &= 0
\end{aligned} \tag{2-75}$$

Hence, using Eq. (2-47), we find that the generalized constraint forces are

$$C_1 = 0$$
$$C_2 = 0 \qquad\qquad (2\text{-}76)$$
$$C_3 = \lambda_1$$

It is important in writing the Lagrangian function to assume that the generalized coordinates are independent. Thus, writing the vertical and horizontal velocity components separately, we obtain

$$T = \frac{1}{2} m_1 \dot{x}_1^2 + \frac{1}{2} m_2 \left[\left(\dot{x}_1 - \frac{\dot{x}_2 + \dot{x}_3}{\sqrt{2}} \right)^2 + \left(\frac{\dot{x}_3 - \dot{x}_2}{\sqrt{2}} \right)^2 \right]$$

$$= \frac{1}{2}(m_1 + m_2)\dot{x}_1^2 + \frac{1}{2} m_2 [\dot{x}_2^2 + \dot{x}_3^2 - \sqrt{2}\, \dot{x}_1(\dot{x}_2 + \dot{x}_3)] \qquad (2\text{-}77)$$

$$V = \frac{1}{\sqrt{2}} m_2 g (x_3 - x_2) \qquad\qquad (2\text{-}78)$$

which results in

$$L = \frac{1}{2}(m_1 + m_2)\dot{x}_1^2 + \frac{1}{2} m_2 [\dot{x}_2^2 + \dot{x}_3^2 - \sqrt{2}\, \dot{x}_1(\dot{x}_2 + \dot{x}_3)]$$

$$- \frac{1}{\sqrt{2}} m_2 g (x_3 - x_2) \qquad (2\text{-}79)$$

If we substitute this Lagrangian into Eq. (2-48) we obtain the following equations of motion:

$$(m_1 + m_2)\ddot{x}_1 - \frac{1}{\sqrt{2}} m_2 \ddot{x}_2 - \frac{1}{\sqrt{2}} m_2 \ddot{x}_3 = 0$$

$$- \frac{1}{\sqrt{2}} m_2 \ddot{x}_1 + m_2 \ddot{x}_2 - \frac{1}{\sqrt{2}} m_2 g = 0 \qquad (2\text{-}80)$$

$$- \frac{1}{\sqrt{2}} m_2 \ddot{x}_1 + m_2 \ddot{x}_3 + \frac{1}{\sqrt{2}} m_2 g = \lambda_1$$

At this point we can use Eq. (2-74) to eliminate the terms containing $\ddot{x}_3$ since they are equal to zero. There remain three equations from which to solve for $\ddot{x}_1$, $\ddot{x}_2$, and λ_1. Again

$$\ddot{x}_1 = \frac{m_2 g}{2m_1 + m_2}$$

$$\ddot{x}_2 = \sqrt{2}\, \frac{(m_1 + m_2)g}{2m_1 + m_2}$$

The constraint force is

$$C_3 = \lambda_1 = \frac{\sqrt{2}\, m_1 m_2 g}{2m_1 + m_2} \qquad (2\text{-}81)$$

and is compressive in nature, indicating that there is no tendency for the blocks to separate.

Example 2-4. A particle of mass m can slide without friction on the inside of a small tube which is bent in the form of a circle of radius r. The tube rotates about a vertical diameter with a constant angular velocity ω, as shown in Fig. 2-4. Write the differential equation of motion.

The kinetic and potential energies of the particle are

$$T = \tfrac{1}{2}mr^2(\dot{\theta}^2 + \omega^2 \sin^2 \theta) \tag{2-82}$$

$$V = mgr \cos \theta \tag{2-83}$$

resulting in

$$L = \tfrac{1}{2}mr^2(\dot{\theta}^2 + \omega^2 \sin^2 \theta) - mgr \cos \theta \tag{2-84}$$

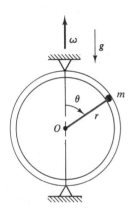

This is a *rheonomic* system because any set of equations giving the inertial Cartesian coordinates of the particle in terms of its single generalized coordinate θ must involve time explicitly. Note, however, that in this case the Lagrangian L is not an explicit function of time.

The tube is a moving constraint which does work on the particle in an *actual* displacement. Nevertheless, no work is done by the constraint forces in a *virtual* displacement, and the constraint is classed as a workless constraint.

Since the only generalized force acting on the system is derivable from a potential function, we can use the standard form of Lagrange's equation given by Eq. (2-27). We find that

Fig. 2-4. A particle in a whirling tube.

$$\frac{d}{dt}\left(\frac{\partial L}{\partial \dot{\theta}}\right) = mr^2\ddot{\theta}$$

$$\frac{\partial L}{\partial \theta} = mr^2\omega^2 \sin \theta \cos \theta + mgr \sin \theta$$

and therefore the equation of motion is

$$mr^2\ddot{\theta} - mr^2\omega^2 \sin \theta \cos \theta - mgr \sin \theta = 0 \tag{2-85}$$

This system will be discussed further in Sec. 2-3.

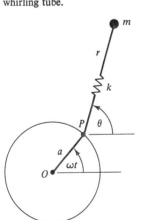

Example 2-5. A particle of mass m is connected by a massless spring of stiffness k and unstressed length r_0 to a point P which is moving along a circular path of radius a at a uniform angular rate ω (Fig. 2-5). Assuming that the particle moves without friction on a horizontal plane, find the differential equations of motion.

Fig. 2-5. A particle attached by a massless spring to a moving point.

This is a *rheonomic* system with two degrees of freedom corresponding to the independent generalized coordinates r and θ. Let us obtain the kinetic energy with the aid of Eq. (1-119) which, for this case of a single particle, can be written in the form

$$T = \tfrac{1}{2}m\dot{r}_p^2 + \tfrac{1}{2}m\dot{\rho}^2 + \dot{r}_p\cdot m\dot{\rho} \tag{2-86}$$

where $\dot{r}_p$ is the absolute velocity of the point P, and $\dot{\rho}$ is the velocity of the particle relative to P. We see that

$$\dot{r}_p^2 = a^2\omega^2$$
$$\dot{\rho}^2 = \dot{r}^2 + r^2\dot{\theta}^2$$

Hence,

$$T_2 = \tfrac{1}{2}m\dot{\rho}^2 = \tfrac{1}{2}m(\dot{r}^2 + r^2\dot{\theta}^2) \tag{2-87}$$

$$T_1 = \dot{r}_p\cdot m\dot{\rho} = ma\omega[\dot{r}\sin(\theta - \omega t) + r\dot{\theta}\cos(\theta - \omega t)] \tag{2-88}$$

$$T_0 = \tfrac{1}{2}m\dot{r}_p^2 = \tfrac{1}{2}ma^2\omega^2 \tag{2-89}$$

Also,

$$V = \tfrac{1}{2}k(r - r_0)^2 \tag{2-90}$$

Therefore, the Lagrangian function is

$$L = \tfrac{1}{2}m[\dot{r}^2 + r^2\dot{\theta}^2 + 2a\omega\dot{r}\sin(\theta - \omega t) + 2a\omega r\dot{\theta}\cos(\theta - \omega t) + a^2\omega^2]$$
$$- \tfrac{1}{2}k(r - r_0)^2 \tag{2-91}$$

We obtain the r equation from

$$\frac{d}{dt}\left(\frac{\partial L}{\partial \dot{r}}\right) = m\ddot{r} - ma\omega^2\cos(\theta - \omega t) + ma\omega\dot{\theta}\cos(\theta - \omega t)$$

$$\frac{\partial L}{\partial r} = mr\dot{\theta}^2 + ma\omega\dot{\theta}\cos(\theta - \omega t) - k(r - r_0)$$

which, after substitution into Lagrange's equation, yields

$$m\ddot{r} - mr\dot{\theta}^2 - ma\omega^2\cos(\theta - \omega t) + k(r - r_0) = 0 \tag{2-92}$$

Also, we obtain

$$\frac{d}{dt}\left(\frac{\partial L}{\partial \dot{\theta}}\right) = mr^2\ddot{\theta} + 2mr\dot{r}\dot{\theta} + ma\dot{r}\omega\cos(\theta - \omega t)$$
$$- ma r\omega\dot{\theta}\sin(\theta - \omega t) + ma r\omega^2\sin(\theta - \omega t)$$

$$\frac{\partial L}{\partial \theta} = ma\dot{r}\omega\cos(\theta - \omega t) - ma r\omega\dot{\theta}\sin(\theta - \omega t)$$

and the θ equation is

$$mr^2\ddot{\theta} + 2mr\dot{r}\dot{\theta} + ma r\omega^2\sin(\theta - \omega t) = 0 \tag{2-93}$$

Example 2-6. Two particles are connected by a rigid massless rod of length l which rotates in a horizontal plane with a constant angular velocity ω (Fig. 2-6). Knife-edge supports at the two particles prevent either particle from having a velocity component along the rod, but the particles can slide without friction in a direction perpendicular to the rod. Find the differential equations of motion. Solve for x, y, and the constraint force as functions of time if the center of mass is initially at the origin and has a velocity v_0 in the positive y direction.

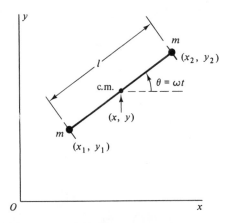

Fig. 2-6. A nonholonomic rheonomic system.

This is a *nonholonomic rheonomic system*. Let us express the configuration initially in terms of the Cartesian coordinates (x_1, y_1) and (x_2, y_2) of the individual particles. There are two independent equations of holonomic constraint, namely,

$$(x_2 - x_1)^2 + (y_2 - y_1)^2 = l^2 \tag{2-94}$$

and

$$y_2 - y_1 = (x_2 - x_1) \tan \omega t \tag{2-95}$$

expressing the given length and orientation of the rod. In addition, there is the nonholonomic constraint equation

$$(\dot{x}_1 + \dot{x}_2) \cos \omega t + (\dot{y}_1 + \dot{y}_2) \sin \omega t = 0 \tag{2-96}$$

which restricts the velocity of the center of the rod to a direction which is perpendicular to the rod, as was shown in the discussion preceding Eq. (1-17).

We have been using four Cartesian coordinates and three independent equations of constraint, indicating that the system has one degree of freedom. It is more convenient, however, to choose generalized coordinates in such a manner that there are no holonomic constraints, and only one nonholonomic constraint. This can be accomplished by choosing the Cartesian coordinates (x, y) of the center of mass as the generalized coordinates. The transformation equations are

$$x_1 = x - \tfrac{1}{2}l \cos \omega t$$
$$y_1 = y - \tfrac{1}{2}l \sin \omega t$$
$$x_2 = x + \tfrac{1}{2}l \cos \omega t \tag{2-97}$$
$$y_2 = y + \tfrac{1}{2}l \sin \omega t$$

From Eqs. (2-96) and (2-97), we see that the nonholonomic constraint equation becomes

$$\dot{x} \cos \omega t + \dot{y} \sin \omega t = 0 \tag{2-98}$$

The total kinetic energy of the system is

$$T = \tfrac{1}{2}m(\dot{x}_1^2 + \dot{y}_1^2 + \dot{x}_2^2 + \dot{y}_2^2) \tag{2-99}$$

or, substituting from Eq. (2-97),

$$T = m(\dot{x}^2 + \dot{y}^2) + \tfrac{1}{4}ml^2\omega^2 \tag{2-100}$$

This result can be obtained directly by adding the translational and rotational kinetic energies, noting that the total mass is $2m$ and the moment of inertia about the center of mass is $ml^2/2$.

The potential energy V is zero for this system, so we can write Eq. (2-48) in the form

$$\frac{d}{dt}\left(\frac{\partial T}{\partial \dot{q}_i}\right) - \frac{\partial T}{\partial q_i} = \lambda_1 a_{1i} \tag{2-101}$$

where, from the nonholonomic constraint equation of (2-98), we see that

$$a_{11} = \cos \omega t$$
$$a_{12} = \sin \omega t \tag{2-102}$$

The differential equations of motion are obtained by substituting from Eqs. (2-100) and (2-102) into Eq. (2-101), the result being

$$2m\ddot{x} = \lambda_1 \cos \omega t$$
$$2m\ddot{y} = \lambda_1 \sin \omega t \tag{2-103}$$

These two equations and the constraint equation of (2-98) must now be solved for x, y, and λ_1. From Eq. (2-103) we obtain

$$\ddot{y} = \ddot{x} \tan \omega t \tag{2-104}$$

and, substituting for $\tan \omega t$ from Eq. (2-98), we find that

$$\frac{d}{dt}(\dot{x}^2 + \dot{y}^2) = 0 \tag{2-105}$$

Next, integrating and using the initial conditions, we see that

$$\dot{x}^2 + \dot{y}^2 = v_0^2 \tag{2-106}$$

indicating that the center of mass moves with a constant speed v_0. Since the direction of the motion is always perpendicular to the rod, we have

$$\dot{x} = -v_0 \sin \omega t$$
$$\dot{y} = v_0 \cos \omega t$$

(2-107)

Integrating again, we obtain

$$x = \frac{v_0}{\omega}(\cos \omega t - 1)$$
$$y = \frac{v_0}{\omega}\sin \omega t$$

(2-108)

From Eqs. (2-103) and (2-107), the Lagrange multiplier is found to be

$$\lambda_1 = -2mv_0\omega$$

(2-109)

It can be seen that the system travels in a circular path of radius v_0/ω at a constant speed v_0. The centripetal constraint force exerted on the system is of magnitude $2mv_0\omega$ and is represented by $-\lambda_1$. The generalized constraint forces, obtained with the aid of Eq. (2-47), are

$$C_1 = -2mv_0\omega \cos \omega t$$
$$C_2 = -2mv_0\omega \sin \omega t$$

(2-110)

and are directed along the positive x and y axes, respectively.

2-3. INTEGRALS OF THE MOTION

Much of the emphasis thus far in this chapter has been on obtaining the differential equations of motion by using the Lagrangian method. We showed that if the configuration of a holonomic system is specified by n independent generalized coordinates, the equations of motion consist, in general, of n second-order nonlinear differential equations with time as the independent variable.

Any general analytical solution of the differential equations of motion contains $2n$ constants of integration which are usually evaluated from the $2n$ initial conditions. One method of expressing the general solution is to obtain $2n$ independent functions of the form

$$f_j(q, \dot{q}, t) = \alpha_j \qquad (j = 1, 2, \ldots, 2n)$$

(2-111)

where the α's are arbitrary constants. These $2n$ functions are called the *integrals or constants of the motion*. Each function f_j maintains a constant value α_j as the motion of the system proceeds, the value of α_j depending upon the initial conditions. In principle, these $2n$ equations can be solved for the q's and $\dot{q}$'s as functions of the α's and t, that is, it is possible to find

$$q_i = g_i(\alpha_1, \ldots, \alpha_{2n}, t)$$
$$\dot{q}_i = \dot{g}_i(\alpha_1, \ldots, \alpha_{2n}, t)$$
$$(i = 1, 2, \ldots, n)$$

(2-112)

such that Eq. (2-111) is satisfied for all j.

It is usually not possible to obtain all the f's by any direct process. Nevertheless, one of the principal topics to be discussed in the following chapters is a procedure for the determination of coordinate transformations which will simplify the finding of integrals of the motion. In fact, we shall look for transformations such that each of the new coordinates and momenta is constant, thereby forming the required $2n$ integrals of the motion.

For the present, however, let us consider some more elementary characteristics of dynamical systems which result in obtaining integrals of the motion directly by *quadratures*, that is, in terms of known elementary functions or indefinite integrals of such functions.

Ignorable Coordinates. Consider a *holonomic* system which can be described by the standard form of Lagrange's equations, that is,

$$\frac{d}{dt}\left(\frac{\partial L}{\partial \dot{q}_i}\right) - \frac{\partial L}{\partial q_i} = 0 \qquad (i = 1, 2, \ldots, n) \qquad (2\text{-}113)$$

Suppose that $L(q, \dot{q}, t)$ contains all n $\dot{q}$'s, but some of the q's, say $q_1, q_2, \ldots, q_k$, are missing from the Lagrangian. These k coordinates are called *ignorable* coordinates. Since $\partial L/\partial q_i$ is zero for each ignorable coordinate, it follows that

$$\frac{d}{dt}\left(\frac{\partial L}{\partial \dot{q}_i}\right) = 0 \qquad (i = 1, 2, \ldots, k) \qquad (2\text{-}114)$$

or

$$p_i = \frac{\partial L}{\partial \dot{q}_i} = \beta_i \qquad (i = 1, 2, \ldots, k) \qquad (2\text{-}115)$$

where the β's are constants evaluated from the initial conditions. Hence we find that the generalized momentum corresponding to each ignorable coordinate is constant, that is, it is an integral of the motion.

Example 2-7. Let us consider the *Kepler problem*, that is, the problem of the motion of a particle of unit mass which is attracted by an inverse-square gravitational force to a fixed point O (Fig. 2-7). Using polar coordinates, the kinetic and potential energies are

$$T = \tfrac{1}{2}(\dot{r}^2 + r^2\dot{\theta}^2) \qquad (2\text{-}116)$$

$$V = -\frac{\mu}{r} \qquad (2\text{-}117)$$

where μ is a positive constant known as the *gravitational coefficient*.

The Lagrangian function is

$$L = \frac{1}{2}(\dot{r}^2 + r^2\dot{\theta}^2) + \frac{\mu}{r} \qquad (2\text{-}118)$$

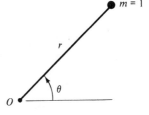

Fig. 2-7. The Kepler problem in terms of polar coordinates.

and, using Eq. (2-113), we find that the r equation of motion is

$$\ddot{r} - r\dot{\theta}^2 + \frac{\mu}{r^2} = 0 \qquad (2\text{-}119)$$

Since θ does not appear explicitly in the Lagrangian function, it is an ignorable coordinate. The θ equation of motion is

$$\frac{d}{dt}(r^2\dot{\theta}) = 0 \qquad (2\text{-}120)$$

or

$$r^2\dot{\theta} = \beta \qquad (2\text{-}121)$$

where β is a constant and is equal to the angular momentum of the particle about the attracting center O.

In this example, we obtain two equations of motion, corresponding to the two degrees of freedom. The principal consequence of the fact that θ is an ignorable coordinate is that one integral of the motion, namely, the constant angular momentum, is obtained immediately. On the other hand, θ is not completely eliminated from the analysis since $\dot{\theta}$ remains in the Lagrangian L.

The Routhian Function. Now let us introduce a procedure which results in the ignorable coordinates being eliminated from consideration as separate degrees of freedom in the Lagrangian formulation of the equations of motion.

Suppose we consider a standard holonomic system whose configuration is given by n independent generalized coordinates, of which the first k are ignorable. In other words, the Lagrangian L is a function of $q_{k+1}, \ldots, q_n$, $\dot{q}_1, \ldots, \dot{q}_n, t$.

Now let us define a *Routhian function* $R(q_{k+1}, \ldots, q_n, \dot{q}_{k+1}, \ldots, \dot{q}_n, \beta_1, \ldots, \beta_k, t)$ as follows:

$$R = L - \sum_{i=1}^{k} \beta_i \dot{q}_i \qquad (2\text{-}122)$$

where we eliminate the $\dot{q}$'s corresponding to the ignorable coordinates by solving the k equations

$$\frac{\partial L}{\partial \dot{q}_i} = \beta_i \qquad (i = 1, 2, \ldots, k) \qquad (2\text{-}123)$$

for $\dot{q}_1, \dot{q}_2, \ldots, \dot{q}_k$ as functions of $q_{k+1}, \ldots, q_n, \dot{q}_{k+1}, \ldots, \dot{q}_n, \beta_1, \ldots, \beta_k, t$. These expressions for the $\dot{q}$'s are *linear* in the β's.

Next, let us make an arbitrary variation of all the variables in the Routhian function. We have

$$\delta R = \sum_{i=k+1}^{n} \frac{\partial R}{\partial q_i} \delta q_i + \sum_{i=k+1}^{n} \frac{\partial R}{\partial \dot{q}_i} \delta \dot{q}_i + \sum_{i=1}^{k} \frac{\partial R}{\partial \beta_i} \delta \beta_i + \frac{\partial R}{\partial t} \delta t \qquad (2\text{-}124)$$

where we note that the β's are regarded as variables. A similar variation in the right-hand side of Eq. (2-122) yields

$$\delta L = \sum_{i=k+1}^{n} \frac{\partial L}{\partial q_i} \delta q_i + \sum_{i=1}^{k} \frac{\partial L}{\partial \dot{q}_i} \delta \dot{q}_i + \sum_{i=k+1}^{n} \frac{\partial L}{\partial \dot{q}_i} \delta \dot{q}_i + \frac{\partial L}{\partial t} \delta t \qquad (2\text{-}125)$$

and, using Eq. (2-123),

$$\delta \sum_{i=1}^{k} \beta_i \dot{q}_i = \sum_{i=1}^{k} \frac{\partial L}{\partial \dot{q}_i} \delta \dot{q}_i + \sum_{i=1}^{k} \dot{q}_i \delta \beta_i \qquad (2\text{-}126)$$

which results in

$$\delta \left(L - \sum_{i=1}^{k} \beta_i \dot{q}_i \right) = \sum_{i=k+1}^{n} \frac{\partial L}{\partial q_i} \delta q_i + \sum_{i=k+1}^{n} \frac{\partial L}{\partial \dot{q}_i} \delta \dot{q}_i$$
$$- \sum_{i=1}^{k} \dot{q}_i \delta \beta_i + \frac{\partial L}{\partial t} \delta t \qquad (2\text{-}127)$$

We assume that the varied quantities in Eqs. (2-124) and (2-127) are independent; hence the corresponding coefficients must be equal. Thus,

$$\frac{\partial L}{\partial q_i} = \frac{\partial R}{\partial q_i}$$
$$\qquad\qquad (i = k + 1, \ldots, n) \qquad (2\text{-}128)$$
$$\frac{\partial L}{\partial \dot{q}_i} = \frac{\partial R}{\partial \dot{q}_i}$$

and

$$\dot{q}_i = -\frac{\partial R}{\partial \beta_i} \qquad (i = 1, 2, \ldots, k)$$
$$\qquad\qquad\qquad (2\text{-}129)$$
$$\frac{\partial L}{\partial t} = \frac{\partial R}{\partial t}$$

Now let us substitute from Eq. (2-128) into Lagrange's equations and obtain

$$\frac{d}{dt}\left(\frac{\partial R}{\partial \dot{q}_i}\right) - \frac{\partial R}{\partial q_i} = 0 \qquad (i = k + 1, \ldots, n) \qquad (2\text{-}130)$$

These equations are of the form of Lagrange's equations with the Routhian function used in place of the Lagrangian function. Note, however, that there are only $(n - k)$ second-order equations in the non-ignorable variables. Thus, the Routhian procedure has succeeded in eliminating the ignorable coordinates from the equations of motion. In effect, the number of degrees of freedom has been reduced to $(n - k)$.

Frequently there is no need to solve for the ignorable coordinates. But if Eq. (2-130) has been solved for the $(n - k)$ non-ignorable coordinates, then we can integrate Eq. (2-129) to obtain expressions for the ignorable coordinates, that is,

$$q_i = -\int \frac{\partial R}{\partial \beta_i} dt \qquad (i = 1, 2, \ldots, k) \qquad (2\text{-}131)$$

To illustrate the Routhian method, consider again the Kepler problem of Example 2-7. From Eq. (2-121), we have

$$\dot{\theta} = \frac{\beta}{r^2} \qquad (2\text{-}132)$$

Substituting this expression for $\dot{\theta}$ into the Lagrangian function of Eq. (2-118), we obtain the following Routhian function:

$$R = L - \beta\dot{\theta} = \frac{1}{2}\dot{r}^2 - \frac{\beta^2}{2r^2} + \frac{\mu}{r} \qquad (2\text{-}133)$$

Thus, we see that the system has been reduced to a single degree of freedom. The corresponding equation of motion is obtained from Eq. (2-130).

$$\frac{d}{dt}\left(\frac{\partial R}{\partial \dot{r}}\right) = \ddot{r}$$

$$\frac{\partial R}{\partial r} = \frac{\beta^2}{r^3} - \frac{\mu}{r^2}$$

yielding

$$\ddot{r} - \frac{\beta^2}{r^3} + \frac{\mu}{r^2} = 0 \qquad (2\text{-}134)$$

This result is identical with that found by substituting from Eq. (2-132) into Eq. (2-119), that is, into the r equation obtained by the Lagrangian method.

We have seen that the Routhian function replaces the Lagrangian function for the system after the number of degrees of freedom has been reduced by the ignoration of coordinates. If we look at the Routhian function of Eq. (2-133) from the viewpoint of an observer rotating with the line drawn from the attracting center O to the particle, we see that

$$R = T' - V' \qquad (2\text{-}135)$$

where

$$T' = \tfrac{1}{2}\dot{r}^2$$

$$V' = \frac{\beta^2}{2r^2} - \frac{\mu}{r} \qquad (2\text{-}136)$$

Here T' is the kinetic energy associated with the single degree of freedom. V' is the potential energy arising from the inverse-square gravitational field and from the centrifugal force field due to the angular motion of the particle in its orbit.

Conservative Systems. In Sec. 1-5, we found that a conservative force field has the properties that (1) the generalized force components are obtained from the potential energy function by using

$$Q_i = -\frac{\partial V}{\partial q_i} \qquad (2\text{-}137)$$

where $V(q)$ is a function of the configuration only, and (2) the integral

$$W = \int_A^B \mathbf{Q} \cdot d\mathbf{q} = \sum_{i=1}^n \int_{A_i}^{B_i} Q_i \, dq_i \qquad (2\text{-}138)$$

is independent of the path taken between the given end-points in q-space. If no other forces do work on the system, the total mechanical energy is conserved; hence, the system is called a *conservative system*. In this case, the total energy $E(q, \dot{q}) = T + V$ is an integral of the motion.

It is possible, however, to find an energy-like integral of the motion which is of greater generality. In so doing, we can arrive at a more suitable definition of a conservative system. Now let us define a system to be *conservative* if it meets the following conditions:

1. The standard form of Lagrange's equation (holonomic or nonholonomic) applies.
2. The Lagrangian function L is not an explicit function of time.
3. Any constraint equations can be expressed in the differential form

$$\sum_{i=1}^n a_{ji} \, dq_i = 0 \qquad (j = 1, 2, \ldots, m) \qquad (2\text{-}139)$$

that is, all the coefficients a_{ji} are equal to zero.†

In order to show that the three given conditions are sufficient to ensure the existence of an energy integral, let us consider the case in which a system is described by the standard nonholonomic form of Lagrange's equations, namely,

$$\frac{d}{dt}\left(\frac{\partial L}{\partial \dot{q}_i}\right) - \frac{\partial L}{\partial q_i} = \sum_{j=1}^m \lambda_j a_{ji} \qquad (i = 1, 2, \ldots, n) \qquad (2\text{-}140)$$

where $L(q, \dot{q})$ is not an explicit function of time. The actual constraints may be holonomic or nonholonomic; but, in either event, let us write the m equations of constraint in the form

$$\sum_{i=1}^n a_{ji}\dot{q}_i = 0 \qquad (j = 1, 2, \ldots, m) \qquad (2\text{-}141)$$

where the a's are functions of the q's, and possibly time. Notice, however, that any holonomic constraint functions $\phi_j(q)$ cannot be explicit functions of time because of the assumption that

$$a_{ji} = \frac{\partial \phi_j}{\partial t} = 0 \qquad (2\text{-}142)$$

Now let us consider the total derivative

$$\frac{dL}{dt} = \sum_{i=1}^n \frac{\partial L}{\partial \dot{q}_i}\ddot{q}_i + \sum_{i=1}^n \frac{\partial L}{\partial q_i}\dot{q}_i \qquad (2\text{-}143)$$

---

†By referring to Eq. (1-37), we see that the requirement that a_{ji} be zero for all j is very similar to the requirement that a virtual velocity at any point of the system be a possible velocity. The conditions stated for a conservative system, however, do not require that the system be scleronomic.

But, from Lagrange's equations as given in Eq. (2-140), we have

$$\frac{\partial L}{\partial q_i} = \frac{d}{dt}\left(\frac{\partial L}{\partial \dot{q}_i}\right) - \sum_{j=1}^{m} \lambda_j a_{ji} \tag{2-144}$$

Hence, we obtain from Eqs. (2-143) and (2-144) that

$$\frac{dL}{dt} = \sum_{i=1}^{n} \frac{\partial L}{\partial \dot{q}_i}\ddot{q}_i + \sum_{i=1}^{n} \frac{d}{dt}\left(\frac{\partial L}{\partial \dot{q}_i}\right)\dot{q}_i - \sum_{i=1}^{n}\sum_{j=1}^{m} \lambda_j a_{ji}\dot{q}_i \tag{2-145}$$

The double summation term of Eq. (2-145) is zero as a result of Eq. (2-141). Therefore, we obtain

$$\frac{dL}{dt} = \frac{d}{dt}\left(\sum_{i=1}^{n} \frac{\partial L}{\partial \dot{q}_i}\dot{q}_i\right) \tag{2-146}$$

which can be integrated to give

$$\sum_{i=1}^{n} \frac{\partial L}{\partial \dot{q}_i}\dot{q}_i - L = h \tag{2-147}$$

where h is a constant.

Thus, we have obtained a constant of the motion which is known as the *Jacobi integral* or the *energy integral*. This integral of the motion exists for all conservative systems.

Let us recall that the Lagrangian function can be written in the form

$$L = T_2 + T_1 + T_0 - V \tag{2-148}$$

where the kinetic energy terms are separated in accordance with their degree in the $\dot{q}$'s, as defined by Eqs. (2-6), (2-8), and (2-10). Assuming that V is not a function of the $\dot{q}$'s, we see that

$$\sum_{i=1}^{n} \frac{\partial L}{\partial \dot{q}_i}\dot{q}_i = 2T_2 + T_1 \tag{2-149}$$

and therefore Eq. (2-147) can be written in the form

$$T_2 - T_0 + V = h \tag{2-150}$$

Hence, we confirm that the Jacobi integral has the units of energy. Notice that T_0 and V are both functions of the q's only, for this case of a conservative system. If we group these functions together, we can write

$$T' + V' = h \tag{2-151}$$

where

$$\begin{aligned} T' &= T_2 \\ V' &= V - T_0 \end{aligned} \tag{2-152}$$

We see that, in addition to the original force field represented by V, the potential function V' includes a force field due to T_0. This field is artificial in the sense that it consists of inertial forces arising from the fact that some of the q's are measured relative to a moving reference. T' is the kinetic energy, assuming that any moving constraints or reference frames are held fixed.

In summary, then, the energy $T' + V'$ is constant for any conservative system, but this energy is not always the total energy of the system measured relative to an inertial frame.

Natural Systems. A natural system is a *conservative* system which has the additional properties that (1) it is described by the standard holonomic form of Lagrange's equations and (2) the kinetic energy is expressed as a homogeneous quadratic function of the $\dot{q}$'s, that is,

$$T = T_2 = \tfrac{1}{2} \sum_{i=1}^{n} \sum_{j=1}^{n} m_{ij} \dot{q}_i \dot{q}_j \qquad (2\text{-}153)$$

where the inertial coefficients m_{ij} may be functions of the q's but not of time.

The Jacobi integral is particularly simple for a natural system; it is equal to the *total energy*. We see this by noting first that $T_0 = T_1 = 0$, and therefore Eq. (2-150) becomes

$$T + V = h \qquad (2\text{-}154)$$

indicating that the total energy is conserved.

Although the transformation equations relating the x's and the q's may contain t explicitly for certain conservative systems, this is no longer possible for a natural system. Since $T_0 = 0$, we see from Eq. (2-10) that $\partial x_k / \partial t$ must be zero for all k.

Now let us consider the form of the equations of motion for a natural system. First, since $T_1 = 0$ and T_2 is not an explicit function of time, we find from Eqs. (2-6) and (2-8) that

$$a_i = 0, \qquad \frac{\partial m_{ij}}{\partial t} = 0 \qquad (i, j = 1, 2, \dots, n) \qquad (2\text{-}155)$$

Then, referring to Eq. (2-37), we obtain

$$\sum_{j=1}^{n} m_{ij} \ddot{q}_j + \frac{1}{2} \sum_{j=1}^{n} \sum_{l=1}^{n} \left(\frac{\partial m_{ij}}{\partial q_l} + \frac{\partial m_{il}}{\partial q_j} - \frac{\partial m_{jl}}{\partial q_i} \right) \dot{q}_j \dot{q}_l + \frac{\partial V}{\partial q_i} = 0$$

$$(i = 1, 2, \dots, n) \qquad (2\text{-}156)$$

Comparing Eqs. (2-37) and (2-156), we see that the differential equations of motion for a natural system are considerably simpler than those for a more general conservative system. In particular, the equations describing a natural system contain no linear terms in the $\dot{q}$'s. Consequently, the $\dot{q}$'s appear only as quadratic terms.

It is important to notice that a holonomic conservative system with $T_1 = 0$ and $T_0 \neq 0$ has equations of motion which are very similar to Eq. (2-156) even though it is not considered as a natural system. This similarity results from the fact that if T_0 is viewed as a part of a potential energy $V' = V - T_0$, then the remaining kinetic energy T_2 is quadratic in the $\dot{q}$'s, as in a natural system. Hence, the equations of motion are those of a natural system with V' as the potential energy function and T_2 as its kinetic energy.

A holonomic conservative system with $T_1 \neq 0$ is, in general, a *gyroscopic* system. The presence of T_1 results in terms of the form $\gamma_{ij}\dot{q}_j$ in the equations of motion, where the coefficients γ_{ij} form a skew symmetric matrix, as was shown in Eq. (2-40). Gyroscopic systems will be discussed further in Sec. 3-3.

Example 2-8. Suppose a mass-spring system is attached to a frame which is translating with a uniform velocity v_0, as shown in Fig. 2-8. Let l_0 be the unstressed spring length and use the elongation x as the generalized coordinate. Find the Jacobi integral for the system.

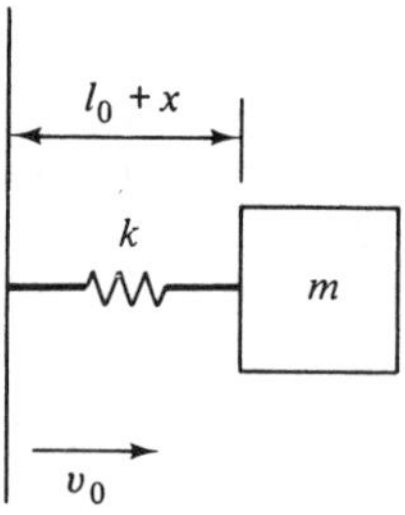

The kinetic energy is

$$T = \tfrac{1}{2}m(v_0 + \dot{x})^2 \tag{2-157}$$

which yields

$$T_2 = \tfrac{1}{2}m\dot{x}^2$$
$$T_1 = mv_0\dot{x} \tag{2-158}$$
$$T_0 = \tfrac{1}{2}mv_0^2$$

Fig. 2-8. A translating mass-spring system.

The potential energy is

$$V = \tfrac{1}{2}kx^2 \tag{2-159}$$

This mass-spring system meets all the conditions of a *holonomic conservative system* since T and V are not explicit functions of time, and the only generalized force Q_x is derivable from V. Although the moving frame does work on the system, resulting in a changing total energy $T + V$, the Jacobi integral exists and is equal to

$$T_2 - T_0 + V = \tfrac{1}{2}m\dot{x}^2 - \tfrac{1}{2}mv_0^2 + \tfrac{1}{2}kx^2 = h \tag{2-160}$$

where h is a constant. T_0 is constant in this example, so we see that $T_2 + V$ is also constant.

Another approach is to notice that the moving frame is a valid inertial reference. Relative to this frame, we have

$$T' = \tfrac{1}{2}m\dot{x}^2$$
$$V' = \tfrac{1}{2}kx^2 \tag{2-161}$$

Since T' is quadratic in $\dot{x}$, we see that we have a *natural system* relative to this reference frame. Hence, $T' + V'$ is constant, that is, the total energy is conserved. Note that this energy is identical with $T_2 + V$ measured relative to the fixed frame.

Example 2-9. Let us consider again the system discussed in Example 2-4. A small tube, bent in the form of a circle of radius r, rotates about a vertical diameter with a constant angular velocity ω. A particle of mass m can slide without friction inside the tube. At any given time, the configuration of the system is specified by the angle θ which is measured from the upward vertical to the line connecting the center O and the particle. Find the Jacobi integral.

Suppose we assume a fixed Cartesian reference frame with its origin at O and with the z axis vertical. The plane of the tube coincides with the xz plane at $t = 0$. The transformation equations relating the generalized coordinate θ and the position (x, y, z) of the particle are the following:

$$x = r \sin \theta \cos \omega t$$
$$y = r \sin \theta \sin \omega t \qquad (2\text{-}162)$$
$$z = r \cos \theta$$

Since these transformation equations contain time explicitly, this is a *rheonomic system*. It is also holonomic and has the same number of degrees of freedom as generalized coordinates, namely, one.

The kinetic and potential energy functions are

$$T = \tfrac{1}{2}m(r^2\dot\theta^2 + r^2\omega^2 \sin^2 \theta) \qquad (2\text{-}163)$$
$$V = mgr \cos \theta \qquad (2\text{-}164)$$

Therefore, the Lagrangian function is

$$L = \tfrac{1}{2}mr^2\dot\theta^2 + \tfrac{1}{2}mr^2\omega^2 \sin^2 \theta - mgr \cos \theta \qquad (2\text{-}165)$$

We see that the Lagrangian L is not an explicit function of time, even though the system is rheonomic. Hence, the system is *conservative*. Its Jacobi integral is

$$T_2 - T_0 + V = \tfrac{1}{2}mr^2\dot\theta^2 - \tfrac{1}{2}mr^2\omega^2 \sin^2 \theta + mgr \cos \theta = h \qquad (2\text{-}166)$$

Now let us consider the Lagrangian function of Eq. (2-165) to be of the form

$$L = T' - V' \qquad (2\text{-}167)$$

where

$$T' = T_2 = \tfrac{1}{2}mr^2\dot\theta^2 \qquad (2\text{-}168)$$

and

$$V' = V - T_0 = mgr \cos \theta - \tfrac{1}{2}mr^2\omega^2 \sin^2 \theta \qquad (2\text{-}169)$$

We see that T' is the kinetic energy of the particle relative to a reference frame which is rotating with the circular tube. The potential energy V' includes the actual gravitational energy plus another term $-T_0$ which accounts for the centrifugal force due to the rotation about the vertical axis.

The result of taking the viewpoint of a rotating observer is that T' and V' are of the form associated with a *natural system;* that is, neither T' nor V' is an explicit function of time, T' is quadratic in $\dot\theta$, and V' is a function of the position θ only. Hence, the total energy relative to this rotating frame is conserved, and we obtain

$$T' + V' = h \qquad (2\text{-}170)$$

in agreement with the Jacobi integral of Eq. (2-166).

V' is plotted as a function of θ in Fig. 2-9 for the case where $\omega^2 > g/r$. Equilibrium points relative to the rotating frame occur at those values of θ for which $dV'/d\theta = 0$, that is, at

$$\theta = 0, \pi, \cos^{-1}\left(\frac{-g}{r\omega^2}\right)$$

The values 0 and π occur at local maxima of V' and therefore are points of unstable equilibrium, while the values $\cos^{-1}(-g/r\omega^2)$ correspond to stable equilibrium points at local minima of V'.

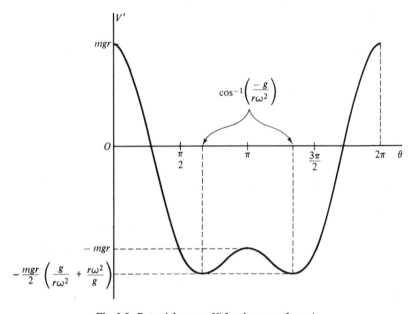

Fig. 2-9. Potential energy V' for the case $\omega^2 > g/r$.

If $0 \leq \omega^2 \leq g/r$, there is no longer a hump in the curve near $\theta = \pi$. In this case, there is a maximum of V' at $\theta = 0$ and a minimum at $\theta = \pi$.

It is interesting to consider the same system using the spherical coordinates θ and ϕ as generalized coordinates. In this case, we obtain

$$T = \tfrac{1}{2}m(r^2\dot{\theta}^2 + r^2\dot{\phi}^2 \sin^2 \theta) \qquad (2\text{-}171)$$

$$V = mgr \cos \theta \qquad (2\text{-}172)$$

The holonomic constraint can be expressed in the differential form

$$d\phi - \omega \, dt = 0 \qquad (2\text{-}173)$$

Here we see that neither T nor V is an explicit function of time and the standard form of Lagrange's equation applies. Nevertheless, the system is *not conservative* in this formulation because $a_{jt} = -\omega \neq 0$. Furthermore, the Jacobi expression ($T + V$ in this case) is not constant because the moving constraint does work on the system.

Finally, let us consider the same system in terms of the Cartesian coordinates of the particle. Here we have

$$T = \tfrac{1}{2}m(\dot{x}^2 + \dot{y}^2 + \dot{z}^2) \qquad (2\text{-}174)$$

$$V = mgz \qquad (2\text{-}175)$$

There are two holonomic constraints, namely,

$$x^2 + y^2 + z^2 = r^2 \qquad (2\text{-}176)$$

$$y = x \tan \omega t \qquad (2\text{-}177)$$

Written in differential form, these constraints are

$$x\,dx + y\,dy + z\,dz = 0 \qquad (2\text{-}178)$$

$$\tan \omega t\,dx - dy + \omega x \sec^2 \omega t\,dt = 0 \qquad (2\text{-}179)$$

Looking at Eq. (2-179) we see that

$$a_{jt} = \omega x \sec^2 \omega t \neq 0$$

Therefore, the system is *not conservative*. Again, the Jacobi expression is the total energy $T + V$ which is not conserved.

In summary, it can be seen that whether a given system is classed as conservative or not may depend upon the coordinates used in its description.

Example 2-10. Let us consider once again the system of Example 2-6. Two particles, each of mass m, are connected by a rigid massless rod of length l (Fig. 2-6). The particles are supported by knife edges placed perpendicular to the rod. Assuming that all motion is confined to the horizontal xy plane, find the Jacobi integral.

Since θ is given explicitly as a function of time, the system is rheonomic. Let us take the coordinates (x, y) of the center of mass as generalized coordinates. We note that the potential energy is zero, so the Lagrangian function is

$$L = T = m(\dot{x}^2 + \dot{y}^2) + \tfrac{1}{4}ml^2\omega^2 \qquad (2\text{-}180)$$

The nonholonomic constraint equation is

$$\cos \omega t\,dx + \sin \omega t\,dy = 0 \qquad (2\text{-}181)$$

which specifies that the velocity of the center of mass must be perpendicular to the rod.

We observe that the Lagrangian L is not an explicit function of time, even though the system is rheonomic and the coefficients in the constraint equation are explicit functions of time. Furthermore, we see that $a_{jt} = 0$. Hence, the system is *conservative*. Its Jacobi integral is

$$T_2 - T_0 + V = m(\dot{x}^2 + \dot{y}^2) - \tfrac{1}{4}ml^2\omega^2 = h \qquad (2\text{-}182)$$

Note that T_0 is a constant in this case and can be omitted from the Lagrangian function and from Jacobi's integral. Thus, we find that the translational kinetic energy is constant, implying that the velocity of the center of mass is constant.

Another approach to this example is to describe the system configuration by using the Cartesian coordinates of its two particles. In this case the Lagrangian function is

$$L = \tfrac{1}{2}m(\dot{x}_1^2 + \dot{y}_1^2 + \dot{x}_2^2 + \dot{y}_2^2) \qquad (2\text{-}183)$$

There are two holonomic equations of constraint, namely,

$$(x_2 - x_1)^2 + (y_2 - y_1)^2 = l^2 \qquad (2\text{-}184)$$

$$(x_2 - x_1)\sin \omega t - (y_2 - y_1)\cos \omega t = 0 \qquad (2\text{-}185)$$

giving the length and orientation of the rod. There is also a nonholonomic constraint which can be described by

$$\cos \omega t \, dx_1 + \sin \omega t \, dy_1 = 0 \qquad (2\text{-}186)$$

expressing the fact that the velocity at the given particle is perpendicular to the rod. A similar equation could be written for the second particle, but it would not be independent of the constraint equations already given.

Equations (2-184) and (2-185) can be combined with (2-186) to yield

$$(x_2 - x_1) \, dx_1 + (y_2 - y_1) \, dy_1 = 0 \qquad (2\text{-}187)$$

$$(x_2 - x_1) \, dx_2 + (y_2 - y_1) \, dy_2 = 0 \qquad (2\text{-}188)$$

Then we see that Eqs. (2-186)—(2-188) represent a set of three independent constraint equations which meet the condition that $a_{jt} = 0$. Hence, the sufficient conditions for a *conservative* system have been met. The existence of the Jacobi integral implies that the total kinetic energy is constant.

It is interesting to note that if the orientation $\theta(t)$ were given as any continuous function of time, the Lagrangian L would, in general, be an explicit function of time, implying that the system is not conservative. Nevertheless, if we assume that the given orientation is enforced by the application of a couple

$$M = \tfrac{1}{2}ml^2\ddot{\theta}$$

then the translational kinetic energy is conserved because no work is done by the constraint forces moving over the path of the center of mass. Of course, the rotational kinetic energy changes as $\dot{\theta}$ changes.

Liouville's System. As we consider the problem of finding the integrals or constants of the motion, a question arises concerning what characteristics of a system make it possible to completely solve for its motion by quadratures, that is, in terms of indefinite integrals, each involving only one variable. The general answer to this question is not known, but we can give examples of some systems which are *separable*, and therefore are capable of being solved by quadratures.

First, we recall that a system having n degrees of freedom requires $2n$ integrals of the motion for a complete solution. For a standard holonomic system, the presence of ignorable coordinates permits a reduction of the effective number of degrees of freedom by using the Routhian procedure. Furthermore, if the system is conservative, the energy integral h is immediately available. Hence, it can be seen that any conservative holonomic system with n degrees of freedom and $(n-1)$ ignorable coordinates can be integrated completely by quadratures. We find that $2(n-1)$ constants are obtained by ignoration of coordinates, and the integral of energy then provides an equation of the form

$$\dot{q} = f(q)$$

which can be integrated to complete the solution. Here q is the last remaining coordinate after all but one have been eliminated by the Routhian procedure.

If a conservative holonomic system does not have a sufficient number of ignorable coordinates to guarantee separability, it may still be separable if it is an *orthogonal system*, that is, a natural system in which T contains only $\dot{q}_i^2$ terms, and no cross-products in the $\dot{q}$'s.

As an example, suppose that

$$T = \tfrac{1}{2} f \sum_{i=1}^{n} \dot{q}_i^2 \tag{2-189}$$

$$V = \frac{1}{f} \sum_{i=1}^{n} v_i(q_i) \tag{2-190}$$

where we define

$$f = \sum_{i=1}^{n} f_i(q_i) > 0 \tag{2-191}$$

We will now show that this system is separable.

Let us use Lagrange's equation in the form

$$\frac{d}{dt}\left(\frac{\partial T}{\partial \dot{q}_i}\right) - \frac{\partial T}{\partial q_i} + \frac{\partial V}{\partial q_i} = 0 \tag{2-192}$$

and we obtain

$$\frac{d}{dt}(f\dot{q}_i) - \frac{1}{2}\frac{\partial f_i}{\partial q_i}\sum_{j=1}^{n}\dot{q}_j^2 + \frac{1}{f}\frac{\partial v_i}{\partial q_i} - \frac{V}{f}\frac{\partial f_i}{\partial q_i} = 0 \tag{2-193}$$

Now, this is a natural system, so it has an energy integral given by

$$T + V = \tfrac{1}{2}f\sum_{j=1}^{n}\dot{q}_j^2 + V = h \tag{2-194}$$

Hence,

$$\frac{1}{2}\sum_{j=1}^{n}\dot{q}_j^2 = \frac{1}{f}(h - V) \tag{2-195}$$

Substituting from Eq. (2-195) into (2-193) and simplifying, we have

$$\frac{d}{dt}(f\dot{q}_i) - \frac{h}{f}\frac{\partial f_i}{\partial q_i} + \frac{1}{f}\frac{\partial v_i}{\partial q_i} = 0 \tag{2-196}$$

Next, we multiply Eq. (2-196) by $2f\dot{q}_i$ and obtain

$$\frac{d}{dt}(f^2\dot{q}_i^2) - 2h\frac{\partial f_i}{\partial q_i}\dot{q}_i + 2\frac{\partial v_i}{\partial q_i}\dot{q}_i = 0$$

or

$$\frac{d}{dt}(f^2\dot{q}_i^2) = 2\frac{d}{dt}(hf_i - v_i) \qquad (2\text{-}197)$$

Integrating, the result is

$$f^2\dot{q}_i^2 = 2[hf_i(q_i) - v_i(q_i) + c_i] \qquad (i = 1, 2, \ldots, n) \qquad (2\text{-}198)$$

where the c's are constants of integration. It can be shown from Eqs. (2-190), (2-194), and (2-198) that

$$\sum_{i=1}^{n} c_i = 0 \qquad (2\text{-}199)$$

Hence the c's and h together comprise n independent constants of the motion.

The remaining n integrals of the motion are obtained by writing Eq. (2-198) in the form

$$\frac{dq_i}{dt} = \frac{\sqrt{2(hf_i - v_i + c_i)}}{f}$$

which implies that

$$\frac{dq_1}{\sqrt{2(hf_1 - v_1 + c_1)}} = \frac{dq_2}{\sqrt{2(hf_2 - v_2 + c_2)}} = \cdots$$

$$= \frac{dq_n}{\sqrt{2(hf_n - v_n + c_n)}} = \frac{dt}{f} = d\tau \qquad (2\text{-}200)$$

where τ is a time-like parameter. Each differential expression is a function of a single q_i, so the problem is reduced to quadratures. Integrating these expressions produces the required n additional constants of the motion.

This system can be generalized rather easily to become a Liouville system. If we replace dq_i by $\sqrt{M_i(q_i)}\,dq_i$, we obtain

$$T = \tfrac{1}{2}f\sum_{i=1}^{n} M_i(q_i)\dot{q}_i^2 \qquad (2\text{-}201)$$

where we assume that $M_i(q_i) > 0$. As before,

$$V = \frac{1}{f}\sum_{i=1}^{n} v_i(q_i) \qquad (2\text{-}202)$$

A natural system having T and V of the form given by Eqs. (2-201) and (2-202) is called a *Liouville system*.

Corresponding to Eq. (2-200) we now have

$$\frac{dq_1}{\sqrt{\phi_1(q_1)}} = \frac{dq_2}{\sqrt{\phi_2(q_2)}} = \cdots = \frac{dq_n}{\sqrt{\phi_n(q_n)}} = \frac{dt}{f} = d\tau \qquad (2\text{-}203)$$

where

$$\phi_i(q_i) = \frac{2}{M_i}(hf_i - v_i + c_i) \qquad (i = 1, 2, \ldots, n) \qquad (2\text{-}204)$$

Using Eqs. (2-191) and (2-203), we obtain

$$\sum_{i=1}^{n} \frac{f_i \, dq_i}{\sqrt{\phi_i(q_i)}} = dt \qquad (2\text{-}205)$$

or

$$\sum_{i=1}^{n} \int \frac{f_i \, dq_i}{\sqrt{\phi_i(q_i)}} = t + \beta_1 \qquad (2\text{-}206)$$

Similarly, taking differences of the indefinite integrals of Eq. (2-203), we have

$$\int \frac{dq_1}{\sqrt{\phi_1(q_1)}} - \int \frac{dq_j}{\sqrt{\phi_j(q_j)}} = \beta_j \qquad (j = 2, 3, \ldots, n) \qquad (2\text{-}207)$$

where the first integral is chosen arbitrarily as a reference. Thus, Eqs. (2-206) and (2-207) provide n independent constants $\beta_1, \beta_2, \ldots, \beta_n$ which, with the previous $(n - 1)$ independent c's and the energy constant h, constitute the required $2n$ independent constants of the motion.

In evaluating the integrals of Eq. (2-206) or (2-207), a question arises concerning the sign of $\sqrt{\phi_j(q_j)}$. Remembering that f is positive, we find from Eq. (2-203) that $\sqrt{\phi_i(q_i)}$ has the same sign as dq_i. This is of particular importance in studying *libration* motions, that is, motions in which one or more q's oscillate between fixed limiting values.

The Liouville system is discussed further in Sec. 5-3.

Example 2-11. Consider again the spherical pendulum of Example 2-1. Reduce the problem to quadratures and obtain the integrals of the motion.

Method 1. Ignoration of coordinates. Using the spherical coordinates θ and ϕ, the expressions for the kinetic and potential energies are

$$T = \tfrac{1}{2}ml^2(\dot{\theta}^2 + \dot{\phi}^2 \sin^2 \theta) \qquad (2\text{-}208)$$

$$V = mgl \cos \theta \qquad (2\text{-}209)$$

where m is the mass of a particle which is suspended by a massless string of length l.

Here we have a conservative holonomic system having two degrees of freedom and one ignorable coordinate. Hence it can be solved completely by quadratures.

First, we see that the Lagrangian function is

$$L = \tfrac{1}{2}ml^2(\dot{\theta}^2 + \dot{\phi}^2 \sin^2 \theta) - mgl \cos \theta \qquad (2\text{-}210)$$

Since ϕ does not appear explicitly, it is an ignorable coordinate, and we have

$$\frac{\partial L}{\partial \dot{\phi}} = ml^2\dot{\phi} \sin^2 \theta = \alpha_\phi \qquad (2\text{-}211)$$

where we now adopt the notation that α_ϕ is the constant generalized momentum conjugate to ϕ. In other words, the angular momentum is conserved about a vertical axis through the support.

The Routhian function is

$$R = L - \alpha_\phi \dot{\phi} = \frac{1}{2} ml^2 \dot{\theta}^2 - \frac{\alpha_\phi^2}{2ml^2 \sin^2 \theta} - mgl \cos \theta \qquad (2\text{-}212)$$

and we note that $\dot{\phi}$ has been eliminated by using Eq. (2-211). Thus, we see that

$$R = T' - V' \qquad (2\text{-}213)$$

where

$$T' = \tfrac{1}{2} ml^2 \dot{\theta}^2 \qquad (2\text{-}214)$$

$$V' = \frac{\alpha_\phi^2}{2ml^2 \sin^2 \theta} + mgl \cos \theta \qquad (2\text{-}215)$$

The form of T' and V' is that of a natural system having one degree of freedom. Hence, we can immediately write the energy integral.

$$T' + V' = \frac{1}{2} ml^2 \dot{\theta}^2 + \frac{\alpha_\phi^2}{2ml^2 \sin^2 \theta} + mgl \cos \theta = h \qquad (2\text{-}216)$$

Solving for $\dot{\theta}$, we obtain

$$\dot{\theta} = \sqrt{\frac{2}{ml^2}(h - mgl \cos \theta - \alpha_\phi^2/2ml^2 \sin^2 \theta)}$$

or

$$\frac{ml^2 \sin \theta \, d\theta}{\sqrt{2ml^2 \sin^2 \theta \, (h - mgl \cos \theta) - \alpha_\phi^2}} = dt \qquad (2\text{-}217)$$

Integrating, we have

$$\int_{\theta_0}^{\theta} \frac{ml^2 \sin \theta \, d\theta}{\sqrt{2ml^2 \sin^2 \theta \, (h - mgl \cos \theta) - \alpha_\phi^2}} = t - t_0 \qquad (2\text{-}218)$$

where $\theta(t_0) = \theta_0$. The motion in θ is usually a libration with $0 < \theta < \pi$. Hence, the sign of the square root should be the same as that of $d\theta$ since each increment dt is positive.

From Eq. (2-211) we obtain

$$d\phi = \frac{\alpha_\phi \, dt}{ml^2 \sin^2 \theta} \qquad (2\text{-}219)$$

and, using Eq. (2-217), we find that

$$\frac{\alpha_\phi \, d\theta}{\sin \theta \sqrt{2ml^2 \sin^2 \theta \, (h - mgl \cos \theta) - \alpha_\phi^2}} = d\phi \qquad (2\text{-}220)$$

Hence,

$$\int_{\theta_0}^{\theta} \frac{\alpha_\phi \, d\theta}{\sin \theta \sqrt{2ml^2 \sin^2 \theta \, (h - mgl \cos \theta) - \alpha_\phi^2}} = \phi - \phi_0 \qquad (2\text{-}221)$$

where $\phi(t_0) = \phi_0$. Again, the sign of the square root is the same as that of $d\theta$. Note that α_ϕ has the same sign as $\dot{\phi}(t_0)$.

Thus, we have obtained the four required constants of the motion, namely, the expressions for α_ϕ, h, t_0, and ϕ_0 given by Eqs. (2-211), (2-216), (2-218), and (2-221).

Method 2. Now consider the spherical pendulum as a Liouville system. Comparing the expressions for T and V given in Eqs. (2-208) and (2-209) with the standard Liouville forms of Eqs. (2-201) and (2-202), we find that

$$f_\theta = ml^2 \sin^2 \theta, \qquad M_\theta = \frac{1}{\sin^2 \theta}, \qquad v_\theta = m^2 g l^3 \sin^2 \theta \cos \theta \qquad (2\text{-}222)$$

and

$$f_\phi = 0, \qquad M_\phi = 1, \qquad v_\phi = 0 \qquad (2\text{-}223)$$

Using Eq. (2-204), we obtain

$$\phi_\theta = 2 \sin^2 \theta \, [ml^2 \sin^2 \theta \, (h - mgl \cos \theta) + c_\theta]$$
$$\phi_\phi = 2c_\phi \qquad\qquad\qquad\qquad\qquad\qquad\qquad\qquad\qquad (2\text{-}224)$$

where, from Eqs. (2-198), (2-199), and (2-211), we have

$$2c_\phi = -2c_\theta = (ml^2\dot{\phi} \sin^2 \theta)^2 = \alpha_\phi^2 \qquad (2\text{-}225)$$

Finally, substituting these expressions into Eqs. (2-206) and (2-207) and writing the results in the form of definite integrals, we obtain

$$\int_{\theta_0}^{\theta} \frac{ml^2 \sin^2 \theta \, d\theta}{\sqrt{\phi_\theta(\theta)}} = t - t_0 \qquad (2\text{-}226)$$

and

$$\int_{\theta_0}^{\theta} \frac{d\theta}{\sqrt{\phi_\theta(\theta)}} = \int_{\phi_0}^{\phi} \frac{d\phi}{\sqrt{2c_\phi}}$$

or

$$\int_{\theta_0}^{\theta} \frac{\alpha_\phi \, d\theta}{\sqrt{\phi_\theta(\theta)}} = \phi - \phi_0 \qquad (2\text{-}227)$$

We see that Eqs. (2-226) and (2-227) agree with Eqs. (2-218) and (2-221) obtained previously.

In Chapter 5, the problem of obtaining integrals of the motion for separable systems will be considered again using the Hamilton-Jacobi method.

2-4. SMALL OSCILLATIONS

One of the principal applications of Lagrange's equations is in the analysis of small motions of a conservative mechanical system about an equilibrium configuration. Although the equations of motion can often be obtained by other means such as the direct application of Newton's laws, the Lagrangian method results in equations of motion in a form which emphasizes the symmetries of the system. Let us consider, then, the Lagrangian approach to the theory of small oscillations. We shall present, however, only an introduction to this rather extensive subject.

Equations of Motion. Suppose we have a *natural system* whose configuration is specified by the n independent generalized coordinates $q_1, q_2, \ldots, q_n$. Let us assume that the q's are measured from a position of equilibrium, and consider *small motions* about this equilibrium position. From Eq. (1-95), we find that if we let the reference value V_0 be zero, the potential energy can be written in the form

$$V = \frac{1}{2} \sum_{i=1}^{n} \sum_{j=1}^{n} \left(\frac{\partial^2 V}{\partial q_i \, \partial q_j} \right)_0 q_i q_j + \cdots \quad (2\text{-}228)$$

Neglecting terms of higher order than the second in the q's, we obtain

$$V = \frac{1}{2} \sum_{i=1}^{n} \sum_{j=1}^{n} k_{ij} q_i q_j \quad (2\text{-}229)$$

where the *stiffness coefficients* are

$$k_{ij} = k_{ji} = \left(\frac{\partial^2 V}{\partial q_i \, \partial q_j} \right)_0 \quad (2\text{-}230)$$

Thus, we see that the potential energy V is a homogeneous quadratic function of the q's for small motions near a position of equilibrium.

Let us assume that the system consists of N particles whose positions are given by the $3N$ Cartesian coordinates $x_1, x_2, \ldots, x_{3N}$. We find from Eqs. (2-7) and (2-153) that the kinetic energy is of the form

$$T = \frac{1}{2} \sum_{i=1}^{n} \sum_{j=1}^{n} m_{ij} \dot{q}_i \dot{q}_j \quad (2\text{-}231)$$

where, for small motions,

$$m_{ij} = m_{ji} = \sum_{k=1}^{3N} m_k \left(\frac{\partial x_k}{\partial q_i} \right)_0 \left(\frac{\partial x_k}{\partial q_j} \right)_0 \quad (2\text{-}232)$$

Also, for the natural system assumed here, the kinetic energy is a *positive definite quadratic function* of the $\dot{q}$'s.

The equations of motion are obtained by using Lagrange's equations. First, we find from Eq. (2-26) that the Lagrangian function is

$$L = \frac{1}{2} \sum_{i=1}^{n} \sum_{j=1}^{n} m_{ij} \dot{q}_i \dot{q}_j - \frac{1}{2} \sum_{i=1}^{n} \sum_{j=1}^{n} k_{ij} q_i q_j \quad (2\text{-}233)$$

Then, using Eq. (2-27), we obtain the following equations:

$$\sum_{j=1}^{n} m_{ij} \ddot{q}_j + \sum_{j=1}^{n} k_{ij} q_j = 0 \qquad (i = 1, 2, \ldots, n) \quad (2\text{-}234)$$

or, in matrix form,

$$\mathbf{m}\ddot{\mathbf{q}} + \mathbf{k}\mathbf{q} = \mathbf{0} \quad (2\text{-}235)$$

Notice that these equations of motion are *linear*, second-order, ordinary differential equations. Also, the **m** and **k** matrices are *constant* and *symmetric*. In general, if Newton's laws are used in obtaining the equations of motion for a system of this sort, the **m** and **k** matrices will not be symmetric. The loss of symmetry in the equations can occur merely by multiplying one of the equations by a constant, or by adding equations. This does not change the nature of the solution, but nevertheless does change the appearance of the equations of motion. An advantage of the Lagrangian method is that the systematic approach preserves the symmetry of the coefficient matrices for those systems in which T and V are adequately represented by quadratic functions of the velocities and displacements, respectively.

Natural Modes. Let us consider a system whose differential equations of motion are given by Eq. (2-234). Assume solutions of the form

$$q_j = A_j C \cos(\omega t + \phi) \qquad (j = 1, 2, \ldots, n) \qquad (2\text{-}236)$$

where the amplitude of the oscillation in q_j is the product of the constants A_j and C. Here C acts as an overall scale factor for the q's, whereas the A's indicate their relative magnitudes.

If we substitute the trial solutions of Eq. (2-236) into Eq. (2-234), we obtain

$$\sum_{j=1}^{n} (-\omega^2 m_{ij} + k_{ij}) A_j C \cos(\omega t + \phi) = 0 \qquad (i = 1, 2, \ldots, n) \qquad (2\text{-}237)$$

The factor $C \cos(\omega t + \phi)$ cannot be zero continuously except for the trivial case in which all the q's remain zero. Therefore, we conclude that

$$\sum_{j=1}^{n} (k_{ij} - \omega^2 m_{ij}) A_j = 0 \qquad (i = 1, 2, \ldots, n) \qquad (2\text{-}238)$$

If the A's are not all zero, then the determinant of their coefficients must vanish, that is,

$$\begin{vmatrix} (k_{11} - \omega^2 m_{11}) & (k_{12} - \omega^2 m_{12}) & \cdots & (k_{1n} - \omega^2 m_{1n}) \\ (k_{21} - \omega^2 m_{21}) & (k_{22} - \omega^2 m_{22}) & & \cdot \\ \cdot & & & \cdot \\ \cdot & & & \cdot \\ \cdot & & & \cdot \\ (k_{n1} - \omega^2 m_{n1}) & \cdots & \cdots & (k_{nn} - \omega^2 m_{nn}) \end{vmatrix} = 0 \qquad (2\text{-}239)$$

The evaluation of this determinant results in an nth-degree algebraic equation in ω^2 which is called the *characteristic equation*. The n roots ω_k^2, where $k = 1, 2, \ldots, n$, are known as *characteristic values* or *eigenvalues*, each being the square of a natural frequency which is usually expressed in rad/sec. It is known from matrix theory that the roots ω_k^2 are all real and finite if T is positive definite and if both **m** and **k** are real symmetric matrices.

If, in addition, V is *positive definite*, then the ω_k^2 are all positive and the motion occurs about a position of stable equilibrium. If V is *positive semidefinite*; that is, if the determinant $|k|$ or any of its principal minors is zero, but none are negative; then at least one of the ω_k^2 is zero and the system is in neutral equilibrium at the reference configuration. Finally, if $|k|$ or any of its principal minors is negative, then at least one of the ω_k^2 is negative and the reference position is one of unstable equilibrium.

Let us summarize the effect of a given root ω_k^2 on the stability of the system as follows:

Case 1: $\omega_k^2 > 0$. The solution for each coordinate contains a term of the form $C_k \cos(\omega_k t + \phi_k)$ in accordance with the assumed solution given by Eq. (2-236). This term is stable, but the overall stability depends upon the other roots as well.

Case 2: $\omega_k^2 = 0$. Here we have a repeated zero root ω_k, corresponding to a term in the solution of the form $(C_k t + D_k)$. This implies that a steady drift can occur in one or more of the coordinates and is characteristic of a neutrally stable system.

Case 3: $\omega_k^2 < 0$. The corresponding pair of roots $\omega_k = \pm i\gamma_k$ is imaginary and the resulting terms of the form $(C_k \cosh \gamma_k t + D_k \sinh \gamma_k t)$ imply an unstable solution.

Returning now to Eq. (2-238), we find that for each ω_k^2 we can write a set of n simultaneous algebraic equations involving the n amplitude coefficients A_j. Because these equations are homogeneous, however, there is no unique solution for the amplitude coefficients, but only for the *ratios* among them. For convenience, let us solve for the amplitude ratios with respect to A_1, that is, let us take $A_1 = 1$. If we arbitrarily eliminate the first equation of (2-238) and use matrix notation in solving the remaining $(n-1)$ equations for the $(n-1)$ unknown amplitudes, we obtain

$$[(k_{ij} - \omega_k^2 m_{ij})]\{A_j^{(k)}\} = \{(\omega_k^2 m_{i1} - k_{i1})\} \qquad (i, j = 2, 3, \ldots, n) \qquad (2\text{-}240)$$

where we have shown a typical element of the matrix in parentheses in each case. The subscripts refer to the original rows and columns. Solving Eq. (2-240) for the amplitudes $A_j^{(k)}$ corresponding to the eigenvalue ω_k^2, we obtain

$$\{A_j^{(k)}\} = [(k_{ij} - \omega_k^2 m_{ij})]^{-1}\{(\omega_k^2 m_{i1} - k_{i1})\} \qquad (i, j = 2, 3, \ldots, n) \qquad (2\text{-}241)$$

where we assume that the *eigenvalues are distinct* and $A_1^{(k)} \neq 0$, thereby ensuring the existence of the inverse matrix. In case $A_1^{(k)} = 0$, another coordinate should be chosen as a reference. In general, one can use any normalization procedure which preserves the correct relative magnitudes of the coordinates.

A complete set of n amplitude coefficients, including the unit reference coefficient, is known as an *eigenvector* or *modal column*. There is an eigenvector $\mathbf{A}^{(k)}$ corresponding to each eigenvalue ω_k^2. This set of amplitude coefficients can be considered to define a *mode shape* associated with the given frequency ω_k. Each natural frequency ω_k with its corresponding eigenvector $\mathbf{A}^{(k)}$ defines a *natural mode of vibration*, sometimes called a *principal mode* or a *normal mode*.

For the natural systems that we are considering, the ω_k^2 are real and the corresponding amplitude ratios are also real. If the system is vibrating in a *single mode* rather than the more general case of a superposition of modes, then all the coordinates execute sinusoidal motion at the same frequency. The relative phase angle between any two coordinates is either $0°$ or $180°$, depending upon whether the particular amplitude ratio is positive or negative.

The *zero frequency modes* are somewhat different physically in that no elastic deformation occurs. For this reason, they are known as *rigid-body modes*. Both the potential and kinetic energies are constant, resulting in uniform translational or rotational motion. The amplitude ratios are calculated in the usual fashion, but are more easily considered as *velocity ratios*.

Principal Coordinates. We have seen that a linear natural system with n degrees of freedom has n eigenvalues or natural frequencies. It is possible for the system to oscillate in a single mode if the initial conditions are properly established; for example, if all the initial velocities are zero and the initial displacements are proportional to the amplitude coefficients $A_j^{(k)}$ of the given mode. In the general case, however, a superposition of modes is required. So, using solutions of the form of Eq. (2-236), we have

$$q_j = \sum_{k=1}^{n} A_{jk} C_k \cos (\omega_k t + \phi_k) \qquad (2\text{-}242)$$

where $A_{jk} \equiv A_j^{(k)}$.

Now let us define the *principal coordinate* U_k corresponding to the kth mode by the equation

$$U_k = C_k \cos (\omega_k t + \phi_k) \qquad (k = 1, 2, \ldots, n) \qquad (2\text{-}243)$$

Then from Eqs. (2-242) and (2-243) we obtain

$$q_j = \sum_{k=1}^{n} A_{jk} U_k \qquad (2\text{-}244)$$

or, in matrix notation,

$$\mathbf{q} = \mathbf{AU} \qquad (2\text{-}245)$$

where the *modal matrix* $\mathbf{A}$ is an $n \times n$ matrix whose columns are the modal columns for the various modes. The columns may be arranged in any order but, by convention, they are usually placed in order of increasing frequency of the corresponding modes.

From Eq. (2-245) we can solve for the principal coordinates. We obtain

$$\mathbf{U} = \mathbf{A}^{-1}\mathbf{q} \tag{2-246}$$

where we assume that the modal columns are linearly independent, a condition which is assured if the corresponding eigenvalues are distinct.

Note that the principal coordinates are generalized coordinates of a particular type. We see from Eq. (2-244) that if only one U_k is nonzero, the q's are proportional to the corresponding A's for that mode. In this case, the principal coordinate U_k oscillates sinusoidally at the frequency ω_k, assuming, of course, that the given mode is stable. Hence, if only one mode is excited, the entire motion is described by using only one principal coordinate. In general, a description of the motion requires all n principal coordinates. Nevertheless, since we are often interested in the transient response within a certain frequency range, the use of coordinates associated specifically with given frequencies may make possible a reduction in the number of coordinates used in the analysis.

Orthogonality of the Eigenvectors. Let us write Eq. (2-238) for the kth natural mode in the form

$$\mathbf{k}\mathbf{A}^{(k)} = \omega_k^2 \mathbf{m}\mathbf{A}^{(k)} \tag{2-247}$$

Similarly, for the lth mode we have

$$\mathbf{k}\mathbf{A}^{(l)} = \omega_l^2 \mathbf{m}\mathbf{A}^{(l)} \tag{2-248}$$

Now premultiply both sides of Eq. (2-247) by the row matrix $\mathbf{A}^{(l)T}$ and premultiply both sides of Eq. (2-248) by $\mathbf{A}^{(k)T}$. The resulting equations are

$$\mathbf{A}^{(l)T}\mathbf{k}\mathbf{A}^{(k)} = \omega_k^2 \mathbf{A}^{(l)T}\mathbf{m}\mathbf{A}^{(k)} \tag{2-249}$$

$$\mathbf{A}^{(k)T}\mathbf{k}\mathbf{A}^{(l)} = \omega_l^2 \mathbf{A}^{(k)T}\mathbf{m}\mathbf{A}^{(l)} \tag{2-250}$$

Since $\mathbf{k}$ and $\mathbf{m}$ are symmetric, the modal columns can be interchanged on both sides of either equation. If we perform this operation on Eq. (2-249), we find that its left-hand side is identical with that of Eq. (2-250). Subtracting Eq. (2-250), we obtain

$$(\omega_k^2 - \omega_l^2)\mathbf{A}^{(k)T}\mathbf{m}\mathbf{A}^{(l)} = 0 \tag{2-251}$$

If the two eigenvalues are *distinct*, that is, if $\omega_k^2 \neq \omega_l^2$, then it follows that

$$\mathbf{A}^{(k)T}\mathbf{m}\mathbf{A}^{(l)} = 0 \qquad (k \neq l) \tag{2-252}$$

This is the *orthogonality* condition for the kth and lth eigenvectors with respect to the inertia matrix $\mathbf{m}$.

Insight into the physical meaning of Eq. (2-252) is obtained by noting that a modal column of amplitude coefficients gives the velocity and the acceleration ratios as well as the displacement ratios for a given mode. So the orthogonality condition can be interpreted as stating that the scalar product of the eigenvector for one mode and the generalized inertia force vector for another mode is zero. This indicates that the two vectors are orthogonal in n-space. In practical terms, there is no inertial coupling between the corresponding principal coordinates.

If one of the modal columns of Eq. (2-252) corresponds to a zero-frequency mode, the orthogonality condition implies that a corresponding momentum component is zero for all the remaining elastic modes. For example, if a system can freely translate in the x direction, a zero-frequency mode will exist corresponding to this uniform translation and it will contain the entire x component of translational momentum. All the remaining elastic modes will have no translational momentum in the x direction. More generally, the *elastic* modes of a body in free space will have zero linear and angular momentum.

Now let us consider the matrix product of Eq. (2-252) for the case where $k = l$. In this instance, the product cannot be zero because it is proportional to the kinetic energy in the given mode. So let us write

$$\mathbf{A}^{(k)^T}\mathbf{m}\mathbf{A}^{(k)} = M_{kk} \qquad (k = 1, 2, \ldots, n) \tag{2-253}$$

where M_{kk} is a positive constant. We will show that M_{kk} is the *generalized mass* or *inertia coefficient* corresponding to the principal coordinate U_k.

Equations (2-252) and (2-253) can be summarized in the equation

$$\mathbf{A}^T\mathbf{m}\mathbf{A} = \mathbf{M} \tag{2-254}$$

where we recall that each column of $\mathbf{A}$ is a modal column. Because of the orthogonality property of the eigenvectors, we see that $\mathbf{M}$ is a *diagonal matrix*.

Now let us consider again the expression for the kinetic energy of the system. From Eqs. (2-231) and (2-245) we have

$$T = \tfrac{1}{2}\dot{\mathbf{q}}^T\mathbf{m}\dot{\mathbf{q}} = \tfrac{1}{2}\dot{\mathbf{U}}^T\mathbf{A}^T\mathbf{m}\mathbf{A}\dot{\mathbf{U}} \tag{2-255}$$

and, using Eq. (2-254), we obtain

$$T = \tfrac{1}{2}\dot{\mathbf{U}}^T\mathbf{M}\dot{\mathbf{U}} \tag{2-256}$$

confirming that $\mathbf{M}$ is the generalized mass matrix associated with the principal coordinates.

We can show that the transformation to principal coordinates also diagonalizes the stiffness matrix $\mathbf{k}$. Let us write Eqs. (2-249) and (2-250) in the form

$$\frac{1}{\omega_k^2}\mathbf{A}^{(l)^T}\mathbf{k}\mathbf{A}^{(k)} = \mathbf{A}^{(l)^T}\mathbf{m}\mathbf{A}^{(k)} \tag{2-257}$$

$$\frac{1}{\omega_l^2}\mathbf{A}^{(k)^T}\mathbf{k}\mathbf{A}^{(l)} = \mathbf{A}^{(k)^T}\mathbf{m}\mathbf{A}^{(l)} \tag{2-258}$$

where we assume that ω_k^2 and ω_l^2 are not zero. Again we note that $\mathbf{k}$ and $\mathbf{m}$ are symmetric; hence the modal columns for the kth and lth modes can be interchanged. Performing this interchange in Eq. (2-257) and subtracting Eq. (2-258), we obtain

$$\left(\frac{1}{\omega_k^2} - \frac{1}{\omega_l^2}\right) \mathbf{A}^{(k)T}\mathbf{k}\mathbf{A}^{(l)} = 0 \tag{2-259}$$

If the modes are *distinct*, we have $\omega_k^2 \neq \omega_l^2$, and it follows that

$$\mathbf{A}^{(k)T}\mathbf{k}\mathbf{A}^{(l)} = 0 \qquad (k \neq l) \tag{2-260}$$

For the case where either ω_k^2 or ω_l^2 is zero, the same result is obtained directly from Eqs. (2-249) and (2-250).

Equation (2-260) represents the orthogonality condition with respect to the stiffness matrix $\mathbf{k}$. It states that the scalar product involving the eigenvector for a given mode and the generalized elastic force vector for another mode is zero. In other words, the principal modes are not elastically coupled because no work is done by the elastic forces of one mode in moving through the displacements of a second mode.

For the case where $k = l$, we can write

$$\mathbf{A}^{(k)T}\mathbf{k}\mathbf{A}^{(k)} = K_{kk} \tag{2-261}$$

where K_{kk} is the *generalized stiffness coefficient* for the kth mode. K_{kk} can be positive, zero, or negative according to whether the kth mode is stable, neutrally stable, or unstable. Equations (2-260) and (2-261) are summarized in the single equation

$$\mathbf{A}^T\mathbf{k}\mathbf{A} = \mathbf{K} \tag{2-262}$$

where $\mathbf{K}$ is an $n \times n$ *diagonal* matrix.

The potential energy V can be written, using Eqs. (2-229) and (2-245), in the form

$$V = \tfrac{1}{2}\mathbf{q}^T\mathbf{k}\mathbf{q} = \tfrac{1}{2}\mathbf{U}^T\mathbf{A}^T\mathbf{k}\mathbf{A}\mathbf{U} \tag{2-263}$$

which can be combined with Eq. (2-262) to yield

$$V = \tfrac{1}{2}\mathbf{U}^T\mathbf{K}\mathbf{U} \tag{2-264}$$

Hence $\mathbf{K}$ is the *generalized stiffness matrix* associated with the principal coordinates.

In summary, the transformation to principal coordinates has resulted in the diagonalization of both the $\mathbf{m}$ and $\mathbf{k}$ matrices. This implies that the natural modes have neither inertial nor elastic coupling and therefore are *independent* for this case of unforced motion.

In order to emphasize further the independence of the natural modes, let us obtain the differential equations of motion in terms of the principal coordinates. From Eqs. (2-256) and (2-264), we find that the Lagrangian function L can be written in the form

$$L = \tfrac{1}{2}\sum_{k=1}^{n} M_{kk}\dot{U}_k^2 - \tfrac{1}{2}\sum_{k=1}^{n} K_{kk}U_k^2 \tag{2-265}$$

Using Lagrange's equation, we obtain

$$M_{kk}\ddot{U}_k + K_{kk}U_k = 0 \qquad (k = 1, 2, \ldots, n) \qquad (2\text{-}266)$$

These differential equations of motion indicate that the free vibrations of the entire system can be described in terms of n independent undamped second-order systems, each system representing a single mode. We note that the natural frequency of each mode is given by

$$\omega_k^2 = \frac{K_{kk}}{M_{kk}} \qquad (k = 1, 2, \ldots, n) \qquad (2\text{-}267)$$

Repeated Roots. In the previous discussion we assumed that the roots are distinct. With this assumption, we found that the eigenvectors are mutually orthogonal with respect to both the **m** and **k** matrices; that is, both matrices are diagonalized by a transformation to principal coordinates.

Now let us consider a *degenerate system* in which the eigenvalues are not all distinct. For example, suppose there is a *double root* such that $\omega_p^2 = \omega_{p+1}^2$. If we use Eq. (2-241) to solve for the modal columns corresponding to the distinct modes, we find that these modes are mutually orthogonal. But if we let $\omega_k^2 = \omega_p^2$ in this equation, we find that the inverse matrix $[(k_{ij} - \omega_p^2 m_{ij})]^{-1}$ does not exist, indicating that the corresponding set of $(n - 1)$ simultaneous algebraic equations is not linearly independent.

In order to avoid this problem, we can choose *two* amplitude coefficients arbitrarily when $\omega_k^2 = \omega_p^2$. As an example, we might take $A_1^{(p)} = 1$, $A_2^{(p)} = 0$. Then we can solve for the $(n - 2)$ remaining amplitude coefficients from the $(n - 2)$ independent equations contained in Eq. (2-240). This results in a modal column $\mathbf{A}^{(p)}$ which is orthogonal to all the modal columns corresponding to the distinct modes, since the assumptions for the derivation of Eq. (2-252) are still valid.

The problem remains, however, of finding a final modal column $\mathbf{A}^{(p+1)}$ such that it is orthogonal to $\mathbf{A}^{(p)}$ and to the other columns. This is accomplished if we set $A_1^{(p+1)} = 1$ and use the $(n - 2)$ independent equations from Eq. (2-240) plus the orthogonality condition

$$\mathbf{A}^{(p)T}\mathbf{m}\mathbf{A}^{(p+1)} = \mathbf{0} \qquad (2\text{-}268)$$

Thus we have $(n - 1)$ equations from which to solve for the $(n - 1)$ remaining amplitude coefficients in $\mathbf{A}^{(p+1)}$.

For the more general case of m repeated roots, a similar procedure is followed. The eigenvector $\mathbf{A}^{(p)}$ has m arbitrary components, and the remaining $(n - m)$ components are obtained from the $(n - m)$ independent equations contained in Eq. (2-240). Each succeeding eigenvector has one fewer arbitrary component, but one more orthogonality condition. Hence, there are sufficient equations to obtain the required components. In this fashion, a complete set of mutually orthogonal eigenvectors is obtained corresponding to the repeated roots.

The result of this procedure is a modal matrix **A** which diagonalizes both **m** and **k**. Note, however, that the amplitude ratios are no longer unique, as they were for the case of distinct roots, since some of the amplitude ratios can now be chosen arbitrarily. In fact, any linear combination of the modal columns corresponding to a repeated root forms another possible modal column for that root. Columns formed in this way, however, are not necessarily mutually orthogonal.

Initial Conditions. The form of the transient solution for a natural system having n degrees of freedom was given in Eq. (2-242), namely,

$$q_j = \sum_{k=1}^{n} A_{jk} C_k \cos(\omega_k t + \phi_k)$$

where we assumed small motions about a position of stable equilibrium. Assuming that the eigenvalues ω_k^2 and the modal matrix **A** have been found, we must now solve for the n C's and the n ϕ's from the $2n$ initial conditions. The orthogonality properties of the natural modes can be used to simplify this process.

Let us assume that the initial q's and $\dot{q}$'s are given. From Eq. (2-242), we see that

$$q_j(0) = \sum_{k=1}^{n} A_{jk} C_k \cos \phi_k \qquad (j = 1, 2, \ldots, n) \qquad (2\text{-}269)$$

and

$$\dot{q}_j(0) = -\sum_{k=1}^{n} A_{jk} C_k \omega_k \sin \phi_k \qquad (j = 1, 2, \ldots, n) \qquad (2\text{-}270)$$

Now multiply each of these equations by $m_{ij} A_{il}$ and sum over i and j. Because of the orthogonality condition of Eqs. (2-252) and (2-253), namely,

$$\sum_{i=1}^{n} \sum_{j=1}^{n} m_{ij} A_{il} A_{jk} = \begin{cases} 0, & k \neq l \\ M_{ll}, & k = l \end{cases} \qquad (2\text{-}271)$$

we obtain from Eq. (2-269) that

$$\sum_{i=1}^{n} \sum_{j=1}^{n} q_j(0) m_{ij} A_{il} = M_{ll} C_l \cos \phi_l$$

or

$$C_l \cos \phi_l = \frac{1}{M_{ll}} \sum_{i=1}^{n} \sum_{j=1}^{n} q_j(0) m_{ij} A_{il} \qquad (2\text{-}272)$$

Similarly, we obtain from Eq. (2-270) that

$$C_l \sin \phi_l = -\frac{1}{\omega_l M_{ll}} \sum_{i=1}^{n} \sum_{j=1}^{n} \dot{q}_j(0) m_{ij} A_{il} \qquad (2\text{-}273)$$

Notice that Eqs. (2-272) and (2-273) enable one to obtain the solutions for C_l and ϕ_l directly, rather than being forced to solve $2n$ simultaneous equations.

Now suppose we consider a system having an eigenvalue $\omega_m^2 = 0$, corresponding to a rigid-body mode. We recall that the solution for q_j contains the term $A_{jm}(C_m t + D_m)$ in this case. If we again use the orthogonality properties of the natural modes, we find that

$$C_m = \frac{1}{M_{mm}} \sum_{i=1}^{n} \sum_{j=1}^{n} \dot{q}_j(0) m_{ij} A_{im} \tag{2-274}$$

$$D_m = \frac{1}{M_{mm}} \sum_{i=1}^{n} \sum_{j=1}^{n} q_j(0) m_{ij} A_{im} \tag{2-275}$$

Example 2-12. In order to illustrate the theory of small oscillations, consider a system consisting of a simple pendulum of length l and mass m which is pivoted at a point O on a block of mass $2m$ (see Fig. 2-10). The block can slide without friction on a horizontal surface. Assuming plane motion, and using x and θ as generalized coordinates, obtain the differential equations of motion and the natural modes. Also obtain the solutions for x and θ as functions of time, assuming the initial conditions are $x(0) = 0$, $\dot{x}(0) = 1$, $\theta(0) = 0.1$, $\dot{\theta}(0) = 0$.

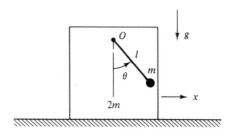

Fig. 2-10. A simple pendulum attached to a sliding block.

First let us obtain an expression for the total kinetic energy T. It is the sum of the individual kinetic energies of the block and the pendulum, or

$$T = m\dot{x}^2 + \tfrac{1}{2}m(\dot{x} + l\dot{\theta})^2 = \tfrac{3}{2}m\dot{x}^2 + ml\dot{x}\dot{\theta} + \tfrac{1}{2}ml^2\dot{\theta}^2 \tag{2-276}$$

where we assume $|\theta| \ll 1$.

The potential energy is entirely gravitational and, again assuming small motions, we can write

$$V = mgl(1 - \cos\theta) \cong \tfrac{1}{2}mgl\theta^2 \tag{2-277}$$

If we next obtain the Lagrangian function $L = T - V$ and substitute into Lagrange's equation, we obtain the following equations of motion:

$$3m\ddot{x} + ml\ddot{\theta} = 0$$
$$ml\ddot{x} + ml^2\ddot{\theta} + mgl\theta = 0 \tag{2-278}$$

Now let $q_1 \equiv x$ and $q_2 \equiv \theta$. Comparing Eq. (2-278) with the standard matrix form of Eq. (2-235), we see that

$$\mathbf{m} = m\begin{bmatrix} 3 & l \\ l & l^2 \end{bmatrix}, \qquad \mathbf{k} = mgl\begin{bmatrix} 0 & 0 \\ 0 & 1 \end{bmatrix}$$

and note that both matrices are constant and symmetric.

In order to obtain the natural modes, let us first assume solutions of the form

$$q_j = A_j C \cos(\omega t + \phi)$$

Substituting into Eq. (2-278), or using Eq. (2-238) directly, we obtain

$$\begin{aligned} -3m\omega^2 A_1 - ml\omega^2 A_2 &= 0 \\ -ml\omega^2 A_1 + (mgl - ml^2\omega^2)A_2 &= 0 \end{aligned} \qquad (2\text{-}279)$$

The characteristic equation is obtained by setting the determinant of the coefficients equal to zero, yielding

$$2l^2\omega^4 - 3gl\omega^2 = 0 \qquad (2\text{-}280)$$

The eigenvalues are

$$\omega_1^2 = 0, \qquad \omega_2^2 = \frac{3g}{2l}$$

From the second equation of (2-279) we have

$$\frac{A_2}{A_1} = \frac{\omega^2}{g - l\omega^2} \qquad (2\text{-}281)$$

Hence we obtain

$$\frac{A_{21}}{A_{11}} = 0, \qquad \frac{A_{22}}{A_{12}} = -\frac{3}{l}$$

and, upon setting $A_{11} = A_{12} = 1$, we find that the modal matrix is

$$\mathbf{A} = \begin{bmatrix} 1 & 1 \\ 0 & -3/l \end{bmatrix}$$

We see that each column of $\mathbf{A}$ is an eigenvector and is associated with a particular ω.

The orthogonality conditions can be checked by using Eqs. (2-254) and (2-262) to obtain $\mathbf{M}$ and $\mathbf{K}$. We find that

$$\mathbf{M} = m\begin{bmatrix} 3 & 0 \\ 0 & 6 \end{bmatrix}, \qquad \mathbf{K} = mgl\begin{bmatrix} 0 & 0 \\ 0 & 9/l^2 \end{bmatrix}$$

Furthermore, we note that

$$\omega_1^2 = \frac{K_{11}}{M_{11}} = 0, \qquad \omega_2^2 = \frac{K_{22}}{M_{22}} = \frac{3g}{2l}$$

in agreement with our previous results.

Now let us obtain the solutions for x and θ as functions of time. These solutions are of the form

$$x = A_{11}(C_1 t + D_1) + A_{12} C_2 \cos(\omega_2 t + \phi_2)$$
$$\theta = A_{21}(C_1 t + D_1) + A_{22} C_2 \cos(\omega_2 t + \phi_2)$$

(2-282)

The constants C_1, C_2, D_1, ϕ_2 are to be determined from the given initial conditions. Instead of finding these constants directly, however, let us use Eqs. (2-274) and (2-275). We obtain

$$C_1 = \frac{1}{M_{11}} \dot{x}(0) m_{11} A_{11} = 1$$

$$D_1 = \frac{1}{M_{11}} \theta(0) m_{12} A_{11} = \frac{l}{30}$$

Similarly, we find from Eqs. (2-272) and (2-273) that

$$C_2 \cos \phi_2 = \frac{\theta(0)}{M_{22}} (m_{12} A_{12} + m_{22} A_{22}) = -\frac{l}{30}$$

$$C_2 \sin \phi_2 = -\frac{\dot{x}(0)}{\omega_2 M_{22}} (m_{11} A_{12} + m_{21} A_{22}) = 0$$

which yields

$$C_2 = \frac{l}{30}, \qquad \phi_2 = \pi$$

Finally, we obtain the solutions

$$x = t + \frac{l}{30}\left(1 - \cos\sqrt{\frac{3g}{2l}}\,t\right)$$

$$\theta = \frac{1}{10} \cos\sqrt{\frac{3g}{2l}}\,t$$

Note that the first mode is a rigid-body mode consisting of a uniform translation in the positive x direction with θ held at zero. The second mode consists of an oscillation of frequency $\sqrt{3g/2l}$ in both x and θ, with the amplitudes and phasing such that the linear momentum in the x direction is always zero.

REFERENCES

1. GOLDSTEIN, H., *Classical Mechanics*. Reading, Mass.: Addison-Wesley Press, Inc., 1950. Lagrange's equations are derived by using virtual work principles. The explanation is brief but well-done.

2. PARS, L. A., *A Treatise on Analytical Dynamics*. London: William Heinemann, 1965. Lagrange's equations are discussed in detail, and alternate derivations are presented in this thorough treatment of the subject.

3. WHITTAKER, E. T., *A Treatise on the Analytical Dynamics of Particles and Rigid Bodies*, 4th ed. New York: Dover Publications, Inc., 1944. The derivation of Lagrange's equations starts with Newton's equations, and the author proceeds directly to a discussion of the various forms and applications of the theory. This is still an excellent reference.

4. MARION, J. B., *Classical Dynamics of Particles and Systems*, 2nd ed. New York: Academic Press, 1970. The author presents a comparison of the Lagrangian and Newtonian viewpoints and gives a good discussion of the principal features of the Lagrangian method.

PROBLEMS

2-1. An inverted pendulum consists of a particle of mass m supported by a rigid massless rod of length l. The pivot O has a vertical motion given by $z = A \sin \omega t$. Obtain the Lagrangian function and find the differential equation of motion.

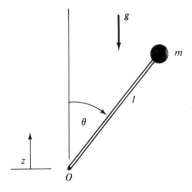

2-2. A particle of mass m is embedded at a distance l from the center of a massless circular disk of radius r. It rolls without slipping down a plane inclined at an angle α with the horizontal. Use the Lagrangian method to write the differential equation of motion for the system.

Fig. P2-1

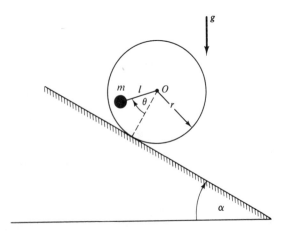

Fig. P2-2

2-3. A uniform rod of mass m and length l moves on the horizontal xy plane. At one end it has a knife-edge constraint which prevents a velocity component perpendicular to the rod at that point.

(a) Write the nonholonomic constraint equation.
(b) Using (x, y, θ) as coordinates, obtain the differential equations of motion.
(c) Show that the Lagrange multiplier λ represents the transverse force at the constraint.

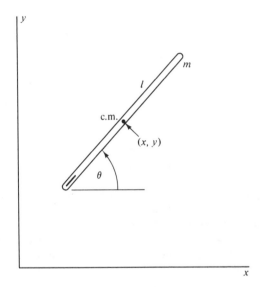

Fig. P2-3

2-4. Given a standard holonomic system for which the T_1 portion of the kinetic energy has the form

$$T_1 = \sum_{j=1}^{n} \frac{\partial f}{\partial q_j} \dot{q}_j = \frac{df}{dt}$$

where $f = f(q_1, \ldots, q_n)$ and is twice differentiable. Show that any T_1 of this form has no influence on the equations of motion.

2-5. A particle of mass m can slide without friction in a straight slot cut in a horizontal turntable. The turntable rotates at a constant angular velocity Ω about a vertical axis through its center at O. The coordinate y represents the position of the particle relative to the turntable and is equal to zero when the spring is unstressed and the particle is at a minimum distance R from the center O. Use the Lagrangian method to find the differential equation of motion. Identify the terms generated by T_2, T_1, T_0, and V. (See p. 98).

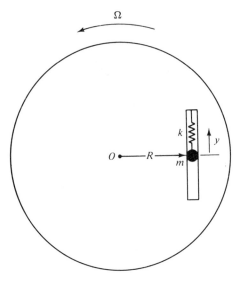

Fig. P2-5

2-6. Consider a natural system with $T = \frac{1}{2}m(q)\dot{q}^2$ and $V = V(q)$. Show that the correct differential equation of motion is obtained by equating to zero the time derivative of the total energy $E = T + V$.

2-7. Given a conservative holonomic system having n degrees of freedom. Show that setting the total time derivative of the energy integral equal to zero gives the same differential equation as multiplying the ith Lagrangian equation by $\dot{q}_i$ and then summing over i.

2-8. Consider the motion of a particle which slides on a smooth fixed wire bent in the form of a helix having a radius R and a constant inclination angle α relative to the horizontal.

    (a) Assuming that the central axis of the helix is vertical, find the time required for the particle to slide through a vertical distance H after it is released from rest.

    (b) Find the corresponding time for the case in which the helix rotates about its vertical axis with a constant angular velocity Ω and the particle is again released with zero velocity relative to the helix.

2-9. A particle of mass m can slide on a smooth rigid wire having the form $y = 3x^2$, where gravity acts in the direction of the negative y axis.

    (a) Use the Lagrangian method to obtain the equations of motion.

    (b) Assuming the initial conditions $\dot{y}(0) = 0$, $y(0) = y_0$, find the maximum constraint force during the resulting motion.

2-10. The position of a particle of mass m is given by the Cartesian coordinates (x, y, z). Assuming a potential energy function $V = \frac{1}{2}k(x^2 + y^2 + z^2)$ and a constraint described by the equation $2\dot{x} + 3\dot{y} + 4\dot{z} + 5 = 0$, find:

(a) the differential equations of motion,
(b) the velocity of the moving constraint.

2-11. A uniform rod of mass m and length l is freely pivoted at a fixed point O. Use the Routhian method to eliminate the coordinate ϕ and obtain the differential equation of motion for θ.

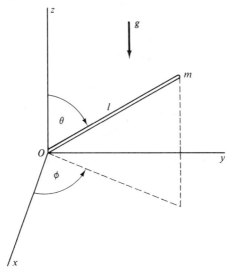

Fig. P2-11

2-12. Use (r, ϕ) as coordinates and employ the Lagrangian method to obtain the differential equations of motion. Show that these equations can be reduced to a

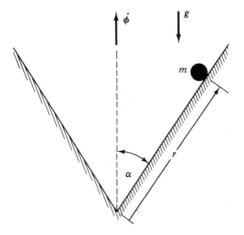

Fig. P2-12

differential equation for r which is identical
with that obtained by the Routhian method.

2-13. A small tube of mass m is bent in
the form of a circle of radius r and is pivoted
about a fixed point O on its circumference. A
particle of mass m can slide without friction
inside the tube.

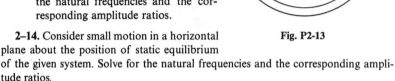

(a) Use the Lagrangian method to obtain
the differential equations for plane
motion.
(b) Now assume small motions and obtain
the natural frequencies and the corresponding amplitude ratios.

Fig. P2-13

2–14. Consider small motion in a horizontal
plane about the position of static equilibrium
of the given system. Solve for the natural frequencies and the corresponding amplitude ratios.

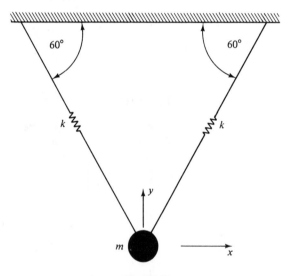

Fig. P2-14

2-15. Consider the small planar transverse vibrations of three particles connected
by an elastic rod. Assume that the potential energy is $V = \frac{1}{2}k(y_1 - 2y_2 + y_3)^2$,
where the constant k is proportional to the bending stiffness. Solve for the
natural frequencies of the system and a corresponding set of orthogonal modal
columns.

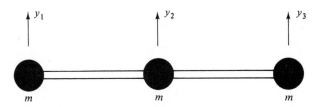

Fig. P2-15

3

SPECIAL APPLICATIONS OF

LAGRANGE'S EQUATIONS

This chapter presents some special applications of Lagrange's equations. These examples serve to illustrate the diversity of problems which can be analyzed by the Lagrangian method. Also, they permit a further discussion of some of the basic topics encountered in classical dynamics, such as virtual work, the representation of constraints, and the use of energy methods.

3-1. RAYLEIGH'S DISSIPATION FUNCTION

The standard form of Lagrange's equation given in Eq. (2-27) is applicable to holonomic systems whose generalized forces are derivable from a potential function $V(q, t)$ in accordance with Eq. (2-24), namely,

$$Q_i = -\frac{\partial V}{\partial q_i}$$

If, in addition, there are applied forces which are functions of the $\dot{q}$'s, these forces are frequently represented by Q_i', and the equations of motion take the form

$$\frac{d}{dt}\left(\frac{\partial L}{\partial \dot{q}_i}\right) - \frac{\partial L}{\partial q_i} = Q_i' \qquad (i = 1, 2, \ldots, n) \tag{3-1}$$

as we found in Eq. (2-29).

Another approach, however, can be applied to systems in which each Q_i' is of the form

$$Q_i' = -\sum_{j=1}^{n} c_{ij}(q, t)\dot{q}_j \tag{3-2}$$

where the c's are known as *damping coefficients* and form a real, symmetric matrix. These generalized friction forces are *dissipative* in nature and result in a loss of energy whenever any Q_i' is nonzero.

Now let us define *Rayleigh's dissipation function* $F(q, \dot{q}, t)$ as follows:

$$F = \frac{1}{2}\sum_{i=1}^{n} \sum_{j=1}^{n} c_{ij}\dot{q}_i\dot{q}_j \tag{3-3}$$

We see from Eqs. (3-1)–(3-3) that the equations of motion can be written in the form

$$\frac{d}{dt}\left(\frac{\partial L}{\partial \dot{q}_i}\right) - \frac{\partial L}{\partial q_i} + \frac{\partial F}{\partial \dot{q}_i} = 0 \qquad (i = 1, 2, \ldots, n) \qquad (3\text{-}4)$$

We assume, of course, that the given friction forces are the only generalized forces which are not derivable from the potential function V.

The rate at which the friction forces do work on the system is

$$\sum_{i=1}^{n} Q_i' \dot{q}_i = -\sum_{i=1}^{n}\sum_{j=1}^{n} c_{ij}\dot{q}_i\dot{q}_j = -2F \qquad (3\text{-}5)$$

Hence, the dissipation function is equal to half the instantaneous rate of dissipation of the total mechanical energy. Since the rate of energy dissipation must be positive or zero at all times, it follows that F is either a positive definite function, or a positive semidefinite function, of the $\dot{q}$'s. Furthermore, we find that the *value* of F is invariant with respect to a coordinate transformation because the rate of energy dissipation is independent of the coordinates used to describe the configuration.

A common application of Rayleigh's dissipation function is in the analysis of small oscillations of a *natural system* to which damping has been added. For example, suppose we consider a system having a Lagrangian function

$$L = \tfrac{1}{2}\sum_{i=1}^{n}\sum_{j=1}^{n} m_{ij}\dot{q}_i\dot{q}_j - \tfrac{1}{2}\sum_{i=1}^{n}\sum_{j=1}^{n} k_{ij}q_iq_j \qquad (3\text{-}6)$$

where the m_{ij} and k_{ij} are constants, and the corresponding **m** and **k** matrices are symmetric.

Using Eqs. (3-3), (3-4), and (3-6), we find that the differential equations of motion are

$$\sum_{j=1}^{n} (m_{ij}\ddot{q}_j + c_{ij}\dot{q}_j + k_{ij}q_j) = 0 \qquad (i = 1, 2, \ldots, n) \qquad (3\text{-}7)$$

or, in matrix form,

$$\mathbf{m}\ddot{q} + \mathbf{c}\dot{q} + \mathbf{k}q = 0 \qquad (3\text{-}8)$$

Now let us assume solutions of the form

$$q_j = A_j C e^{\lambda t} \qquad (3\text{-}9)$$

and substitute into the equations of motion. After canceling the common factor $Ce^{\lambda t}$, we obtain

$$(\lambda^2\mathbf{m} + \lambda\mathbf{c} + \mathbf{k})\mathbf{A} = 0 \qquad (3\text{-}10)$$

where **A** is a column matrix. These n algebraic equations have a nontrivial solution if, and only if, the determinant of the coefficients is zero, that is,

$$|(\lambda^2\mathbf{m} + \lambda\mathbf{c} + \mathbf{k})| = 0 \qquad (3\text{-}11)$$

This is the characteristic equation for the system and is of degree $2n$ in λ. Since the coefficients are real, the $2n$ roots either are real or occur in complex conjugate pairs. Making the usual assumption that $A_1^{(k)} = 1$ for each root

λ_k, we can solve for the corresponding modal column $\mathbf{A}^{(k)}$ by using Eq. (3-10). These amplitudes are real for real roots and are complex, in general, for complex roots.

The general solution for the free vibrations is obtained by superimposing solutions corresponding to each of the $2n$ roots. Thus we have

$$q_j = \sum_{k=1}^{2n} A_{jk} C_k e^{\lambda_k t} \qquad (j = 1, 2, \ldots, n) \tag{3-12}$$

where $A_{jk} \equiv A_j^{(k)}$ and we assume that the roots are distinct. The $2n$ C's are real, or occur in complex conjugate pairs, and can be determined from the $2n$ initial conditions.

Now let us consider the effect on the stability of the solutions arising from the addition of damping to the given natural system. Let us assume first that F and V are *both positive definite*. We see that the total energy $T + V$ decreases continuously due to the damping, except when $\dot{\mathbf{q}} = 0$ or, equivalently, when $T = 0$. Since the only equilibrium position is at $\mathbf{q} = 0$, corresponding to a minimum potential energy $V = 0$, we see that the system must approach the condition $T = V = 0$ as t approaches infinity. This type of stability is known as *asymptotic stability*. For this case of a linear system, it implies that all the roots have negative real parts. By contrast, all the roots of the original undamped system are imaginary.

If we now assume that F is positive definite but V is not, we find that asymptotic stability does not result. In fact, if the undamped system is unstable, it will remain unstable after the addition of damping. In other words, each positive real root remains positive. This occurs because V does not have a minimum value at $\mathbf{q} = 0$ in this instance.

Finally, if V is positive definite but F is positive semidefinite, the system may not be asymptotically stable because it may be possible to find a continuing motion for which the energy dissipation is zero. For example, there may be a pair of imaginary roots, corresponding to an undamped sinusoidal oscillation, even though the system contains dissipative elements.

3-2. IMPULSIVE MOTION

The study of impulsive motion involves an analysis of the response of mechanical systems to forces of very large magnitude and short duration. These *impulsive forces* often arise as the result of an impact or collision, but one can imagine other sources such as an explosion or the sudden application of a constraint. If the mechanical system under consideration includes rigid bodies or other rigid constraints, the application of known impulsive external forces will usually result in unknown impulsive constraint forces. As we have observed, however, the Lagrangian method often allows one to calculate the motion of a system without having to solve for the constraint forces. Hence the use of Lagrange's equations is particularly convenient in the analysis of the impulsive motion of systems with rigid constraints.

Impulse and Momentum. Before we proceed with the Lagrangian approach, let us introduce some of the necessary notation and dynamical theory. Suppose we consider a system of N particles whose positions, relative to a point O fixed in an inertial frame, are given by $\mathbf{r}_1, \mathbf{r}_2, \ldots, \mathbf{r}_N$. The basic equation of motion for the center of mass is

$$\mathbf{F} = \dot{\mathbf{p}} \tag{3-13}$$

where $\mathbf{F}$ is the *total external force* acting on the system, and the *total linear momentum* is

$$\mathbf{p} = \sum_{i=1}^{N} m_i \dot{\mathbf{r}}_i = m \dot{\mathbf{r}}_c \tag{3-14}$$

Here m is the total mass of the system and $\mathbf{r}_c$ is the position of its center of mass.

Let us integrate both sides of Eq. (3-13) with respect to time over the interval t_1 to t_2. We obtain

$$\int_{t_1}^{t_2} \mathbf{F} \, dt = \int_{t_1}^{t_2} \dot{\mathbf{p}} \, dt = \mathbf{p}(t_2) - \mathbf{p}(t_1)$$

or

$$\hat{\mathbf{F}} = \mathbf{p}_2 - \mathbf{p}_1 \tag{3-15}$$

where the *total impulse* $\hat{\mathbf{F}}$ of the external forces is

$$\hat{\mathbf{F}} = \int_{t_1}^{t_2} \mathbf{F} \, dt \tag{3-16}$$

Thus we have the *principle of linear impulse and momentum: The change in the total linear momentum of a system during a given time interval is equal to the total impulse of the external forces acting over the same interval.*

Let us consider next the basic rotational equation for the system, namely,

$$\mathbf{M} = \dot{\mathbf{H}} \tag{3-17}$$

where $\mathbf{M}$ is the *moment* of the external forces and $\mathbf{H}$ is the *total angular momentum* given by Eq. (1-123) or Eq. (1-126). The reference point is the same for both $\mathbf{M}$ and $\mathbf{H}$ and is assumed to be either (1) fixed relative to an inertial frame or (2) at the center of mass.

Notice that the form of Eq. (3-17) is similar to Eq. (3-13). If we integrate Eq. (3-17) with respect to time over the interval t_1 to t_2, we obtain

$$\hat{\mathbf{M}} = \mathbf{H}_2 - \mathbf{H}_1 \tag{3-18}$$

where the *angular impulse* is

$$\hat{\mathbf{M}} = \int_{t_1}^{t_2} \mathbf{M} \, dt \tag{3-19}$$

Eq. (3-18) is a statement of the *principle of angular impulse and momentum: The change in the total angular momentum of a system during a given time interval is equal to the total angular impulse of the external forces acting over the same interval, provided that the reference point for* $\hat{\mathbf{M}}$ *and* H *either is fixed in an inertial frame or is taken at the center of mass.*

It is important to notice that the principles expressed in Eqs. (3-15) and (3-18) are valid for time intervals that are arbitrarily large and for forces and moments which are not impulsive in nature. But these principles are also valid for the case of impulsive forces and moments which are applied over very short intervals of time, and it is with this latter case that we are primarily concerned. So let us assume that the sudden changes $\Delta\mathbf{p} = \mathbf{p}_2 - \mathbf{p}_1$ and $\Delta\mathbf{H} = \mathbf{H}_2 - \mathbf{H}_1$ occur during the short interval $\Delta t = t_2 - t_1$. Then Eqs. (3-15) and (3-18) can be written as follows:

$$\Delta\mathbf{p} = \hat{\mathbf{F}} \tag{3-20}$$

$$\Delta\mathbf{H} = \hat{\mathbf{M}} \tag{3-21}$$

If these vector equations are expressed in terms of the corresponding scalar components, we obtain a set of *algebraic* equations which are *linear in the velocity changes* at various points of the system. These velocity changes are finite, so the actual velocities remain finite, resulting in a negligible change in the *configuration* of the system during the interval Δt. It follows that the inertia coefficients (masses and moments of inertia) can be considered as *constants* during this interval. Furthermore, any *finite forces* such as gravitational and elastic forces are small compared with the impulsive applied forces and can be neglected in obtaining $\hat{\mathbf{F}}$ and $\hat{\mathbf{M}}$. Note, however, that $\hat{\mathbf{F}}$ and $\hat{\mathbf{M}}$ may include impulsive constraint forces.

Lagrangian Method. Having considered some of the results which can be derived directly from Newton's laws of motion, let us now turn to the Lagrangian approach to impulsive motion. Suppose the configuration of a mechanical system is specified by the *independent* generalized coordinates $q_1, q_2, \ldots, q_n$. From Eq. (2-23) we can write

$$\frac{d}{dt}\left(\frac{\partial T}{\partial \dot{q}_i}\right) - \frac{\partial T}{\partial q_i} = Q_i \qquad (i = 1, 2, \ldots, n) \tag{3-22}$$

where the Q's are due to the *applied forces.*

Now let us assume that impulsive forces are applied to the system during the interval Δt. From Eq. (1-138) we recall that the generalized momentum is given by

$$p_i = \frac{\partial T}{\partial \dot{q}_i} \tag{3-23}$$

Hence, if we integrate both sides of Eq. (3-22) over the interval Δt, we obtain

$$\int_{t_1}^{t_1+\Delta t} \left[\dot{p}_i - \frac{\partial T}{\partial q_i} \right] dt = \int_{t_1}^{t_1+\Delta t} Q_i \, dt$$

or

$$\Delta p_i = \hat{Q}_i \qquad (i = 1, 2, \ldots, n) \tag{3-24}$$

where Δp_i is the change in p_i during Δt, and where the *generalized impulse* $\hat{Q}_i$ is given by

$$\hat{Q}_i = \int_{t_1}^{t_1+\Delta t} Q_i \, dt \tag{3-25}$$

The term $-\partial T/\partial q_i$ in the integrand can be neglected because it is finite at time t_1, and therefore

$$\lim_{\Delta t \to 0} \int_{t_1}^{t_1+\Delta t} \frac{\partial T}{\partial q_i} \, dt = 0 \tag{3-26}$$

Now recall that the q's and t are essentially constant during the interval Δt; hence we obtain from Eq. (2-30) that

$$\Delta p_i = \sum_{j=1}^{n} m_{ij} \, \Delta \dot{q}_j \tag{3-27}$$

where we assume that the functions $m_{ij}(q, t)$ and $a_i(q, t)$ are continuous. Then, from Eqs. (3-24) and (3-27), we find that

$$\sum_{j=1}^{n} m_{ij} \, \Delta \dot{q}_j = \hat{Q}_i \qquad (i = 1, 2, \ldots, n) \tag{3-28}$$

Here we have n algebraic equations from which to solve for the n $\Delta \dot{q}$'s. Using matrix notation, we can write

$$\Delta \dot{\mathbf{q}} = \mathbf{m}^{-1} \hat{\mathbf{Q}} \tag{3-29}$$

The inertia matrix $\mathbf{m}$ is positive definite, so its inverse exists. Note also that the equations are linear, so the superposition principle applies; that is, the change $\Delta \dot{\mathbf{q}}$ due to $\hat{\mathbf{Q}}_1 + \hat{\mathbf{Q}}_2$ is equal to the sum of the $\Delta \mathbf{q}$'s due to $\hat{\mathbf{Q}}_1$ and $\hat{\mathbf{Q}}_2$ applied separately.

The generalized impulse $\hat{\mathbf{Q}}_i$ is found in a manner similar to that used in obtaining Q_i. For example, suppose a holonomic system of N particles is described in terms of the Cartesian coordinates $x_1, x_2, \ldots, x_{3N}$; and the corresponding impulses $\hat{F}_1, \hat{F}_2, \ldots, \hat{F}_{3N}$ are applied to the system during the small interval Δt. Then, using Eqs. (1-73), (3-16), and (3-25), we obtain

$$\hat{Q}_i = \sum_{k=1}^{3N} \frac{\partial x_k}{\partial q_i} \hat{F}_k \tag{3-30}$$

where the transformation equations relating the x's and q's are of the form

$$x_k = x_k(q, t) \qquad (k = 1, 2, \ldots, 3N) \tag{3-31}$$

If the q's are *independent*, and if the constraints on the x's are workless, then the constraint impulses will not contribute to the $\hat{Q}$'s.

Ordinary Constraints. Now let us broaden the discussion of impulsive motion to include systems for which the q's are *not independent*. Whether the constraints are holonomic or nonholonomic, we will express the m constraint equations in the form

$$\sum_{i=1}^{n} a_{ji}\dot{q}_i + a_{jt} = 0 \qquad (j = 1, 2, \ldots, m) \tag{3-32}$$

where the a's are continuous functions of the q's and t. Lagrange's equations for this system can be written in the form

$$\frac{d}{dt}\left(\frac{\partial T}{\partial \dot{q}_i}\right) - \frac{\partial T}{\partial q_i} = Q_i + C_i \qquad (i = 1, 2, \ldots, n) \tag{3-33}$$

where the C's are generalized constraint forces given by

$$C_i = \sum_{j=1}^{m} \lambda_j a_{ji} \tag{3-34}$$

and the λ's are Lagrange multipliers. The Q's are generalized forces associated with the *applied forces*, as in Eq. (1-73).

Now suppose we consider the virtual work expression

$$\delta W = \sum_{i=1}^{n}\left[\frac{d}{dt}\left(\frac{\partial T}{\partial \dot{q}_i}\right) - \frac{\partial T}{\partial q_i} - Q_i - C_i\right]\delta q_i \tag{3-35}$$

From Eq. (3-33), it is clear that $\delta W = 0$ for any set of δq's since all of the coefficients are zero. If, however, we assume that the constraints are *workless* and the δq's conform to the *instantaneous constraints*, then

$$\sum_{i=1}^{n} C_i \,\delta q_i = 0 \tag{3-36}$$

and we obtain

$$\sum_{i=1}^{n}\left[\frac{d}{dt}\left(\frac{\partial T}{\partial \dot{q}_i}\right) - \frac{\partial T}{\partial q_i} - Q_i\right]\delta q_i = 0 \tag{3-37}$$

This important result is a generalization of the Lagrangian form of d'Alembert's principle given in Eq. (1-63).

In the development of Eq. (3-37), we have not assumed that the applied forces or the constraint forces are impulsive in nature. But now let us assume that impulsive forces are applied to the system during the interval Δt, beginning at time t_1. Then, integrating Eq. (3-37) with respect to time over the interval Δt, we obtain

$$\sum_{i=1}^{n}(\Delta p_i - \hat{Q}_i)\delta q_i = 0 \tag{3-38}$$

where we use the same limiting approximations as were used previously in finding Eq. (3-24).

It is convenient at this point to think in terms of *virtual velocities* rather than virtual displacements. So let us consider a set of virtual velocities where each component u_i is proportional to the corresponding δq_i, but is not necessarily small. With the aid of Eq. (3-27), we can write Eq. (3-38) in the form

$$\sum_{i=1}^{n} \left[\sum_{j=1}^{n} m_{ij}(\dot{q}_j - \dot{q}_{j0}) - \hat{Q}_i \right] u_i = 0 \qquad (3\text{-}39)$$

where

$$\Delta \dot{q}_j = \dot{q}_j - \dot{q}_{j0} \qquad (3\text{-}40)$$

and the u's meet the *instantaneous* constraint conditions, namely,

$$\sum_{i=1}^{n} a_{ji}u_i = 0 \qquad (j = 1, 2, \ldots, m) \qquad (3\text{-}41)$$

The $\dot{q}$'s, however, conform to the actual constraint conditions given in Eq. (3-32).

To summarize this method, let us assume that the configuration and the velocities $\dot{q}_{j0}$ are given at time t_1, as well as the generalized impulse components $\hat{Q}_i$. We can choose $(n - m)$ *independent* sets of virtual velocity components u_i which meet the instantaneous constraint conditions of Eq. (3-41). Hence, we obtain $(n - m)$ equations having the form of Eq. (3-39) plus the m constraint equations given by Eq. (3-32). These n equations can be solved for the n $\dot{q}$'s, that is, the velocity components immediately after the impulses have been applied. The q's are assumed to remain unchanged during the small interval Δt.

An alternative approach, which applies to constrained systems with applied impulses, is the *Lagrange multiplier method*. We integrate Eq. (3-33) with respect to time over the interval Δt and obtain

$$\Delta p_i = \hat{Q}_i + \hat{C}_i \qquad (i = 1, 2, \ldots, n) \qquad (3\text{-}42)$$

where the $\hat{C}$'s are *generalized constraint impulses* given by

$$\hat{C}_i = \sum_{k=1}^{m} \lambda_k a_{ki} \qquad (3\text{-}43)$$

The λ's are Lagrange multipliers which are *impulsive* in this instance. Using Eqs. (3-27), (3-42), and (3-43), we find that

$$\sum_{j=1}^{n} m_{ij}(\dot{q}_j - \dot{q}_{j0}) = \hat{Q}_i + \sum_{k=1}^{m} \lambda_k a_{ki} \qquad (i = 1, 2, \ldots, n) \qquad (3\text{-}44)$$

If we consider the actual constraint equations given in Eq. (3-32) and the generalized impulse and momentum relations of Eq. (3-44), we have a total of $(n + m)$ algebraic equations from which to solve for the n $\dot{q}$'s and the m λ's. Then, knowing the λ's, the constraint impulses are readily found by using Eq. (3-43).

Impulsive Constraints. So far we have assumed that the coefficients a_{ji} and a_{jt} are continuous functions of the q's and t. Now let us consider the case of *impulsive constraints* where one or more of the a's may be *discon-*

tinuous at a given time t_1. This allows for the sudden appearance of a constraint, or a sudden change in its motion. For example, the sudden application of a fixed constraint might be represented by a change in the coefficients a_{ji} from zero to nonzero values. On the other hand, a discontinuity in the coefficient a_{ji} would represent a change in the velocity of a moving constraint. *Unilateral* or *inequality* constraints can also be represented in this manner, provided that the time of the switching is known.

At this point, let us emphasize the difference between *impulsive constraints* and *constraint impulses*. An impulsive constraint is a suddenly applied constraint which is represented by a discontinuous constraint equation. On the other hand, constraint impulses are constraint forces of an impulsive nature which may arise as a result of impulsive constraints or of applied impulses $\hat{Q}_i$.

Now let us assume that any impulsive constraints are *workless*; that is, no work is done by the constraint forces in moving through an arbitrary virtual displacement which is consistent with the instantaneous constraints, either before or after the discontinuity. With this assumption, the preceding theory can be generalized easily to include impulsive constraints.

Suppose, then, that any discontinuities in the a's occur at time t_1; and any impulses $\hat{Q}_i$ are applied during the small interval Δt immediately after t_1. Equations (3-32), (3-39), (3-41), and (3-44) are still valid, but a question arises concerning which of the discontinuous values of the a's should be chosen. It can be seen that the initial velocities $\dot{q}_{j0}$ must conform to Eq. (3-32), where the a's are evaluated at $t = t_1-$, that is, just before the application of the impulsive constraints. On the other hand, the a's are evaluated at $t = t_1+$ in the expressions involving the $\dot{q}$'s, u's, and λ's in Eqs. (3-32), (3-41), and (3-44), respectively.

Energy Considerations. At the outset, one should observe that the sudden application of a constraint or an impulse nearly always changes the kinetic energy of a system because, in general, the $\dot{q}$'s are suddenly changed.

Now suppose we have a system whose kinetic energy function is *quadratic* in the $\dot{q}$'s. The change in the kinetic energy during the period Δt is

$$T - T_0 = \tfrac{1}{2} \sum_{i=1}^{n} \sum_{j=1}^{n} m_{ij} \dot{q}_i \dot{q}_j - \tfrac{1}{2} \sum_{i=1}^{n} \sum_{j=1}^{n} m_{ij} \dot{q}_{i0} \dot{q}_{j0} \qquad (3\text{-}45)$$

From Eq. (3-23) we see that

$$p_i = \sum_{j=1}^{n} m_{ij} \dot{q}_j \qquad (3\text{-}46)$$

and therefore

$$T - T_0 = \tfrac{1}{2} \sum_{i=1}^{n} p_i \dot{q}_i - \tfrac{1}{2} \sum_{i=1}^{n} p_{i0} \dot{q}_{i0} \qquad (3\text{-}47)$$

But, noting that $m_{ij} = m_{ji}$, we obtain from Eq. (3-46) that

$$\sum_{i=1}^{n} p_i \dot{q}_{i0} = \sum_{i=1}^{n} p_{i0} \dot{q}_i \qquad (3\text{-}48)$$

Hence, we find from Eqs. (3-47) and (3-48) that

$$T - T_0 = \tfrac{1}{2} \sum_{i=1}^{n} (p_i - p_{i0})(\dot{q}_i + \dot{q}_{i0}) \qquad (3\text{-}49)$$

If $\hat{Q}_i$ is the generalized impulse component due to the *applied* forces and $\hat{C}_i$ is the corresponding generalized impulse component due to the *constraint* forces, then we see from Eq. (3-42) that

$$p_i - p_{i0} = \hat{Q}_i + \hat{C}_i \qquad (i = 1, 2, \ldots, n) \qquad (3\text{-}50)$$

and we obtain

$$T - T_0 = \sum_{i=1}^{n} (\hat{Q}_i + \hat{C}_i)\left(\frac{\dot{q}_i + \dot{q}_{i0}}{2}\right) \qquad (3\text{-}51)$$

Note that each impulse is multiplied by the *mean value* of the corresponding $\dot{q}_i$, taken before and after the impulse.

The impulses $\hat{C}_i$ can be eliminated from the analysis in the special case where the a's are *continuous* at time t_1 and $a_{jt} = 0$ for all j. Then

$$\sum_{i=1}^{n} a_{ji} \dot{q}_i = 0 \qquad (j = 1, 2, \ldots, m) \qquad (3\text{-}52)$$

both before and after the impulse, and, using Eq. (3-34), we obtain

$$\sum_{i=1}^{n} \hat{C}_i \dot{q}_i = \sum_{i=1}^{n} \sum_{j=1}^{m} \lambda_j a_{ji} \dot{q}_i = 0 \qquad (3\text{-}53)$$

Similarly, we have

$$\sum_{i=1}^{n} \hat{C}_i \dot{q}_{i0} = 0 \qquad (3\text{-}54)$$

with the result that Eq. (3-51) reduces to

$$T - T_0 = \tfrac{1}{2} \sum_{i=1}^{n} \hat{Q}_i (\dot{q}_i + \dot{q}_{i0}) \qquad (3\text{-}55)$$

Now let us define the *kinetic energy of relative motion* as follows:

$$K = \tfrac{1}{2} \sum_{i=1}^{n} \sum_{j=1}^{n} m_{ij} (\dot{q}_i - \dot{q}_{i0})(\dot{q}_j - \dot{q}_{j0}) \qquad (3\text{-}56)$$

Here we substitute *relative* velocities for the actual velocities in the kinetic energy T which again is assumed to be quadratic in the $\dot{q}$'s.

In order to illustrate the significance of the kinetic energy of relative motion, consider a system having impulsive constraints and assume that $\hat{Q}_i = 0$. Let us suppose that the constraints are given by

$$\sum_{i=1}^{n} a_{ji} \dot{q}_i = 0 \qquad (j = 1, 2, \ldots, m) \qquad (3\text{-}57)$$

where the a's are *discontinuous* at time t_1.† We can write the basic equation (3-39) in the form

$$\sum_{i=1}^{n} \sum_{j=1}^{n} m_{ij}(\dot{q}_j - \dot{q}_{j0})\dot{q}_i = 0 \qquad (3\text{-}58)$$

where we let $u_i = \dot{q}_i$ since actual $\dot{q}$'s meet the conditions on the virtual velocities given by Eq. (3-41). Here we evaluate the a's in Eq. (3-57) at time t_1+.

Using Eq. (3-58) and recalling that $m_{ij} = m_{ji}$, we obtain

$$\sum_{i=1}^{n} \sum_{j=1}^{n} m_{ij}\dot{q}_i\dot{q}_j = \sum_{i=1}^{n} \sum_{j=1}^{n} m_{ij}\dot{q}_i\dot{q}_{j0} = \sum_{i=1}^{n} \sum_{j=1}^{n} m_{ij}\dot{q}_{i0}\dot{q}_j \qquad (3\text{-}59)$$

Then, from Eqs. (3-56) and (3-59), we have

$$K = \tfrac{1}{2}\sum_{i=1}^{n} \sum_{j=1}^{n} m_{ij}\dot{q}_{i0}\dot{q}_{j0} - \tfrac{1}{2}\sum_{i=1}^{n} \sum_{j=1}^{n} m_{ij}\dot{q}_i\dot{q}_j = T_0 - T \qquad (3\text{-}60)$$

where we again note that T_0 is the initial kinetic energy and T is its value after the application of the impulsive constraint. Thus, the *energy lost* because of the sudden appearance of a *fixed constraint* is equal to the kinetic energy of relative motion.

As a second illustration, let us consider a system in which the a_{ji} coefficients are continuous, but the a_{jt} coefficients change from zero to nonzero values at time t_1, corresponding to the sudden start of a moving constraint. In this case, the final $\dot{q}$'s are not possible virtual velocities since they do not conform to Eq. (3-41). The initial velocities, however, do meet these conditions at time t_1. So let us take $u_i = \dot{q}_{i0}$ and, again assuming $\hat{Q}_i = 0$, we find that Eq. (3-39) results in

$$\sum_{i=1}^{n} \sum_{j=1}^{n} m_{ij}(\dot{q}_j - \dot{q}_{j0})\dot{q}_{i0} = 0 \qquad (3\text{-}61)$$

or

$$\sum_{i=1}^{n} \sum_{j=1}^{n} m_{ij}\dot{q}_{i0}\dot{q}_{j0} = \sum_{i=1}^{n} \sum_{j=1}^{n} m_{ij}\dot{q}_{i0}\dot{q}_j = \sum_{i=1}^{n} \sum_{j=1}^{n} m_{ij}\dot{q}_i\dot{q}_{j0} \qquad (3\text{-}62)$$

Hence, from Eqs. (3-56) and (3-62), we obtain

$$K = \tfrac{1}{2}\sum_{i=1}^{n} \sum_{j=1}^{n} m_{ij}\dot{q}_i\dot{q}_j - \tfrac{1}{2}\sum_{i=1}^{n} \sum_{j=1}^{n} m_{ij}\dot{q}_{i0}\dot{q}_{j0} = T - T_0 \qquad (3\text{-}63)$$

which implies that the *increase* in the kinetic energy of a system, due to the sudden start of a *moving constraint*, is equal to the kinetic energy of relative motion.

Quasi-Coordinates. The standard forms of Lagrange's equations apply for cases where the q's are *true coordinates*; that is, the q's specify the con-

---

†More precisely, Eq. (3-57) gives the orientation of a differential element of a constraint surface in q-space. This element may suddenly appear, or may suddenly change its orientation, but it cannot have a normal velocity component since this would require that $a_{jt} \neq 0$.

figuration of the system and each q_i can be obtained by integrating the corresponding $\dot{q}_i$ with respect to time. Furthermore, the transformation equations (1-8) give the x's as *explicit* functions of the q's and t. Thus, we obtain expressions for the $\dot{x}$'s as functions of $(q, \dot{q}, t)$, and, similarly, we find that the kinetic energy is of the form $T = T(q, \dot{q}, t)$.

It is possible, however, in certain common situations, to obtain a correct expression for the kinetic energy T which is not of this form. For example, the rotational kinetic energy of a rigid body, expressed in terms of the Cartesian body-axis components of the angular velocity $\boldsymbol{\omega}$, is

$$T = \frac{1}{2} \sum_{i=1}^{3} \sum_{j=1}^{3} I_{ij}\, \omega_i\, \omega_j \tag{3-64}$$

where the ω's are *not* the time derivatives of true coordinates. In other words, the integral of ω_j with respect to time does not result in a parameter which helps specify the configuration.

Although the ω's are not considered as $\dot{q}$'s, they can be expressed as linear functions of the $\dot{q}$'s using equations of the form

$$\omega_j = \sum_{i=1}^{n} A_{ji}\dot{q}_i + A_{jt} \tag{3-65}$$

where the A's are functions of the q's and t. If we associate a *quasi-coordinate* θ_j with each ω_j such that

$$d\theta_j = \omega_j\, dt \tag{3-66}$$

we obtain

$$d\theta_j = \sum_{i=1}^{n} A_{ji}\, dq_i + A_{jt}\, dt \tag{3-67}$$

The characteristic that distinguishes the θ's from true coordinates is the fact that the expressions having the form of Eq. (3-67) are *not integrable*; hence, the θ's cannot be expressed as *explicit* functions of the q's and t.

Returning to the example of rigid body rotation, we can see from Fig. 3-1 that the body-axis (x, y, z) components of the angular velocity can be expressed in terms of the Euler angle (ψ, θ, ϕ) rates as follows:

$$\begin{aligned}
\omega_x &= \dot{\phi} - \dot{\psi} \sin \theta \\
\omega_y &= \dot{\theta} \cos \phi + \dot{\psi} \cos \theta \sin \phi \\
\omega_z &= \dot{\psi} \cos \theta \cos \phi - \dot{\theta} \sin \phi
\end{aligned} \tag{3-68}$$

These equations are of the form of Eq. (3-65) where $A_{jt} = 0$. In terms of quasi-coordinates, we have

$$\begin{aligned}
d\theta_x &= -\sin \theta \, d\psi + d\phi \\
d\theta_y &= \cos \theta \sin \phi \, d\psi + \cos \phi \, d\theta \\
d\theta_z &= \cos \theta \cos \phi \, d\psi - \sin \phi \, d\theta
\end{aligned} \tag{3-69}$$

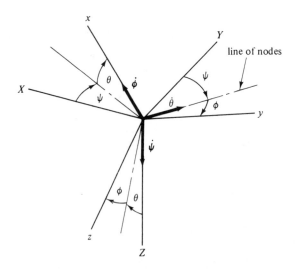

Fig. 3-1. Eulerian angles and angular velocity components.

Using the criteria given in Eq. (1-20) or Eq. (1-23), we see that none of these equations is integrable. In particular, we note that

$$\frac{\partial A_{ji}}{\partial q_k} \neq \frac{\partial A_{jk}}{\partial q_i} \tag{3-70}$$

for at least one pair of coefficients in each equation.

Even though quasi-coordinates cannot be used with the standard forms of Lagrange's equations, we can use the corresponding ω's in the analysis of impulsive motion. Suppose, for example, that the kinetic energy of a system of rotating bodies is given by

$$T = \tfrac{1}{2} \sum_{i=1}^{n} \sum_{j=1}^{n} I_{ij} \omega_i \omega_j \tag{3-71}$$

we can treat the ω's like $\dot{q}$'s in the sense that we define the ith component of the generalized momentum according to the equation

$$p_i = \frac{\partial T}{\partial \omega_i} = \sum_{j=1}^{n} I_{ij} \, \omega_j \tag{3-72}$$

The generalized impulse $\hat{Q}_i$ associated with the quasi-coordinate θ is defined in much the same fashion as it was previously. Note, however, that Eq. (3-30) does not apply because the x's cannot be expressed as functions of the θ's. So let us make use of the idea of virtual work to obtain each component $\hat{Q}_i$ due to the *applied* forces. We see that

$$\sum_{i=1}^{n} \hat{Q}_i \, \delta\theta_i = \sum_{k=1}^{3N} \hat{F}_k \, \delta x_k \tag{3-73}$$

where the $\delta\theta$'s and δx's correspond to the same virtual displacement. By taking all the $\delta\theta$'s except $\delta\theta_i$ equal to zero, we can solve for each $\hat{Q}_i$ in turn.

Now let us write Eq. (3-39) in the form

$$\sum_{i=1}^{n}\left[\sum_{j=1}^{n} I_{ij}(\omega_j - \omega_{j0}) - \hat{Q}_i\right]u_i = 0 \tag{3-74}$$

where, as before, the u's are *any set* of virtual velocities which meet the instantaneous constraints of Eq. (3-41). The ω's conform to the actual constraints, namely,

$$\sum_{i=1}^{n} a_{ji}\omega_i + a_{jt} = 0 \qquad (j = 1, 2, \ldots, m) \tag{3-75}$$

where the a's are evaluated at time t_1+. A similar constraint equation applies to the ω_0's, and the a's are evaluated at time t_1- in this case.

Also, Eq. (3-44) can be written in the form

$$\sum_{j=1}^{n} I_{ij}(\omega_j - \omega_{j0}) = \hat{Q}_i + \sum_{k=1}^{m} \lambda_k a_{ki} \qquad (i = 1, 2, \ldots, n) \tag{3-76}$$

where the a's are evaluated at time t_1+.

The principal results for impulsive constrained motion given in Eqs. (3-39), (3-44), (3-74), and (3-76) apply to systems in which the kinetic energy is expressed as a function of the $\dot{q}$'s or the ω's. The same approach, however, is valid for systems in which T contains a mixture of $\dot{q}$'s and ω's.

Example 3-1. Four rigid uniform rods, each of mass m and length l are joined by pin joints at their ends to form a rhombus $ABCD$ (Fig. 3-2). Ini-

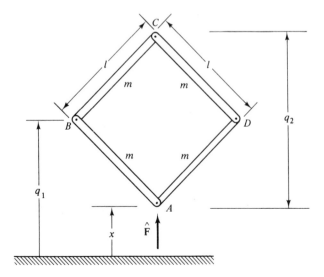

Fig. 3-2. A square framework of uniform rods.

tially the system is motionless in a square configuration with point C directly above point A. (a) If point A is given a vertical impulse $\hat{\mathbf{F}}$, solve for its velocity immediately after the impulse. (b) Suppose that the system falls with velocity v_0 and point A strikes the floor inelastically. Solve for $\dot{q}_1$ and $\dot{q}_2$ immediately after the impact.

First, let us obtain an expression for the total kinetic energy. It is equal to the sum of the translational kinetic energy due to the motion of the center of mass plus the kinetic energy arising from the motion of the rods relative to the center of mass. If we concentrate on the latter portion first, we find that each rod has an angular velocity

$$\omega = \frac{1}{\sqrt{2}\,l}\dot{q}_2$$

and a translational velocity

$$v = \frac{1}{2\sqrt{2}}\dot{q}_2$$

Hence, adding the two portions, we find that the total kinetic energy of the system is

$$T = 2m\dot{q}_1^2 + 4\left(\frac{1}{2}mv^2 + \frac{ml^2}{24}\omega^2\right) = 2m\dot{q}_1^2 + \frac{1}{3}m\dot{q}_2^2 \qquad (3\text{-}77)$$

where we note that the moment of inertia of each rod about its center of mass is $ml^2/12$. The generalized momenta are

$$p_1 = \frac{\partial T}{\partial \dot{q}_1} = 4m\dot{q}_1$$

$$p_2 = \frac{\partial T}{\partial \dot{q}_2} = \frac{2}{3}m\dot{q}_2 \qquad (3\text{-}78)$$

and the inertia matrix is

$$\mathbf{m} = \begin{bmatrix} 4m & 0 \\ 0 & \frac{2}{3}m \end{bmatrix} \qquad (3\text{-}79)$$

The displacement of the point of application A of the impulse $\hat{\mathbf{F}}$ is

$$x = q_1 - \tfrac{1}{2}q_2 \qquad (3\text{-}80)$$

Hence, from Eq. (3-30) we obtain

$$\hat{Q}_1 = \hat{F}, \qquad \hat{Q}_2 = -\tfrac{1}{2}\hat{F} \qquad (3\text{-}81)$$

The generalized coordinates are independent, so Eq. (3-28) applies and yields

$$4m\,\Delta\dot{q}_1 = \hat{F}, \qquad \Delta\dot{q}_1 = \frac{1}{4m}\hat{F}$$

$$\frac{2}{3}m\,\Delta\dot{q}_2 = -\frac{1}{2}\hat{F}, \qquad \Delta\dot{q}_2 = -\frac{3}{4m}\hat{F} \qquad (3\text{-}82)$$

For the case where the system is initially motionless, we find that the velocity of point A immediately after the impulse is

$$\dot{x} = \dot{q}_1 - \frac{1}{2}\dot{q}_2 = \frac{5}{8m}\hat{F} \tag{3-83}$$

Now let us consider part (b). The initial generalized velocities are

$$\dot{q}_{10} = -v_0, \qquad \dot{q}_{20} = 0$$

and the final velocity of point A is

$$\dot{x} = \dot{q}_1 - \tfrac{1}{2}\dot{q}_2 = 0 \tag{3-84}$$

From Eq. (3-82) we obtain

$$\Delta\dot{q}_2 = -3\,\Delta\dot{q}_1$$

or

$$3(\dot{q}_1 + v_0) + \dot{q}_2 = 0 \tag{3-85}$$

Then, solving Eqs. (3-84) and (3-85), we find that the final velocities are

$$\dot{q}_1 = -\tfrac{2}{5}v_0, \qquad \dot{q}_2 = -\tfrac{6}{5}v_0$$

Using Eq. (3-82) we can solve for the impulse $\hat{F}$ applied at point A at the time of impact.

$$\hat{F} = 4m\,\Delta\dot{q}_1 = \tfrac{8}{5}mv_0$$

Another method of analyzing part (b) is to consider the inelastic impact as the sudden application of an impulsive constraint given by Eq. (3-84). Using this approach, we have

$$\hat{Q}_1 = \hat{Q}_2 = 0$$

since the constraint impulse is *internal*. Thus, Eq. (3-39) takes the form

$$\sum_{i=1}^{2}\sum_{j=1}^{2} m_{ij}(\dot{q}_j - \dot{q}_{j0})u_i = 0 \tag{3-86}$$

where the virtual velocities in this case meet the same constraint condition as the actual velocities, namely,

$$u_1 - \tfrac{1}{2}u_2 = 0 \tag{3-87}$$

Let us choose $u_1 = 1$, $u_2 = 2$. Then Eq. (3-86) becomes

$$4m(\dot{q}_1 + v_0) + \tfrac{4}{3}m\dot{q}_2 = 0 \tag{3-88}$$

which is equivalent to Eq. (3-85). Hence, from Eqs. (3-84) and (3-88) we obtain the same results as before.

Finally, let us consider the kinetic energy lost in the impact. Using the kinetic energy expression of Eq. (3-77), we obtain

$$T_0 - T = 2mv_0^2 - (\tfrac{18}{25}mv_0^2 + \tfrac{12}{25}mv_0^2) = \tfrac{4}{5}mv_0^2$$

The same result is found by recalling that the lost energy due to the sudden appearance of a fixed constraint is equal to the kinetic energy of relative motion. Using Eq. (3-56), we find that

$$K = 2m(\tfrac{2}{3}v_0)^2 + \tfrac{1}{3}m(-\tfrac{5}{3}v_0)^2 = \tfrac{4}{3}mv_0^2$$

A further check can be obtained by multiplying the constraint impulse by the mean value of the velocity at point A. Adapting Eq. (3-51) to this case of a single coordinate x, we obtain

$$T - T_0 = \tfrac{1}{2}\hat{F}(\dot{x} + \dot{x}_0) = \tfrac{1}{2}(\tfrac{8}{3}mv_0)(-v_0) = -\tfrac{4}{3}mv_0^2$$

where the negative sign indicates that energy is lost.

Example 3-2. A rigid bar can rotate freely about a fixed pivot O (Fig. 3-3) and has a moment of inertia I about this point. A particle of mass m

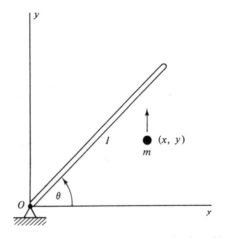

Fig. 3-3. The impact of a particle and a freely pivoted bar.

strikes the bar inelastically at time t_1, and slides along the bar after the impact. Solve for the velocities $\dot{x}, \dot{y}, \dot{\theta}$ after impact if the initial conditions are

$$x(t_1) = 1 \text{ m} \qquad y(t_1) = 1 \text{ m} \qquad \theta(t_1) = \pi/4$$
$$\dot{x}(t_1-) = 0 \qquad \dot{y}(t_1-) = 1 \text{ m/sec} \qquad \dot{\theta}(t_1-) = -1 \text{ rad/sec}$$

Let $m = 1$ kg and $I = 10$ kg m²

First method: The kinetic energy of the system is

$$T = \tfrac{1}{2}m(\dot{x}^2 + \dot{y}^2) + \tfrac{1}{2}I\dot{\theta}^2 = \tfrac{1}{2}(\dot{x}^2 + \dot{y}^2) + 5\dot{\theta}^2 \qquad (3\text{-}89)$$

Hence the inertia matrix is

$$\mathbf{m} = \begin{bmatrix} 1 & 0 & 0 \\ 0 & 1 & 0 \\ 0 & 0 & 10 \end{bmatrix} \qquad (3\text{-}90)$$

The *applied* impulses are zero so Eq. (3-39) takes the form

$$\sum_{i=1}^{3} \sum_{j=1}^{3} m_{ij}(\dot{q}_j - \dot{q}_{j0})u_i = 0 \tag{3-91}$$

or

$$\dot{x}u_1 + (\dot{y} - 1)u_2 + 10(\dot{\theta} + 1)u_3 = 0 \tag{3-92}$$

The equation of constraint after the impact is found by noting that

$$x \tan \theta = y$$

or

$$\tan \theta \, \dot{x} - \dot{y} + x \sec^2 \theta \, \dot{\theta} = 0 \tag{3-93}$$

For the configuration at time t_1, this equation becomes

$$\dot{x} - \dot{y} + 2\dot{\theta} = 0 \tag{3-94}$$

Similarly, the virtual velocities must satisfy the instantaneous constraint at time t_1+, namely,

$$u_1 - u_2 + 2u_3 = 0 \tag{3-95}$$

Let us take two independent sets of virtual velocities which are solutions of Eq. (3-95), for example,

$$
\begin{array}{ccc}
u_1 = 1 & & u_1 = 0 \\
u_2 = 0 & \text{and} & u_2 = 1 \\
u_3 = -\tfrac{1}{2} & & u_3 = \tfrac{1}{2}
\end{array}
$$

Substituting these virtual velocities into Eq. (3-92), we obtain

$$\dot{x} - 5\dot{\theta} = 5, \qquad \dot{y} + 5\dot{\theta} = -4 \tag{3-96}$$

The velocities immediately after impact are found by solving Eqs. (3-94) and (3-96). They are

$$\dot{x} = \tfrac{5}{4}\,\text{m/sec}, \qquad \dot{y} = -\tfrac{1}{4}\,\text{m/sec}, \qquad \dot{\theta} = -\tfrac{3}{4}\,\text{rad/sec}$$

Second method: Another approach is to use Eq. (3-44). In this example the $\hat{Q}$'s are zero, and we have

$$\sum_{j=1}^{n} m_{ij}(\dot{q}_j - \dot{q}_{j0}) = \sum_{k=1}^{m} \lambda_k a_{ki} \tag{3-97}$$

where the $\dot{q}$'s satisfy the actual constraint equations at time t_1+, namely,

$$\sum_{i=1}^{n} a_{ki}\dot{q}_i + a_{kt} = 0 \qquad (k = 1, 2, \ldots, m) \tag{3-98}$$

Substituing the values of the coefficients, Eq. (3-97) yields

$$
\begin{aligned}
\dot{x} &= \lambda \\
\dot{y} - 1 &= -\lambda \\
10\dot{\theta} + 10 &= 2\lambda
\end{aligned}
$$

The constraint equation is

$$\dot{x} - \dot{y} + 2\dot{\theta} = 0$$

as before. Solving these four equations, we obtain

$$\dot{x} = \tfrac{5}{4}\,\text{m/sec}, \qquad \dot{y} = -\tfrac{1}{4}\,\text{m/sec}, \qquad \dot{\theta} = -\tfrac{3}{4}\,\text{rad/sec}, \qquad \lambda = \tfrac{5}{4}\,\text{N sec}$$

in agreement with our earlier results.

The constraint impulse acting on the particle in a direction normal to the bar is

$$\frac{1}{\sqrt{2}}(\hat{C}_1 - \hat{C}_2) = \frac{\lambda}{\sqrt{2}}(a_{11} - a_{12}) = \frac{5}{2\sqrt{2}}\,\text{N sec}$$

Example 3-3. An ellipsoid of revolution (Fig. 3-4) having mass m, semimajor axis a, and semiminor axis b is lying motionless on a horizontal sur-

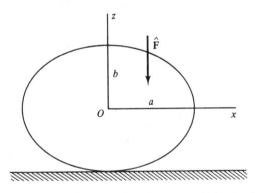

Fig. 3-4. An ellipsoid on a horizontal surface.

face, as shown. It is struck by a vertical impulse $\hat{F}$ which is applied at a position such that a unit angular impulse is applied about the x and y axes, where the x axis is the axis of symmetry. Assuming that the ellipsoid rolls without slipping, solve for the translational and rotational velocities immediately after the impulse.

First method: Let us write the kinetic energy using a body-axis coordinate system with its origin O at the center of mass. For the given orientation, we have

$$T = \tfrac{1}{2}m(\dot{x}^2 + \dot{y}^2) + \tfrac{1}{2}I_a\omega_x^2 + \tfrac{1}{2}I_t(\omega_y^2 + \omega_z^2) \qquad (3\text{-}99)$$

where I_a is the moment of inertia about the axis of symmetry (x axis) and I_t is the moment of inertia about a transverse axis in the yz plane. Here the translational and rotational velocities are absolute, but the components are taken in the directions of the body axes at the instant of the impulse. Furthermore, the generalized velocities are associated with quasi-coordinates.

If we adopt the velocity sequence $\dot{x}, \dot{y}, \omega_x, \omega_y, \omega_z$, we obtain the following components of generalized momentum and applied impulse:

$$p_1 = \frac{\partial T}{\partial \dot{x}} = m\dot{x} \qquad \hat{Q}_1 = 0$$

$$p_2 = \frac{\partial T}{\partial \dot{y}} = m\dot{y} \qquad \hat{Q}_2 = 0$$

$$p_3 = \frac{\partial T}{\partial \omega_x} = I_a\omega_x \qquad \hat{Q}_3 = 1$$

$$p_4 = \frac{\partial T}{\partial \omega_y} = I_t\omega_y \qquad \hat{Q}_4 = 1$$

$$p_5 = \frac{\partial T}{\partial \omega_z} = I_t\omega_z \qquad \hat{Q}_5 = 0$$

The rolling constraint can be expressed in terms of two nonholonomic constraint equations, namely,

$$\dot{x} - b\omega_y = 0, \qquad \dot{y} + b\omega_x = 0 \tag{3-100}$$

and the corresponding nonzero a's are

$$a_{11} = 1, \qquad a_{14} = -b$$
$$a_{22} = 1, \qquad a_{23} = b$$

Since there are five generalized velocities and two equations of constraint, we can find three independent sets of virtual velocities. Let us choose the following:

$$
\begin{array}{lll}
u_1 = 1 & u_1 = 0 & u_1 = 1 \\
u_2 = 0 & u_2 = 1 & u_2 = 0 \\
u_3 = 0 & u_3 = -1/b & u_3 = 0 \\
u_4 = 1/b & u_4 = 0 & u_4 = 1/b \\
u_5 = 0 & u_5 = 0 & u_5 = 1
\end{array}
$$

Noting that the inertia matrix is diagonal, and using each column of u's in turn, we obtain from Eq. (3-39) or (3-74) that

$$m\dot{x} + \frac{I_t\omega_y - 1}{b} = 0$$

$$m\dot{y} - \frac{I_a\omega_x - 1}{b} = 0 \tag{3-101}$$

$$m\dot{x} + \frac{I_t\omega_y - 1}{b} + I_t\omega_z = 0$$

Solving Eq. (3-101), we obtain the velocities immediately after the impulse, namely,

$$\dot{x} = \frac{b}{I_t + mb^2}, \qquad \omega_x = \frac{1}{I_a + mb^2}, \qquad \omega_z = 0$$

$$\dot{y} = \frac{-b}{I_a + mb^2}, \qquad \omega_y = \frac{1}{I_t + mb^2}$$

Second method: Now let us use the Lagrange multiplier method, as given by Eq. (3-44) or (3-76). We obtain the following five equations of impulse and momentum:

$$m\dot{x} = \lambda_1$$
$$m\dot{y} = \lambda_2$$
$$I_a \omega_x - 1 = b\lambda_2 \qquad\qquad (3\text{-}102)$$
$$I_t \omega_y - 1 = -b\lambda_1$$
$$I_t \omega_z = 0$$

These equations, plus the constraint equations of (3-100), can be solved for the five velocities and two Lagrange multipliers. The results agree with those obtained previously, with the addition of

$$\lambda_1 = \frac{mb}{I_t + mb^2}, \qquad \lambda_2 = \frac{-mb}{I_a + mb^2}$$

The constraint impulses obtained from Eq. (3-43) are

$$\hat{C}_1 = \lambda_1 a_{11} = \frac{mb}{I_t + mb^2}$$

$$\hat{C}_2 = \lambda_2 a_{22} = \frac{-mb}{I_a + mb^2}$$

$$\hat{C}_3 = \lambda_2 a_{23} = \frac{-mb^2}{I_a + mb^2}$$

$$\hat{C}_4 = \lambda_1 a_{14} = \frac{-mb^2}{I_t + mb^2}$$

$$\hat{C}_5 = 0$$

Third method: Perhaps the most direct way of obtaining the motion of the ellipsoid immediately after the impulse is to use Eq. (3-21) which can be written in the form

$$\Delta H_i = \hat{M}_i \qquad\qquad (3\text{-}103)$$

The contact point of the ellipsoid with the horizontal surface is instantaneously motionless, since we assume no slipping. Taking this point as the origin of a fixed Cartesian frame having its x, y, z axes parallel to the original body axes at the time of the impulse, we have the following principal moments of inertia:

$$I_{xx} = I_a + mb^2, \qquad I_{yy} = I_t + mb^2, \qquad I_{zz} = I_t$$

where we note that the x and y axes have each been translated a distance b. The constraint impulse has no moment about the origin; hence the applied angular impulse components are

$$\hat{M}_x = 1, \qquad \hat{M}_y = 1, \qquad \hat{M}_z = 0$$

Then, from Eq. (3-103) we obtain

$$I_{xx}\omega_x = (I_a + mb^2)\omega_x = 1$$
$$I_{yy}\omega_y = (I_t + mb^2)\omega_y = 1$$
$$I_{zz}\omega_z = I_t\omega_z = 0$$

which yield the same values of ω_x, ω_y, and ω_z as we obtained previously. The values of $\dot{x}$ and $\dot{y}$ are then obtained from Eq. (3-100).

3-3. GYROSCOPIC SYSTEMS

Gyroscopic Forces. Let us consider once again the explicit form of the equations of motion which arise by applying Lagrange's equations to a holonomic system. As given previously in Eq. (2-39), we obtain

$$\sum_{j=1}^{n} m_{ij}\ddot{q}_j + \sum_{j=1}^{n}\sum_{l=1}^{n} [jl, i]\dot{q}_j\dot{q}_l + \sum_{j=1}^{n} \gamma_{ij}\dot{q}_j$$
$$+ \sum_{j=1}^{n} \frac{\partial m_{ij}}{\partial t}\dot{q}_j + \frac{\partial a_i}{\partial t} - \frac{\partial T_0}{\partial q_i} + \frac{\partial V}{\partial q_i} = 0 \qquad (i = 1, 2, \ldots, n) \quad (3\text{-}104)$$

The terms $\gamma_{ij}\dot{q}_j$ are known as *gyroscopic terms*, and a system whose equations of motion contain gyroscopic terms is known as a *gyroscopic system*. The *gyroscopic coefficients* γ_{ij} are *skew-symmetric* and are given by

$$\gamma_{ij} = -\gamma_{ji} = \frac{\partial a_i}{\partial q_j} - \frac{\partial a_j}{\partial q_i} \qquad (3\text{-}105)$$

where the a's are the coefficients of the $\dot{q}$'s in T_1. If the equations of motion are obtained from a Routhian function R, then the a's are the coefficients in R_1, the portion of R which is linear in the $\dot{q}$'s.

Now let us consider how gyroscopic terms arise. If we assume a standard Lagrangian system, we recall from Eq. (2-9) that

$$a_i = \sum_{k=1}^{3N} m_k \frac{\partial x_k}{\partial q_i}\frac{\partial x_k}{\partial t}$$

where the x's give the inertial positions of the particles comprising the system.

It is apparent that a necessary condition for a_i to be nonzero is that one or more of the transformation equations

$$x_k = x_k(q, t) \tag{3-106}$$

be an *explicit* function of both q_i and t, as might occur with a moving constraint, for example. Furthermore, we see from Eq. (3-105) that some of the a's must be explicit functions of the q's if we are to obtain gyroscopic terms in the equations of motion.

Expressed in physical terms, gyroscopic forces are most likely to occur if (1) the system contains a rotating body (or reference frame) and the configuration is expressed relative to this body, or if (2) the system has ignorable coordinates and the Routhian procedure is used.

An important characteristic of gyroscopic terms is that *coupling* of the motions in two or more coordinates is always involved. From Eq. (3-104) we see that the gyroscopic force conjugate to q_i is

$$Q_i = -\sum_{j=1}^{n} \gamma_{ij} \dot{q}_j \tag{3-107}$$

Since $\gamma_{ii} = 0$, it follows that Q_i may be due to all the $\dot{q}$'s except $\dot{q}_i$, thereby providing a coupling effect.

The rate at which the gyroscopic forces do work on the system is

$$\sum_{i=1}^{n} Q_i \dot{q}_i = -\sum_{i=1}^{n} \sum_{j=1}^{n} \gamma_{ij} \dot{q}_i \dot{q}_j = 0 \tag{3-108}$$

where we note once again the skew symmetry of the γ's. In other words, the quadratic form $\dot{\mathbf{q}}^T \boldsymbol{\gamma} \dot{\mathbf{q}}$ is *identically zero*, implying that the gyroscopic forces do no work on the system, regardless of the motion. The contrast between gyroscopic and dissipative forces is emphasized by recalling that the Rayleigh dissipation function $F = \frac{1}{2} \dot{\mathbf{q}}^T \mathbf{c} \dot{\mathbf{q}}$ is a *positive definite* (or semidefinite) quadratic form; consequently, any motion of a dissipative element results in a loss of energy. Also, the damping coefficient matrix $\mathbf{c}$ is *symmetric* rather than skew symmetric.

To illustrate how the Routhian procedure can result in gyroscopic terms in the equations of motion, let us consider a system in which q_1 is *ignorable*. Suppose the original Lagrangian function is of the form

$$L = \frac{1}{2} \sum_{i=1}^{n} \sum_{j=1}^{n} m_{ij} \dot{q}_i \dot{q}_j + \sum_{i=1}^{n} a_i \dot{q}_i + T_0 - V \tag{3-109}$$

Then the generalized momentum p_1 is a *constant* of the motion, that is,

$$p_1 = \sum_{j=1}^{n} m_{1j} \dot{q}_j + a_1 = \beta_1 \tag{3-110}$$

and we obtain

$$\dot{q}_1 = \frac{1}{m_{11}} \left(\beta_1 - a_1 - \sum_{j=2}^{n} m_{1j} \dot{q}_j \right) \tag{3-111}$$

The Routhian function in this case is

$$R = L - \beta_1 \dot{q}_1$$

$$= \tfrac{1}{2}m_{11}\dot{q}_1^2 + \sum_{j=2}^{n} m_{1j}\dot{q}_1\dot{q}_j + \tfrac{1}{2}\sum_{i=2}^{n}\sum_{j=2}^{n} m_{ij}\dot{q}_i\dot{q}_j$$

$$+ a_1\dot{q}_1 + \sum_{j=2}^{n} a_j\dot{q}_j - \beta_1\dot{q}_1 + T_0 - V \qquad (3\text{-}112)$$

If we substitute for $\dot{q}_1$ from Eq. (3-111) and consider only that portion R_1 which is linear in the $\dot{q}$'s (since we are interested primarily in gyroscopic forces), we obtain

$$R_1 = \left(\frac{\beta_1 - a_1}{m_{11}}\right)\sum_{j=2}^{n} m_{1j}\dot{q}_j + \sum_{j=2}^{n} a_j\dot{q}_j \qquad (3\text{-}113)$$

This equation shows that linear terms in the $\dot{q}$'s appear in the Routhian function, in general, even if these terms are entirely absent in the original Lagrangian function. If more than one coordinate is ignored, a repeated application of this procedure indicates that additional gyroscopic terms occur. Notice, however, that if only one non-ignorable coordinate remains, there are no gyroscopic terms in the equations of motion because, as we showed previously, gyroscopic forces always involve coupling of the motions in two or more coordinates.

Small Motions. Now let us consider the small vibrations of a conservative gyroscopic system about a reference condition of *steady motion* due to moving constraints or ignored coordinates. Let us choose the q's such that, at the reference position, all the non-ignorable q's and $\dot{q}$'s are zero, and the ignorable $\dot{q}$'s are constant. Furthermore, the reference position meets the following equilibrium conditions:

$$\left(\frac{\partial(V - T_0)}{\partial q_i}\right)_0 = \left(\frac{\partial V'}{\partial q_i}\right)_0 = 0 \qquad (i = 1, 2, \ldots, n) \qquad (3\text{-}114)$$

where, in accordance with Eq. (2-152), V' represents inertial forces as well as applied forces occurring in the reference motion. If we expand V' in a Taylor series about the origin in a manner similar to Eq. (2-228), we obtain

$$V' = \tfrac{1}{2}\sum_{i=1}^{n}\sum_{j=1}^{n} k_{ij}q_i q_j \qquad (3\text{-}115)$$

where

$$k_{ij} = k_{ji} = \left(\frac{\partial^2 V'}{\partial q_i \, \partial q_j}\right)_0 \qquad (3\text{-}116)$$

and the higher-order terms in the q's are neglected. If the Routhian procedure is used, R_0 (that portion of R not containing $\dot{q}$'s) replaces $-V'$.

Now let us evaluate the coefficients m_{ij} and a_i at the reference configuration and assume that none are explicit functions of time. Then, from Eq.

(3-104), we see that the *linearized* equations of motion are of the form

$$\sum_{j=1}^{n} m_{ij}\ddot{q}_j + \sum_{j=1}^{n} \gamma_{ij}\dot{q}_j + \sum_{j=1}^{n} k_{ij}q_j = 0 \qquad (i = 1, 2, \ldots, n) \qquad (3\text{-}117)$$

where the coefficients are constant. The corresponding Lagrangian function is

$$L = \tfrac{1}{2}\sum_{i=1}^{n}\sum_{j=1}^{n} m_{ij}\dot{q}_i\dot{q}_j + \sum_{i=1}^{n}\sum_{j=1}^{n} g_{ij}\dot{q}_i q_j - \tfrac{1}{2}\sum_{i=1}^{n}\sum_{j=1}^{n} k_{ij}q_i q_j \qquad (3\text{-}118)$$

where

$$g_{ij} = \frac{\partial a_i}{\partial q_j} \qquad (3\text{-}119)$$

and, from Eq. (3-105),

$$\gamma_{ij} = g_{ij} - g_{ji} \qquad (3\text{-}120)$$

Notice that the matrices **m** and **k** are symmetric, but **g** is not.

Next let us assume solutions of the form

$$q_j = A_j C e^{\lambda t} \qquad (3\text{-}121)$$

and substitute into Eq. (3-117). After canceling the common factor $Ce^{\lambda t}$, we have

$$\lambda^2 \sum_{j=1}^{n} m_{ij}A_j + \lambda \sum_{j=1}^{n} \gamma_{ij}A_j + \sum_{j=1}^{n} k_{ij}A_j = 0 \qquad (3\text{-}122)$$

or, in matrix form,

$$(\lambda^2\mathbf{m} + \lambda\boldsymbol{\gamma} + \mathbf{k})\mathbf{A} = 0 \qquad (3\text{-}123)$$

The *characteristic equation* for the system is obtained by setting the determinant of the coefficients of the A's equal to zero, that is,

$$|(\lambda^2\mathbf{m} + \lambda\boldsymbol{\gamma} + \mathbf{k})| = 0 \qquad (3\text{-}124)$$

This yields a polynomial of degree $2n$ in λ. Because of the skew symmetry of $\boldsymbol{\gamma}$, however, and the symmetry of **m** and **k**, it can be shown that all *odd* powers of λ are missing from the characteristic equation. Hence we actually obtain a polynomial of nth degree in λ^2, as would be the case without the gyroscopic terms.

The coefficients in the characteristic equation are real, implying that the n eigenvalues λ_k^2 either are real or occur in complex conjugate pairs. For each root λ_k, we can solve for the corresponding modal column $\mathbf{A}^{(k)}$ by using Eq. (3-123). As a result of the gyroscopic effects, however, the amplitude ratios are *not real*, in general. Thus, for a stable system, the motion corresponding to a single negative eigenvalue λ_k^2 consists of oscillatory motions of the various q's which have the same frequency but *do not move in phase*.

The general solution for the free motion is obtained by superposition. Assuming distinct λ's, we find that

$$q_j = \sum_{k=1}^{2n} A_{jk}C_k e^{\lambda_k t} \qquad (j = 1, 2, \ldots, n) \qquad (3\text{-}125)$$

as in Eq. (3-12). The $2n$ C's are determined from the $2n$ initial conditions, and either are real or occur in complex conjugate pairs. Because $\mathbf{m}$, $\boldsymbol{\gamma}$, and $\mathbf{k}$ are real, we can see from Eq. (3-123) that the modal column corresponding to λ_k^*, the complex conjugate of λ_k, is $\mathbf{A}^{(k)*}$.

Gyroscopic Stability. In the discussion of the small oscillations of a conservative system in Sec. 2-4 we found that the motion is stable if V is positive definite, it is neutrally stable if V is positive semidefinite, and it is unstable if V is indefinite, negative definite, or negative semidefinite.

Now let us consider a definition of stability which is based upon the response of the *non-ignored coordinates* to an *infinitesimal* disturbance from a reference equilibrium position at the origin of configuration space. If the response of the system remains infinitesimal; that is, if $|q_i| < \epsilon \ll 1$ for all the non-ignored q's for all future values of t, then the system is *stable*. Otherwise, it is unstable.

This definition of stability differs from the previous definitions in that a neutrally stable system is now considered to be *unstable* because any term which is linear in t will ultimately become larger than ϵ. Also note that we are considering *stability in the small*; that is, we are concerned only with motions in the immediate vicinity of the reference configuration at the origin of q-space. For linear systems, of course, the stability analysis of small motions applies to large motions as well.

Now let us consider the effect of the addition of gyroscopic terms, as in Eq. (3-117), upon the stability of the linearized system. Assume first that the reference condition is *stable* without the gyroscopic terms; that is, V' is *positive definite*. Furthermore, the system is conservative and there is a Jacobi integral

$$T_2 + V' = h \tag{3-126}$$

Assuming an infinitesimal disturbance from the reference condition, both T_2 and V' are initially small and positive; hence h is also small and positive. Since T_2 and V' are positive definite, it follows that both must remain small in the ensuing motion, confirming the assumed stability of the nongyroscopic system. We note, furthermore, that the addition of gyroscopic terms does not change the Jacobi integral since the gyroscopic forces do no work. Therefore, an initially stable system remains stable with the addition of gyroscopic terms.

Next assume that V' is *negative definite*, corresponding to an unstable system without the gyroscopic terms. In this case, energy considerations must allow for the instability, but do not require it. Hence, a more detailed analysis of the equations of motion is necessary to determine the stability. It can be shown from the characteristic equation, however, that if the system has an even number of degrees of freedom after ignoring coordinates, then gyroscopic stabilization of the system is always possible. A familiar case of this type is the gyroscopic stabilization of a vertical top.

If V' is *indefinite*, that is, if it has both positive and negative regions in the vicinity of the origin, gyroscopic stabilization is sometimes possible. In this case it is necessary for the system to have an even number of unstable modes.

The remaining possible forms for V', namely, *positive semidefinite* and *negative semidefinite*, result in systems which cannot be gyroscopically stabilized. A special case of this sort occurs when V' is identically zero.

Now let us consider the effect of *damping* upon a gyroscopic system. Let us suppose, for example, that the Rayleigh dissipation function associated with the system is *positive definite* in the non-ignored $\dot{q}$'s, implying that no motion of these coordinates is possible without the loss of energy. The ignored coordinates, however, are assumed to remain undamped. If V' is *positive definite*, the addition of damping converts a stable system to an *asymptotically stable* system. This result occurs because the only position at which the system can remain without dissipation of energy is the minimum of V' at the origin.

On the other hand, if V' is *negative definite*, *negative semidefinite*, or *indefinite*, the presence of damping will result in an *unstable* system, regardless of any gyroscopic effects. Here the damping is associated with a continuing energy dissipation and an increasingly negative total energy, resulting ultimately in one or more of the q_i's being no longer infinitesimal.

Finally, if V' is *zero* or *positive semidefinite*, the system remains *unstable* with damping. Note, however, that this analysis applies to the linearized system, that is, V' is assumed to be a quadratic form in the non-ignored q's. The inclusion of higher-order terms may also influence the stability.

Example 3-4. Consider a gyroscopic system having two degrees of freedom and whose equations of motion are

$$m_{11}\ddot{q}_1 + m_{12}\ddot{q}_2 + \gamma_{12}\dot{q}_2 + k_{11}q_1 + k_{12}q_2 = 0$$
$$m_{12}\ddot{q}_1 + m_{22}\ddot{q}_2 - \gamma_{12}\dot{q}_1 + k_{12}q_1 + k_{22}q_2 = 0 \tag{3-127}$$

where the coefficients m_{ij} and k_{ij} are symmetric and γ_{ij} is skew-symmetric. Let us obtain conditions on these coefficients which will ensure stability.

If we assume exponential solutions having the form of Eq. (3-121), the characteristic equation is

$$\begin{vmatrix} (m_{11}\lambda^2 + k_{11}) & (m_{12}\lambda^2 + \gamma_{12}\lambda + k_{12}) \\ (m_{12}\lambda^2 - \gamma_{12}\lambda + k_{12}) & (m_{22}\lambda^2 + k_{22}) \end{vmatrix} = 0$$

or

$$(m_{11}m_{22} - m_{12}^2)\lambda^4 + (m_{11}k_{22} + m_{22}k_{11} - 2m_{12}k_{12} + \gamma_{12}^2)\lambda^2$$
$$+ (k_{11}k_{22} - k_{12}^2) = 0 \tag{3-128}$$

The system is stable if the eigenvalues λ_k^2 are negative real and distinct, resulting in nonrepeated imaginary roots λ_k. We recall that the coefficient $(m_{11}m_{22} - m_{12}^2)$ must be positive because the kinetic energy is positive definite for any real system. Hence the system is stable if

$$k_{11}k_{22} - k_{12}^2 > 0$$

$$(3\text{-}129)$$

$$(m_{11}k_{22} + m_{22}k_{11} - 2m_{12}k_{12} + \gamma_{12}^2) > 2\sqrt{(m_{11}m_{22} - m_{12}^2)(k_{11}k_{22} - k_{12}^2)}$$

These results follow directly from the solution for the roots of the characteristic equation.

Assuming that the system is *gyroscopic* ($\gamma_{12} \neq 0$), it can be shown that repeated imaginary roots λ_k occur only if V is negative definite, in which case the system is unstable. Also, repeated zero roots result in an unstable solution. Furthermore, if λ_k^2 is complex or positive real, the solution is clearly unstable. Hence the conditions of (3-129) are, in fact, the *necessary and sufficient conditions* for the *stability* of a gyroscopic system with two degrees of freedom.

Now let us illustrate the nature of the motion of this linear gyroscopic system with the aid of numerical examples.

Case 1: Positive definite V. Let the inertia and stiffness coefficients be as follows:

$$m_{11} = 2 \qquad k_{11} = 1$$
$$m_{22} = 4 \qquad k_{22} = 8$$
$$m_{12} = 1 \qquad k_{12} = -1$$

where we assume a consistent set of units. In this case Eq. (3-123) becomes

$$(2\lambda^2 + 1)A_1 + (\lambda^2 + \gamma_{12}\lambda - 1)A_2 = 0$$
$$(\lambda^2 - \gamma_{12}\lambda - 1)A_1 + (4\lambda^2 + 8)A_2 = 0$$

$$(3\text{-}130)$$

The characteristic equation is

$$7\lambda^4 + (22 + \gamma_{12}^2)\lambda^2 + 7 = 0 \qquad (3\text{-}131)$$

which yields

$$\lambda_{1,2}^2 = \tfrac{1}{14}[-(22 + \gamma_{12}^2) \pm \sqrt{(22 + \gamma_{12}^2)^2 - 196}]$$

Since both eigenvalues λ^2 are negative real, the system is *stable* for all values of the gyroscopic coefficient γ_{12}, that is, for all rotation speeds of the gyroscopic coupling element.

Now let us choose $\gamma_{12} = 1$. Substituting into the expression for the eigenvalues, we obtain

$$\lambda_1^2 = -0.339, \qquad \lambda_1 = i0.583$$
$$\lambda_2^2 = -2.95, \qquad \lambda_2 = i1.716$$

The amplitude ratios can be found from Eq. (3-130). In general,

$$\frac{A_2}{A_1} = -\frac{\lambda^2 - \lambda - 1}{4\lambda^2 + 8}$$

which yields

$$\frac{A_{21}}{A_{11}} = 0.202 + i0.0877$$

$$\frac{A_{22}}{A_{12}} = -1.043 - i0.453$$

The first complex ratio indicates that in the first mode of oscillation q_2 leads q_1 by $\tan^{-1}(0.0877/0.202) = 23.5°$ and the amplitude ratio

$$\frac{|q_2|}{|q_1|} = \sqrt{0.202^2 + 0.0877^2} = 0.220$$

Similarly, in the second mode q_2 lags q_1 by $156.5°$ and the ratio of the amplitudes is 1.137. The motion in both modes is sinusoidal, the frequency of the first being 0.583 rad/sec and the second 1.716 rad/sec.

Case 2: Negative definite V. Now suppose the inertia and stiffness coefficients are as follows:

$$m_{11} = 2 \quad k_{11} = -1$$
$$m_{22} = 4 \quad k_{22} = -8$$
$$m_{12} = 1 \quad k_{12} = 1$$

The amplitude coefficient equations are

$$(2\lambda^2 - 1)A_1 + (\lambda^2 + \gamma_{12}\lambda + 1)A_2 = 0$$
$$(\lambda^2 - \gamma_{12}\lambda + 1)A_1 + (4\lambda^2 - 8)A_2 = 0 \tag{3-132}$$

and the characteristic equation is

$$7\lambda^4 + (\gamma_{12}^2 - 22)\lambda^2 + 7 = 0 \tag{3-133}$$

Thus we obtain

$$\lambda_{1,2}^2 = \tfrac{1}{14}[-(\gamma_{12}^2 - 22) \pm \sqrt{(\gamma_{12}^2 - 22)^2 - 196}]$$

From Eq. (3-129) we see that, for stability, we must have $\gamma_{12}^2 > 36$.
Let us take $\gamma_{12} = 7$. Then we obtain

$$\lambda_1^2 = -0.280, \quad \lambda_1 = i0.529$$
$$\lambda_2^2 = -3.58, \quad \lambda_2 = i1.891$$

The amplitude ratios are

$$\frac{A_2}{A_1} = -\frac{\lambda^2 - 7\lambda + 1}{4\lambda^2 - 8}$$

yielding

$$\frac{A_{21}}{A_{11}} = 0.0790 - i0.406$$

$$\frac{A_{22}}{A_{12}} = -0.1155 - i0.594$$

Here is an example of a system which has two unstable modes when the gyroscopic coefficient is zero. It is gyroscopically stabilized for $|\gamma_{12}| > 6$, however, and its free motion consists of sinusoidal oscillations.

Case 3: Indefinite V. Now let the inertia and stiffness coefficients be as follows:

$$m_{11} = 2 \qquad k_{11} = -1$$
$$m_{22} = 4 \qquad k_{22} = 8$$
$$m_{12} = 1 \qquad k_{12} = 1$$

The characteristic equation is

$$\begin{vmatrix} (2\lambda^2 - 1) & (\lambda^2 + \gamma_{12}\lambda + 1) \\ (\lambda^2 - \gamma_{12}\lambda + 1) & (4\lambda^2 + 8) \end{vmatrix} = 0$$

or

$$7\lambda^4 + (10 + \gamma_{12}^2)\lambda^2 - 9 = 0 \qquad (3\text{-}134)$$

From the first condition of (3-129) we see that the system is *unstable*, regardless of the value of the gyroscopic constant γ_{12}. Specifically, we find that the eigenvalues λ_1^2 and λ_2^2 are real and of opposite signs, resulting in an unstable term of the form $e^{\lambda t}$ in the solution, where λ is positive real.

Example 3-5. Consider the motion of a top (Fig. 3-5) whose configuration is expressed in terms of Eulerian angles. Obtain the differential equations of motion using the Routhian procedure and show the nature of small motions near a reference condition of steady precession.

Let the moment of inertia about the axis of symmetry be I_a, and let the transverse moment of inertia about the fixed point O be I_t. The axial component of the angular velocity $\boldsymbol{\omega}$ is called the *total spin* Ω and is given by

$$\Omega = \dot{\phi} - \dot{\psi} \sin \theta \qquad (3\text{-}135)$$

The transverse component of the angular velocity is the vector sum of the orthogonal components $\dot{\theta}$ and $\dot{\psi} \cos \theta$. Hence the total kinetic energy is

$$T = \tfrac{1}{2}I_a(\dot{\phi} - \dot{\psi} \sin \theta)^2 + \tfrac{1}{2}I_t(\dot{\theta}^2 + \dot{\psi}^2 \cos^2 \theta) \qquad (3\text{-}136)$$

The potential energy is entirely gravitational and is given by

$$V = mgl \sin \theta \qquad (3\text{-}137)$$

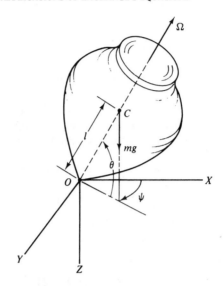

Fig. 3-5. A top with a fixed point O.

Thus the Lagrangian function is

$$L = \tfrac{1}{2}I_a(\dot{\phi} - \dot{\psi} \sin \theta)^2 + \tfrac{1}{2}I_t(\dot{\theta}^2 + \dot{\psi}^2 \cos^2 \theta) - mgl \sin \theta \quad (3\text{-}138)$$

Since L is not an explicit function of ϕ and ψ, these coordinates are *ignorable*. The Routhian procedure, however, does not *require* us to treat both coordinates as ignorable. So let us consider first the consequences of ignoring just one of these coordinates.

Case 1: Ignore ϕ only. We note that the axial component of angular momentum is constant, that is,

$$p_\phi = \frac{\partial L}{\partial \dot{\phi}} = I_a(\dot{\phi} - \dot{\psi} \sin \theta) = \beta_\phi \quad (3\text{-}139)$$

or

$$\dot{\phi} = \dot{\psi} \sin \theta + \frac{\beta_\phi}{I_a} \quad (3\text{-}140)$$

Then, using Eq. (2-122) we find that the Routhian function is

$$R = L - \beta_\phi \dot{\phi}$$
$$= \frac{\beta_\phi^2}{2I_a} + \frac{1}{2}I_t(\dot{\theta}^2 + \dot{\psi}^2 \cos^2 \theta) - mgl \sin \theta - \beta_\phi\left(\dot{\psi} \sin \theta + \frac{\beta_\phi}{I_a}\right)$$

or

$$R = \tfrac{1}{2}I_t(\dot{\theta}^2 + \dot{\psi}^2 \cos^2 \theta) - \beta_\phi\dot{\psi} \sin \theta - mgl \sin \theta \quad (3\text{-}141)$$

where we have omitted the constant $-\beta_\phi^2/2I_a$ since it does not influence the motion.

The equations of motion are obtained by using R as the Lagrangian function in the standard form of Lagrange's equations. We find that

$$\frac{\partial R}{\partial \dot\psi} = I_t \dot\psi \cos^2 \theta - \beta_\phi \sin \theta, \qquad \frac{\partial R}{\partial \psi} = 0$$

which results in the following ψ equation of motion:

$$I_t \ddot\psi \cos^2 \theta - 2I_t \dot\psi \dot\theta \sin \theta \cos \theta - \beta_\phi \dot\theta \cos \theta = 0 \qquad (3\text{-}142)$$

In a similar fashion, we obtain

$$\frac{\partial R}{\partial \dot\theta} = I_t \dot\theta$$

$$\frac{\partial R}{\partial \theta} = - I_t \dot\psi^2 \sin \theta \cos \theta - \beta_\phi \dot\psi \cos \theta - mgl \cos \theta$$

and the θ equation is

$$I_t \ddot\theta + I_t \dot\psi^2 \sin \theta \cos \theta + \beta_\phi \dot\psi \cos \theta + mgl \cos \theta = 0 \qquad (3\text{-}143)$$

The gyroscopic terms are $-\beta_\phi \dot\theta \cos \theta$ in the ψ equation and $\beta_\phi \dot\psi \cos \theta$ in the θ equation. Notice that these terms are linear in the $\dot q$'s and have equal and opposite coefficients.

To determine the steady precession rate, let us set $\ddot\theta = 0$ in Eq. (3-143) and solve for $\dot\psi$. We obtain

$$\dot\psi = \frac{-\beta_\phi}{2I_t \sin \theta}\left(1 \pm \sqrt{1 - \frac{4I_t mgl \sin \theta}{\beta_\phi^2}}\right) \qquad (3\text{-}144)$$

Note that there are *two* possible steady precession rates, provided that the square root is real, that is, for

$$\beta_\phi^2 > 4I_t mgl \sin \theta$$

or

$$\Omega^2 > \frac{4I_t mgl}{I_a^2} \sin \theta \qquad (3\text{-}145)$$

In other words, the magnitude of the total spin must be large enough to provide gyroscopic stabilization. For the common case of *large spin* in which $I_a^2 \Omega^2 \gg 4I_t mgl \sin \theta$, the slower precession rate is the one usually observed and is approximately

$$\dot\psi = -\frac{mgl}{\beta_\phi} = -\frac{mgl}{I_a \Omega} \qquad (3\text{-}146)$$

In order to obtain the linearized equations for small motions near the reference condition of steady precession, it is convenient to apply *elementary perturbation theory* to the nonlinear equations given in (3-142) and (3-143). In general, if we consider the Taylor expansion of a function $F(\alpha, \beta, \gamma, \ldots)$

about a reference value $F_0(\alpha_0, \beta_0, \gamma_0, \ldots)$, we can write

$$F = F_0 + \delta F = F_0 + \left(\frac{\partial F}{\partial \alpha}\right)_0 \delta\alpha + \left(\frac{\partial F}{\partial \beta}\right)_0 \delta\beta + \left(\frac{\partial F}{\partial \gamma}\right)_0 \delta\gamma + \cdots$$
$$+ \frac{1}{2!}\left[\left(\frac{\partial^2 F}{\partial \alpha^2}\right)_0 (\delta\alpha)^2 + 2\left(\frac{\partial^2 F}{\partial \alpha\, \partial \beta}\right)_0 \delta\alpha\, \delta\beta + \left(\frac{\partial^2 F}{\partial \beta^2}\right)_0 (\delta\beta)^2 + \cdots\right] + \cdots$$

$$(3\text{-}147)$$

where $\alpha = \alpha_0 + \delta\alpha$, $\beta = \beta_0 + \delta\beta$, and so on. If we assume small perturbations and neglect all terms involving products of these small quantities, we obtain the *linear approximation*

$$\delta F = \left(\frac{\partial F}{\partial \alpha}\right)_0 \delta\alpha + \left(\frac{\partial F}{\partial \beta}\right)_0 \delta\beta + \left(\frac{\partial F}{\partial \gamma}\right)_0 \delta\gamma + \cdots \qquad (3\text{-}148)$$

Now let $F = 0$ represent the differential equation to be linearized, and consider each coordinate q and its derivatives as separate variables in the expansion. Noting that $F_0 = 0$, Eq. (3-142) becomes

$$I_t \cos^2 \theta_0\, \delta\ddot{\psi} - (I_t\dot{\psi}_0 \sin 2\theta_0 + \beta_\phi \cos \theta_0)\, \delta\dot{\theta} = 0 \qquad (3\text{-}149)$$

In a similar fashion, Eq. (3-143) results in

$$I_t\, \delta\ddot{\theta} + (I_t\dot{\psi}_0 \sin 2\theta_0 + \beta_\phi \cos \theta_0)\delta\dot{\psi}$$
$$+ (I_t\dot{\psi}_0^2 \cos 2\theta_0 - \beta_\phi\dot{\psi}_0 \sin \theta_0 - mgl \sin \theta_0)\delta\theta = 0 \qquad (3\text{-}150)$$

But from Eq. (3-143) we see that, since $\ddot{\theta}_0 = 0$,

$$I_t\dot{\psi}_0^2 \sin \theta_0 + \beta_\phi\dot{\psi}_0 + mgl = 0 \qquad (3\text{-}151)$$

where we assume that $\cos \theta_0 \neq 0$. Hence, from Eqs. (3-150) and (3-151), we obtain

$$I_t\, \delta\ddot{\theta} + (I_t\dot{\psi}_0 \sin 2\theta_0 + \beta_\phi \cos \theta_0)\delta\dot{\psi} + I_t\dot{\psi}_0^2 \cos^2 \theta_0\, \delta\theta = 0 \qquad (3\text{-}152)$$

Equations (3-149) and (3-152) constitute the linearized equations for small perturbations from the steady reference motion of the system. The second term in each equation represents the gyroscopic coupling.

Next let us make the further assumption that $\theta_0 = 0$. Then the equations of motion are

$$I_t\, \delta\ddot{\psi} - \beta_\phi\, \delta\dot{\theta} = 0$$
$$I_t\, \delta\ddot{\theta} + I_t\dot{\psi}_0^2\, \delta\theta + \beta_\phi\, \delta\dot{\psi} = 0 \qquad (3\text{-}153)$$

where the reference precessional rate is

$$\dot{\psi}_0 = -\frac{mgl}{\beta_\phi} \qquad (3\text{-}154)$$

Another method of obtaining these equations is to substitute $\psi = \dot{\psi}_0 + \delta\psi$, $\theta = \delta\theta$, $\dot{\theta} = \delta\dot{\theta}$ into the Routhian function of Eq. (3-141), retaining only quadratic terms in the small quantities since linear or constant terms will not contribute to the equations of motion if the reference motion is steady. Using this procedure, we obtain

$$R = \tfrac{1}{2}I_t[(\delta\psi)^2 + (\delta\dot{\theta})^2] - \beta_\phi\,\delta\psi\,\delta\theta - \tfrac{1}{2}I_t\psi_0^2\,(\delta\theta)^2 \qquad (3\text{-}155)$$

which is of the form of the Lagrangian function given in Eq. (3-118). Then, using Lagrange's equations and noting that the operations of differentiation and variation can be interchanged, that is, $d/dt\,(\delta q) = \delta\dot{q}$, we obtain Eq. (3-153) once again.

Now let us assume that the solutions of the equations of motion are of the form

$$\delta\psi = A_1 e^{\lambda t}, \qquad \delta\theta = A_2 e^{\lambda t} \qquad (3\text{-}156)$$

Substituting into Eq. (3-153), we have

$$\begin{aligned} I_t\lambda^2 A_1 - \beta_\phi\lambda A_2 &= 0 \\ \beta_\phi\lambda A_1 + I_t(\lambda^2 + \psi_0^2)A_2 &= 0 \end{aligned} \qquad (3\text{-}157)$$

The characteristic equation, obtained by setting the determinant of the coefficients equal to zero, is

$$I_t^2\lambda^4 + (I_t^2\psi_0^2 + \beta_\phi^2)\lambda^2 = 0 \qquad (3\text{-}158)$$

which yields the eigenvalues

$$\lambda_{1,2}^2 = 0, \qquad -(\psi_0^2 + \beta_\phi^2/I_t^2)$$

From the second equation of (3-157) we see that the amplitude ratio is

$$\frac{A_2}{A_1} = -\frac{\beta_\phi\lambda}{I_t(\lambda^2 + \psi_0^2)} \qquad (3\text{-}159)$$

Hence we obtain

$$\frac{A_{21}}{A_{11}} = 0, \qquad \frac{A_{22}}{A_{12}} = i\sqrt{1 + \left(\frac{I_t\psi_0}{\beta_\phi}\right)^2}$$

The first ratio shows that the solution for $\delta\theta$ contains no *secular* term, that is, no linear term in t associated with the repeated zero root. It may, however, contain a constant term, as we shall see. The factor i in the second ratio indicates that $\delta\theta$ leads $\delta\psi$ by 90° in the sinusoidal motion.

If we use the notation

$$\omega_0 = \sqrt{\psi_0^2 + (\beta_\phi/I_t)^2} = \sqrt{\psi_0^2 + (I_a\Omega/I_t)^2} \qquad (3\text{-}160)$$

we can write the solutions for $\delta\psi$ and $\delta\theta$ as follows:

$$\begin{aligned} \delta\psi &= C_1 + C_2 t + A_1 \sin\,(\omega_0 t + \alpha) \\ \delta\theta &= -\frac{I_a\Omega}{I_t\psi_0^2}C_2 + A_1\sqrt{1 + \left(\frac{I_t\psi_0}{I_a\Omega}\right)^2}\cos\,(\omega_0 t + \alpha) \end{aligned} \qquad (3\text{-}161)$$

where the constants C_1, C_2, A_1, α are determined from the initial conditions. In accordance with the assumption of small motion, A_1 and C_2 must be small, resulting in $\delta\theta \ll 1$ and $\delta\dot{\psi} \ll 1$. $\delta\psi$ need not remain small, however, since it does not appear explicitly in the differential equations of motion.

Case 2: Ignore ϕ and ψ. The generalized momentum conjugate to ψ is

$$p_\psi = \frac{\partial L}{\partial \dot{\psi}} = -I_a(\dot{\phi} - \dot{\psi}\sin\theta)\sin\theta + I_t\dot{\psi}\cos^2\theta = \beta_\psi \qquad (3\text{-}162)$$

and, using Eq. (3-139), we obtain

$$\dot{\psi} = \frac{\beta_\psi + \beta_\phi \sin\theta}{I_t \cos^2\theta} \qquad (3\text{-}163)$$

Also, from Eqs. (3-140) and (3-163) we find that

$$\dot{\phi} = \left(\frac{\beta_\psi + \beta_\phi \sin\theta}{I_t \cos^2\theta}\right)\sin\theta + \frac{\beta_\phi}{I_a} \qquad (3\text{-}164)$$

The Routhian function is obtained by substituting these expressions for $\dot{\psi}$ and $\dot{\phi}$ into Eq. (2-122). Thus,

$$\begin{aligned} R &= L - \beta_\phi\dot{\phi} - \beta_\psi\dot{\psi} \\ &= \frac{1}{2}I_t\dot{\theta}^2 - \frac{(\beta_\psi + \beta_\phi \sin\theta)^2}{2I_t \cos^2\theta} - mgl\sin\theta \end{aligned} \qquad (3\text{-}165)$$

where we again neglect constant terms. The equation of motion which results from using this Routhian function in Lagrange's equation is

$$I_t\ddot{\theta} + \frac{(\beta_\psi + \beta_\phi \sin\theta)^2}{I_t \cos^3\theta}\sin\theta + \frac{\beta_\phi(\beta_\psi + \beta_\phi \sin\theta)}{I_t \cos\theta} + mgl\cos\theta = 0 \qquad (3\text{-}166)$$

The same equation would have been obtained if we had substituted the expression for $\dot{\psi}$ given in Eq. (3-163) into the θ equation of (3-143). Note that Eq. (3-166) contains no gyroscopic terms since there is only one degree of freedom remaining.

The form of the Routhian function of Eq. (3-165) is that of a *natural system* with

$$\begin{aligned} T' &= \tfrac{1}{2}I_t\dot{\theta}^2 \\ V' &= \frac{(\beta_\psi + \beta_\phi \sin\theta)^2}{2I_t \cos^2\theta} + mgl\sin\theta \end{aligned} \qquad (3\text{-}167)$$

Thus, we can use conservation of energy to obtain the Jacobi integral

$$T' + V' = \frac{1}{2}I_t\dot{\theta}^2 + \frac{(\beta_\psi + \beta_\phi \sin\theta)^2}{2I_t \cos^2\theta} + mgl\sin\theta = h \qquad (3\text{-}168)$$

Hence

$$\dot{\theta}^2 = \frac{2h}{I_t} - \left(\frac{\beta_\psi + \beta_\phi \sin\theta}{I_t \cos\theta}\right)^2 - \frac{2mgl}{I_t}\sin\theta \qquad (3\text{-}169)$$

This equation, giving $\dot{\theta}^2$ as an explicit function of θ, could have been obtained directly by integrating Eq. (3-166). If we let $\dot{\theta}$ equal zero we obtain an equation whose roots are the *turning points* defining the limits of the nutational motion in θ.

Now let us return to Eq. (3-166) and find the linear approximation for small motions about a reference value $\theta_0 = 0$. In this case the linear approximation is obtained by assuming that $\sin\theta \cong \theta$ and $\cos\theta \cong 1$. Then the resulting equation is

$$\ddot{\theta} + \left(\frac{\beta_\psi^2 + \beta_\phi^2}{I_t^2}\right)\theta = 0 \qquad (3\text{-}170)$$

which yields the nutational frequency

$$\omega_0 = \frac{1}{I_t}\sqrt{\beta_\psi^2 + \beta_\phi^2} = \sqrt{\dot{\psi}_0^2 + (I_a\Omega/I_t)^2} \qquad (3\text{-}171)$$

in agreement with the result previously given in Eq. (3-160).

In considering systems such as this, a question may arise concerning the proper initial values which should be assigned to the ignored $\dot{q}$'s. This question is easily answered if we recall that the ignored $\dot{q}$'s can be expressed in terms of the β's and the remaining q's by equations similar to Eqs. (3-163) and (3-164). Furthermore, for the case of small perturbations about a reference condition of steady motion, the variation of these equations provides expressions for the ignored $\delta\dot{q}$'s in terms of the remaining δq's $\delta\dot{q}$'s and the constant β's.

3-4. VELOCITY-DEPENDENT POTENTIALS

In discussing the applications of the standard form of Lagrange's equations given in Eq. (2-27), we previously assumed that all the applied forces can be obtained from the potential energy V which is a function of the q's and possibly t, but is not a function of the $\dot{q}$'s. This restriction is not necessary, however. Let us recall the basic form of Lagrange's equations from Eq. (2-23), namely,

$$\frac{d}{dt}\left(\frac{\partial T}{\partial \dot{q}_i}\right) - \frac{\partial T}{\partial q_i} = Q_i \qquad (i = 1, 2, \ldots, n) \qquad (3\text{-}172)$$

If we now assume that Q_i is obtained from a *velocity-dependent potential function* $U(q, \dot{q}, t)$ in accordance with the equation

$$Q_i = \frac{d}{dt}\left(\frac{\partial U}{\partial \dot{q}_i}\right) - \frac{\partial U}{\partial q_i} \qquad (3\text{-}173)$$

we see that

$$\frac{d}{dt}\left(\frac{\partial L}{\partial \dot{q}_i}\right) - \frac{\partial L}{\partial q_i} = 0 \qquad (i = 1, 2, \ldots, n) \qquad (3\text{-}174)$$

where

$$L = T - U \qquad (3\text{-}175)$$

Electromagnetic Forces. As an example of a velocity-dependent potential, consider the *electromagnetic force* acting on a charged particle. Using rationalized mks units, the force on a particle having a charge e and velocity $\mathbf{v}$ is

$$\mathbf{F} = e(\mathbf{E} + \mathbf{v} \times \mathbf{B}) \tag{3-176}$$

where $\mathbf{E}$ is the electric field intensity and $\mathbf{B}$ is the magnetic induction vector. $\mathbf{E}$ and $\mathbf{B}$ are obtained from a *scalar potential* ϕ and a *vector potential* $\mathbf{A}$ in accordance with the equations

$$\mathbf{E} = -\nabla\phi - \frac{\partial \mathbf{A}}{\partial t} \tag{3-177}$$

$$\mathbf{B} = \nabla \times \mathbf{A} \tag{3-178}$$

where ϕ and $\mathbf{A}$ are, in general, functions of position and time. Then, using Eqs. (3-176)—(3-178), we obtain

$$\mathbf{F} = e\left(-\nabla\phi - \frac{\partial \mathbf{A}}{\partial t} + \mathbf{v} \times \nabla \times \mathbf{A}\right) \tag{3-179}$$

In order to show how this force can be obtained from a function U, let us assume that the position of the particle is given by the Cartesian coordinates (x, y, z). The term $-e\nabla\phi$ is easily explained since $e\phi$ is analogous to the ordinary potential energy function V.

Next let us consider the x component of the term $\mathbf{v} \times \nabla \times \mathbf{A}$. If we designate the Cartesian components of $\mathbf{A}$ by A_x, A_y, A_z, we see that

$$(\mathbf{v} \times \nabla \times \mathbf{A})_x = v_y\left(\frac{\partial A_y}{\partial x} - \frac{\partial A_x}{\partial y}\right) - v_z\left(\frac{\partial A_x}{\partial z} - \frac{\partial A_z}{\partial x}\right)$$

$$= v_y\frac{\partial A_y}{\partial x} + v_z\frac{\partial A_z}{\partial x} + v_x\frac{\partial A_x}{\partial x} - v_x\frac{\partial A_x}{\partial x} - v_y\frac{\partial A_x}{\partial y} - v_z\frac{\partial A_x}{\partial z} \tag{3-180}$$

where $v_x(\partial A_x/\partial x)$ has been added and subtracted on the right-hand side. But $\mathbf{A}$ is a function of position and time, so

$$\frac{dA_x}{dt} = \frac{\partial A_x}{\partial x}\dot{x} + \frac{\partial A_x}{\partial y}\dot{y} + \frac{\partial A_x}{\partial z}\dot{z} + \frac{\partial A_x}{\partial t}$$

$$= v_x\frac{\partial A_x}{\partial x} + v_y\frac{\partial A_x}{\partial y} + v_z\frac{\partial A_x}{\partial z} + \frac{\partial A_x}{\partial t} \tag{3-181}$$

Hence

$$(\mathbf{v} \times \nabla \times \mathbf{A})_x = \frac{\partial}{\partial x}(\mathbf{v} \cdot \mathbf{A}) - \frac{dA_x}{dt} + \frac{\partial A_x}{\partial t} \tag{3-182}$$

Then, from Eqs. (3-179) and (3-182), we find that the x component of the electromagnetic force is

$$F_x = e\left[-\frac{\partial \phi}{\partial x} + \frac{\partial}{\partial x}(\mathbf{v} \cdot \mathbf{A}) - \frac{dA_x}{dt}\right] \tag{3-183}$$

Next we observe that

$$\frac{dA_x}{dt} = \frac{d}{dt}\left[\frac{\partial}{\partial \dot{x}}(\mathbf{v}\cdot\mathbf{A})\right] \tag{3-184}$$

and find that if we take

$$U = e(\phi - \mathbf{v}\cdot\mathbf{A}) \tag{3-185}$$

we obtain

$$F_x = \frac{d}{dt}\left(\frac{\partial U}{\partial \dot{x}}\right) - \frac{\partial U}{\partial x}$$

in agreement with Eq. (3-173). Similar expressions occur for F_y and F_z.

Thus we see that the electromagnetic forces on a particle are represented by the velocity-dependent potential function given by Eq. (3-185). The Lagrangian function is

$$L = T - U = \tfrac{1}{2}m(\dot{x}^2 + \dot{y}^2 + \dot{z}^2) - e\phi + e(A_x\dot{x} + A_y\dot{y} + A_z\dot{z})$$
$$= \tfrac{1}{2}mv^2 - e(\phi - \mathbf{v}\cdot\mathbf{A}) \tag{3-186}$$

where m is the mass of the particle.

Now let us consider the generalized momentum components associated with a charged particle moving in an electromagnetic field. From Eqs. (1-137) and (3-186) we obtain

$$p_x = \frac{\partial L}{\partial \dot{x}} = m\dot{x} + eA_x \tag{3-187}$$

or, more generally,

$$\mathbf{p} = m\mathbf{v} + e\mathbf{A} \tag{3-188}$$

This rather surprising result indicates that a portion of the *momentum* is associated with the *electromagnetic field*. If L is not an explicit function of x, that is, if x is ignorable, then it is the momentum p_x given by Eq. (3-187) (rather than the mechanical momentum $m\dot{x}$) which is conserved.

The *total energy* of the particle is

$$T + e\phi = \tfrac{1}{2}mv^2 + e\phi \tag{3-189}$$

where $e\phi$ is the potential energy associated with the electric field. The magnetic force $\mathbf{v}\times\mathbf{B}$ does no work on the system because it is perpendicular to the velocity; hence $\mathbf{A}$ does not enter into the energy expression. If ϕ and $\mathbf{A}$ are not explicit functions of time, the system is *conservative* and the total energy is constant.

Now consider an isolated system of two particles, each of mass m and charge e (Fig. 3-6). Suppose we choose an inertial reference frame which translates with the center of mass O. We wish to find the interaction forces between the two particles as each moves in the electromagnetic field of the other.

First we note that the velocities of the two particles are equal and opposite relative to the center of mass. Then, using Eq. (3-176) or (3-179), it is

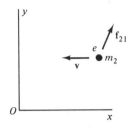

Fig. 3-6. Two interacting charged particles.

apparent that the interaction forces are equal and opposite, or

$$\mathbf{f}_{12} = -\mathbf{f}_{21} \tag{3-190}$$

a result that also follows from symmetry considerations. The symmetry does not require that the two forces be directed along the line connecting the particles, however, and in general they are not collinear.

A consequence of the fact that the electromagnetic interaction forces are not collinear is that the total angular momentum, as defined in Eq. (1-126), is no longer conserved, even for an isolated system with no external forces. This occurs because the interaction forces exert a nonzero moment about O, as can be seen in Fig. 3-6.

Gyroscopic Forces. There are similarities in the mathematical representations of gyroscopic and electromagnetic forces. As an example, consider a gyroscopic system whose Lagrangian function has the form

$$L = T_2 + T_1 - V' \tag{3-191}$$

where

$$V'(x, y, z, t) = V - T_0 \tag{3-192}$$

The gyroscopic forces arise from T_1 which, in general, has the form

$$T_1 = P\dot{x} + Q\dot{y} + R\dot{z} \tag{3-193}$$

where P, Q, and R are each functions of (x, y, z, t). If we define a vector $\mathbf{G}$ given by

$$\mathbf{G} = P\mathbf{i} + Q\mathbf{j} + R\mathbf{k} \tag{3-194}$$

and let

$$\mathbf{v} = \dot{x}\mathbf{i} + \dot{y}\mathbf{j} + \dot{z}\mathbf{k} \tag{3-195}$$

we obtain

$$T_1 = \mathbf{v} \cdot \mathbf{G} \tag{3-196}$$

Now place the terms arising from T_1 on the right-hand side of the equation and consider them as additional forces. We find that the x component of these forces is

$$\frac{\partial T_1}{\partial x} - \frac{d}{dt}\left(\frac{\partial T_1}{\partial \dot{x}}\right) = \mathbf{v} \cdot \left(\frac{\partial P}{\partial x}\mathbf{i} + \frac{\partial Q}{\partial x}\mathbf{j} + \frac{\partial R}{\partial x}\mathbf{k}\right) - \dot{P}$$

$$= (\mathbf{v} \times \nabla \times \mathbf{G})_x - \frac{\partial P}{\partial t} \qquad (3\text{-}197)$$

where $\mathbf{v} \times \nabla \times \mathbf{G}$ is the *gyroscopic force*. It is analogous to the force $e\mathbf{v} \times \nabla \times \mathbf{A}$ in the electromagnetic example.

The x equation of motion is

$$\frac{d}{dt}\left(\frac{\partial T_2}{\partial \dot{x}}\right) - \frac{\partial T_2}{\partial x} = -\frac{\partial V'}{\partial x} - \frac{\partial P}{\partial t} + (\mathbf{v} \times \nabla \times \mathbf{G})_x \qquad (3\text{-}198)$$

Comparing the results for the gyroscopic and electromagnetic systems, we see from Eqs. (3-179) and (3-198) that V' corresponds to $e\phi$ and $\mathbf{G}$ corresponds to $e\mathbf{A}$. Furthermore, the Lagrangian function for the gyroscopic system is

$$L = T_2 - U' \qquad (3\text{-}199)$$

where the velocity-dependent potential is

$$U' = V' - T_1 = V' - \mathbf{v} \cdot \mathbf{G} \qquad (3\text{-}200)$$

which is of the same form as in Eq. (3-185).

Finally, we note that if L is not an explicit function of time, the system is conservative and its Jacobi integral is

$$T_2 + V' = T_2 - T_0 + V = h \qquad (3\text{-}201)$$

Example 3-6. A particle of mass m and charge e moves under the influence of uniform electric and magnetic fields which are mutually orthogonal. Relative to a fixed Cartesian frame, these fields are $\mathbf{E} = E\mathbf{j}$ and $\mathbf{B} = B\mathbf{k}$. Find the equations of motion and the path of the particle if it is initially at rest at the origin O (Fig. 3-7).

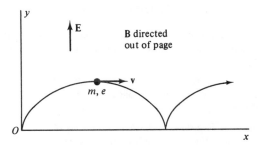

Fig. 3-7. The path of a charged particle.

The scalar and vector potentials which yield the correct fields with the aid of Eqs. (3-177) and (3-178) are

$$\phi = -Ey \tag{3-202}$$

$$\mathbf{A} = \tfrac{1}{2}B(-y\mathbf{i} + x\mathbf{j}) \tag{3-203}$$

The Lagrangian function is obtained by using Eq. (3-186), and results in

$$L = \tfrac{1}{2}m(\dot{x}^2 + \dot{y}^2 + \dot{z}^2) + eEy + \tfrac{1}{2}eB(x\dot{y} - y\dot{x}) \tag{3-204}$$

Then, employing the standard form of Lagrange's equation, we obtain the following equations of motion:

$$m\ddot{x} - eB\dot{y} = 0$$
$$m\ddot{y} + eB\dot{x} = eE \tag{3-205}$$
$$m\ddot{z} = 0$$

It is clear from the third equation and the initial conditions $z(0) = 0$, $\dot{z}(0) = 0$, that all the motion is confined to the xy plane. Assuming, then, solutions for x and y having the form $e^{\lambda t}$, we obtain the characteristic equation

$$m^2\lambda^4 + e^2B^2\lambda^2 = 0 \tag{3-206}$$

which yields the eigenvalues

$$\lambda_{1,2}^2 = 0, \qquad -\frac{e^2B^2}{m^2}$$

The solution for the path of the particle is

$$x = \frac{E}{B}t - \frac{mE}{eB^2}\sin\frac{eB}{m}t$$
$$y = \frac{mE}{eB^2}\left(1 - \cos\frac{eB}{m}t\right) \tag{3-207}$$

This represents a cycloid having cusps which lie on the x axis and are separated by a distance $2\pi mE/eB^2$. This distance is just the average x velocity E/B multiplied by the period $2\pi m/eB$ of the sinusoidal terms.

REFERENCES

1. PARS, L. A., *A Treatise on Analytical Dynamics.* London: Heinemann, 1965. There is an excellent presentation of the application of Lagrangian methods to impulsive motion.

2. WHITTAKER, E. T., *A Treatise on the Analytical Dynamics of Particles and Rigid Bodies,* 4th ed. New York: Dover Publications, Inc., 1944. This book is particularly helpful in its explanation of quasi-coordinates.

3. CORBEN, H. C., and P. STEHLE, *Classical Mechanics,* 2nd ed. New York: John Wiley and Sons, Inc., 1960. The application of Lagrange's equations to systems with electromagnetic forces is particularly well-done.

4. GOLDSTEIN, H., *Classical Mechanics.* Reading, Mass.: Addison-Wesley Press,

Inc., 1950. This text is very helpful, particularly in its explanations of velocity-dependent potentials and the dissipation function.

PROBLEMS

3-1. Two rods, each of mass m and length l, are connected by a pin at their ends, as shown. The system is motionless until a transverse impulse $\hat{F}_1$ is applied at the coordinate x_1. Solve for the initial values of $\dot{x}_1$, $\dot{x}_2$, $\dot{x}_3$. Show that the kinetic energy just after the impulse is equal to $\frac{1}{2}\hat{F}_1\dot{x}_1$.

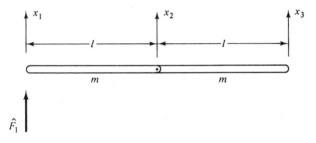

Fig. P3-1.

3-2. Solve Example 3-2 for the case in which the particle sticks to the rod after the impact, assuming the same initial conditions. Also find the energy lost in the impact.

3-3. A rigid rod of mass m and length l translates with velocity v_0 parallel to the x axis. Then it hits a smooth wall inelastically; that is, after impact the end

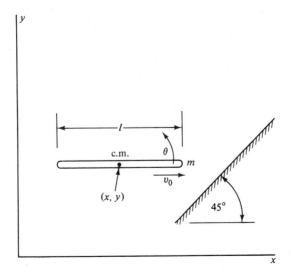

Fig. P3-3.

slides without friction along the wall. Assuming that θ and $\dot{\theta}$ are zero before impact, find the values of $\dot{x}$, $\dot{y}$, and $\dot{\theta}$ immediately after impact. Evaluate the impulse applied to the rod by the wall.

3-4. A rigid rod of mass m and length l is motionless with $\theta = 45°$ when, at $t = 0$, the vertical wall begins to move in the positive x direction at a constant speed v_0. Assuming frictionless surfaces, solve for the initial values of $\dot{x}$, $\dot{y}$, and $\dot{\theta}$. Find the impulse exerted by the wall on the rod at $t = 0$.

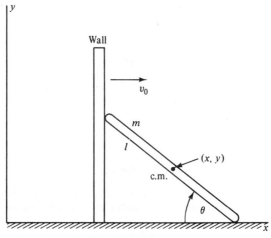

Fig. P3-4.

3-5. For the previous problem of the rod and the moving wall, use only a single generalized coordinate θ in the analysis. Obtain an expression for p_θ and solve for the initial value of $\dot{\theta}$.

3-6. Let xyz be a principal axis system with the origin at the center of mass of a rigid body having a mass m and principal moments of inertia I_{xx}, I_{yy}, and I_{zz}. Suppose this body has a velocity $\mathbf{v} = v_0\mathbf{i}$ but is not rotating until the point $(x_0, y_0, 0)$ in the body is suddenly fixed. Assume that the body can rotate freely about this point. Let (v_x, v_y, v_z) and $(\omega_x, \omega_y, \omega_z)$ be the translational and rotational velocity components for the body immediately after the constraint is applied. Write the constraint equations and use either the Lagrange multiplier method or the virtual velocity method to solve for $(v_x, v_y, v_z, \omega_x, \omega_y, \omega_z)$.

3-7. Given a system with nonzero T_1 and more than one degree of freedom, under what conditions on a_i does the Lagrangian method produce no gyroscopic terms in the equations of motion? Give an example of such a system.

3-8. Two massless rods of length l and two particles are connected by pin joints to each other and to a point P on the edge of a disk of radius l which rotates at a constant rate Ω. Assume that all motion occurs in a horizontal plane and let q_1 and q_2 be angles measured relative to the direction of line OP. Obtain T_1 and the

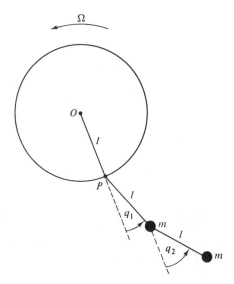

Fig. P3-8.

gyroscopic coefficient γ_{12}. Show that the gyroscopic terms in the equations of motion represent inertial moments due to Coriolis accelerations.

3-9. Consider the motion of a top (Fig. 3-5) relative to a reference frame XYZ which rotates at a constant rate ω_0 about its Z axis. Use the Euler angles (ψ, θ, ϕ) as coordinates (defined relative to the XYZ frame) and obtain the differential equations of motion. Indicate the gyroscopic terms.

3-10. An undamped gyroscopic system has the following **m**, **k**, and γ matrices, namely,

$$\mathbf{m} = \begin{bmatrix} m & 0 & 0 \\ 0 & m & 0 \\ 0 & 0 & m \end{bmatrix}, \quad \mathbf{k} = \begin{bmatrix} 2k & -\frac{1}{2}k & 0 \\ -\frac{1}{2}k & -k & 0 \\ 0 & 0 & -k \end{bmatrix}, \quad \gamma = \begin{bmatrix} 0 & 0 & 0 \\ 0 & 0 & \gamma \\ 0 & -\gamma & 0 \end{bmatrix}$$

where m and k are positive real constants and $\gamma^2 = 5mk$. Obtain the characteristic equation and solve for the roots. Is the system stable?

3-11. For the gyroscopic system of the previous problem, find the range of the values of γ^2 for stability. Note that the borderline case occurs for a repeated negative real eigenvalue λ_k^2.

3-12. The given system consists of masses, springs, and a linear damper having a positive coefficient c.

    (a) Obtain Rayleigh's dissipation function and the differential equations of motion.

    (b) Find the characteristic equation for the system and determine if the system is asymptotically stable.

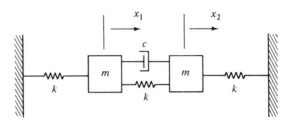

Fig. P3-12.

(c) Now add gyroscopic coupling terms to the differential equations. Assume unit values for m, c, k, and γ, and determine the stability by finding the roots.

3-13. Consider a particle of mass m and charge e whose position is given by the cylindrical coordinates (r, θ, z). Suppose it moves in an axially-symmetric electric field $\mathbf{E} = (E_0/r)\mathbf{e}_r$, and a uniform magnetic field $\mathbf{B} = B_0\mathbf{k}$. Obtain the velocity-dependent potential U and the differential equations of motion.

4

HAMILTON'S EQUATIONS

In the previous chapters we have become accustomed to the equations of motion being expressed in differential form. This procedure emphasizes the *evolution* of the system from instant to instant. On the other hand, one can use a *variational* principle as the basis for the description of a dynamical system. This approach tends to view the motion as a whole and involves a search for the path in configuration space which yields a stationary value (usually a minimum) for a certain integral. The variational principle of most importance in dynamics is Hamilton's principle which was first announced in 1834. Although other variational principles such as the principle of least action preceded Hamilton's principle, none has held the same theoretical interest.

4-1. HAMILTON'S PRINCIPLE

The mathematics of extremum problems is partially covered in ordinary calculus and partially in the calculus of variations. Before we enter into a discussion of Hamilton's principle, let us review briefly some of the mathematical concepts associated with these problems.

Stationary Values of a Function. Consider a function $f(q_1, q_2, \ldots, q_n)$ which is assumed to be continuous through the second partial derivatives. The *first variation* of f at the reference point $\mathbf{q}_0$ is

$$\delta f = \sum_{i=1}^{n} \left(\frac{\partial f}{\partial q_i} \right)_0 \delta q_i \tag{4-1}$$

where the δq's are the variations in the individual q's and can be considered as virtual displacements. The *necessary* and *sufficient* condition that f have a *stationary value* at $\mathbf{q}_0$ is that $\delta f = 0$ for all geometrically possible δq's, where

$$\mathbf{q} = \mathbf{q}_0 + \delta \mathbf{q} \tag{4-2}$$

For the case in which the δq's are *independent* and *reversible*, this implies that

$$\left(\frac{\partial f}{\partial q_i} \right)_0 = 0 \qquad (i = 1, 2, \ldots, n) \tag{4-3}$$

This result is reminiscent of the requirement expressed by Eq. (1-93) for the static equilibrium of a mechanical system, where the function under consideration is the potential energy V.

Now consider the *second variation* of the function f about a stationary point q_0. We have

$$\delta^2 f = \frac{1}{2} \sum_{i=1}^{n} \sum_{j=1}^{n} \left(\frac{\partial^2 f}{\partial q_i \, \partial q_j} \right)_0 \delta q_i \, \delta q_j \tag{4-4}$$

Let us use the notation

$$k_{ij} \equiv \left(\frac{\partial^2 f}{\partial q_i \, \partial q_j} \right)_0 \tag{4-5}$$

where the k's form the elements of the symmetric $n \times n$ matrix $\mathbf{k}$. If the δq's are again independent and reversible, the sufficient condition that q_0 be a local minimum is that the matrix $\mathbf{k}$ be positive definite. Conversely, if $\mathbf{k}$ is negative definite, the point q_0 is a local maximum. If $\mathbf{k}$ is indefinite, q_0 is a saddle point.

The reason that we assume the δq's are reversible is that we wish to exclude the possibility that the reference configuration q_0 lies on the boundary of its allowed domain. This might occur, for example, at a unilateral constraint. One can visualize that the function f might assume an extreme value at a point on this boundary without being stationary, that is, where $\delta f \neq 0$ for certain geometrically allowable δq's. If q_0 is an interior point, however, f takes on a minimum or a maximum value only if it is also stationary.

Constrained Stationary Values. Now let us consider the conditions necessary for a stationary value of the function $f(q_1, q_2, \ldots, q_n)$, subject to the m independent constraint equations

$$\phi_j(q_1, q_2, \ldots, q_n) = 0 \qquad (j = 1, 2, \ldots, m) \tag{4-6}$$

where we assume that the ϕ's are continuous through the second partial derivatives. Once again we have $\delta f = 0$, or

$$\sum_{i=1}^{n} \left(\frac{\partial f}{\partial q_i} \right)_0 \delta q_i = 0 \tag{4-7}$$

where the reference configuration q_0 satisfies Eq. (4-6). The δq's are no longer independent, but conform to the m equations

$$\delta \phi_j = \sum_{i=1}^{n} \left(\frac{\partial \phi_j}{\partial q_i} \right)_0 \delta q_i = 0 \qquad (j = 1, 2, \ldots, m) \tag{4-8}$$

Note that because the δq's are not independent Eq. (4-3) no longer applies, in general.

It is possible, in theory, to use Eq. (4-6) to eliminate m q's in favor of the remaining $(n - m)$ q's and to find the stationary value of the resulting unconstrained function. Usually it is preferable, however, to apply the *Lagrange multiplier method* to problems involving constrained minima or maxima.

Now let us proceed with the Lagrange multiplier method. First multiply each $\delta\phi_j$ by a Lagrange multiplier λ_j. If we sum over j and add Eq. (4-7) we obtain

$$\sum_{i=1}^{n}\left[\left(\frac{\partial f}{\partial q_i}\right)_0 + \sum_{j=1}^{m}\lambda_j\left(\frac{\partial\phi_j}{\partial q_i}\right)_0\right]\delta q_i = 0 \qquad (4\text{-}9)$$

So far one might normally assume that the λ's are arbitrary and the δq's satisfy Eq. (4-8). On the other hand, if we can find a set of m λ's such that the coefficient of each δq_i is zero, that is, if

$$\left(\frac{\partial f}{\partial q_i}\right)_0 + \sum_{j=1}^{m}\lambda_j\left(\frac{\partial\phi_j}{\partial q_i}\right)_0 = 0 \qquad (i=1,2,\ldots,n) \qquad (4\text{-}10)$$

then we can choose the δq's arbitrarily. In other words, they can be considered to be *independent*. We will now show that the required set of m λ's can indeed be found, assuming that the given n equations are consistent.

Let us consider the $m \times n$ matrix **c** where

$$c_{ji} = \left(\frac{\partial\phi_j}{\partial q_i}\right)_0 \qquad (4\text{-}11)$$

If **c** is of rank m, that is, if there is at least one nonzero minor of order m, then the λ's are obtained by solving the corresponding m equations given in Eq. (4-10). But this requirement is met if the m constraints are independent at the reference configuration, as we have previously assumed. Hence we can always find a set of λ's such that the δq's can be considered to be freely variable.

In summary, the problem of finding a stationary value of the function $f(q_1, q_2, \ldots, q_n)$ subject to m constraints of the form $\phi_j(q_1, q_2, \ldots, q_n) = 0$ is solved by finding m λ's and n q's from the $(n + m)$ equations given by Eqs. (4-6) and (4-10).

Another way of viewing the Lagrange multiplier method is to consider the *free variations* of an *augmented function* $F(q_1, \ldots, q_n; \lambda_1, \ldots, \lambda_m)$ where

$$F = f + \sum_{j=1}^{m}\lambda_j\phi_j \qquad (4\text{-}12)$$

and the n q's and m λ's are regarded as independent variables. The necessary and sufficient conditions for F to be stationary are

$$\left(\frac{\partial F}{\partial q_i}\right)_0 = 0 \qquad (i = 1, 2, \ldots, n) \qquad (4\text{-}13)$$

and

$$\left(\frac{\partial F}{\partial \lambda_j}\right)_0 = 0 \qquad (j = 1, 2, \ldots, m) \qquad (4\text{-}14)$$

It can be seen that Eqs. (4-12) and (4-13) lead to Eq. (4-10) while Eqs. (4-12) and (4-14) result in Eq. (4-6). Hence we conclude that an unconstrained stationary value of the augmented function F occurs at the same configuration $\mathbf{q}_0$ as the corresponding constrained stationary value of the original function f.

Example 4-1. Find the stationary values of the function $f = z$, subject to the constraints

$$\phi_1 = x^2 + y^2 + z^2 - 4 = 0 \qquad (4\text{-}15)$$

$$\phi_2 = xy - 1 = 0 \qquad (4\text{-}16)$$

This corresponds to finding the highest and lowest points on the curve formed by the intersection of a sphere and a hyperbolic cylinder.

First we form the augmented function F with the aid of Eq. (4-12).

$$F = z + \lambda_1(x^2 + y^2 + z^2 - 4) + \lambda_2(xy - 1) \qquad (4\text{-}17)$$

Using Eq. (4-13), we obtain

$$\frac{\partial F}{\partial x} = 2x\lambda_1 + y\lambda_2 = 0$$

$$\frac{\partial F}{\partial y} = 2y\lambda_1 + x\lambda_2 = 0 \qquad (4\text{-}18)$$

$$\frac{\partial F}{\partial z} = 1 + 2z\lambda_1 = 0$$

These three equations plus Eqs. (4-15) and (4-16) can be solved for x, y, z, λ_1, λ_2. The result is that there are four stationary points, namely, $(1, 1, \sqrt{2})$, $(-1, -1, \sqrt{2})$, $(1, 1, -\sqrt{2})$, and $(-1, -1, -\sqrt{2})$. It can be seen that the first two are constrained maximum points of f, while the remaining two are constrained minimum points. The Lagrange multipliers are

$$\lambda_1 = -\frac{1}{2z} = \mp \frac{1}{2\sqrt{2}}, \qquad \lambda_2 = \frac{1}{z} = \pm \frac{1}{\sqrt{2}}$$

Stationary Value of a Definite Integral. Thus far we have been concerned with the ordinary minimization (or maximization) problem; that is, we have searched for *points in configuration space* which will yield a stationary value for a given function $f(q_1, q_2, \ldots, q_n)$. Now let us consider the problem of finding the stationary values of a definite integral. In particular, suppose we wish to find the necessary conditions for a stationary value of

$$I = \int_{x_0}^{x_1} f[y(x), y'(x), x] \, dx \qquad (4\text{-}19)$$

where $y'(x) \equiv dy/dx$ and the limits x_0 and x_1 are *fixed*. Furthermore, let us assume that $f(y, y', x)$ has two continuous derivatives in each of its arguments.

Here we are confronted with a problem in the calculus of variations which involves finding a *function* $y^*(x)$, out of some admissible class, which yields a stationary value for the integral I as compared to its value for other neighboring functions. Thus we have

$$y(x) = y*(x) + \delta y(x) \qquad (4\text{-}20)$$

where $\delta y(x)$ is a small variation in y. The quantity I is called a *functional* because its value depends upon the form of the function $y(x)$; that is, it is a function of the function $y(x)$. Note, however, that $f(y, y', x)$ is *given*.

Let us choose functions having two continuous derivatives as the admissible class for $y(x)$. It is convenient to represent the variation δy in the form

$$\delta y = \alpha \eta(x) \qquad (4\text{-}21)$$

where $\eta(x)$ is an arbitrary function having the required smoothness, and α is a small parameter which does not depend upon x (Fig. 4-1). Hence, for

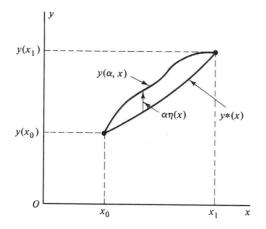

Fig. 4-1. The variation of a curve between fixed end-points.

any given $\eta(x)$, we can consider the varied curve y to be a function of α and x, that is,

$$y(\alpha, x) = y*(x) + \alpha\eta(x) \qquad (4\text{-}22)$$

Now let us make the additional assumption that the variation δy is zero at the end-points, that is,

$$\eta(x_0) = \eta(x_1) = 0 \qquad (4\text{-}23)$$

which implies that $y(x_0)$ and $y(x_1)$ are fixed.

We see that the integral I is a function of α only, for any given $\eta(x)$. Thus, a necessary condition that $y*(x)$ result in a stationary value of I is that its first variation be zero, or

$$\delta I = \left(\frac{dI}{d\alpha}\right)_{\alpha=0} \alpha = 0 \qquad (4\text{-}24)$$

for arbitrary $\eta(x)$ and nonzero α. Since the limits x_0 and x_1 are not dependent on α, we can differentiate under the integral sign. Hence we obtain

$$\frac{dI}{d\alpha} = \int_{x_0}^{x_1} \left(\frac{\partial f}{\partial y} \frac{\partial y}{\partial \alpha} + \frac{\partial f}{\partial y'} \frac{\partial y'}{\partial \alpha} \right) dx$$

$$= \int_{x_0}^{x_1} \left[\frac{\partial f}{\partial y} \eta(x) + \frac{\partial f}{\partial y'} \eta'(x) \right] dx = 0 \qquad (4\text{-}25)$$

Now let us integrate the second term by parts and obtain

$$\int_{x_0}^{x_1} \frac{\partial f}{\partial y'} \eta'(x) \, dx = \left[\frac{\partial f}{\partial y'} \eta(x) \right]_{x_0}^{x_1} - \int_{x_0}^{x_1} \eta(x) \frac{d}{dx} \left(\frac{\partial f}{\partial y'} \right) dx \qquad (4\text{-}26)$$

But we have chosen $\eta(x)$ to be zero at x_0 and x_1. Hence the first term on the right is zero and, from Eqs. (4-25) and (4-26), we find that

$$\int_{x_0}^{x_1} \left[\frac{\partial f}{\partial y} - \frac{d}{dx} \left(\frac{\partial f}{\partial y'} \right) \right] \eta(x) \, dx = 0 \qquad (4\text{-}27)$$

Because $\eta(x)$ is arbitrary except for the restrictions upon continuity and the end conditions, it follows that a necessary condition for the integral to be zero is that

$$\frac{\partial f}{\partial y} - \frac{d}{dx} \left(\frac{\partial f}{\partial y'} \right) = 0 \qquad (4\text{-}28)$$

for any curve $y = y*(x)$ which results in a stationary value of I. This is the *Euler-Lagrange equation*. With the given assumptions, it is both *necessary and sufficient* for a stationary value of the integral I. The sufficiency condition is apparent from the fact that Eq. (4-28) implies that the integral of Eq. (4-27) vanishes, resulting in the variation δI being zero.

The Euler-Lagrange equation was first derived by Euler in 1744 and later was applied to mechanical systems by Lagrange. In most problems in dynamics, the solution of this equation represents a *minimizing* curve for the functional I.

Example 4-2. The *brachistochrone problem* is one of the classical problems of the calculus of variations. We wish to find a curve $y(x)$ between the origin O and the point (x_1, y_1) such that a particle starting from rest at O, and sliding down the curve without friction under the influence of a uniform gravitational field, will reach the end of the curve in a minimum time.

Let us assume that the gravitational force is directed along the positive x axis, as shown in Fig. 4-2. We can obtain an expression for the velocity v as a function of position by using the principle of conservation of energy. Thus we obtain

$$\tfrac{1}{2}mv^2 = mgx$$

or

$$v = \sqrt{2gx} \qquad (4\text{-}29)$$

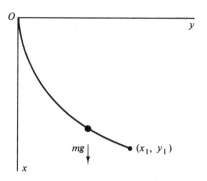

Fig. 4-2. The brachistochrone problem.

The time t required to reach the point (x_1, y_1) is found by noting first that an infinitesimal path element ds is given by

$$ds = \pm \sqrt{1 + y'^2}\, dx \qquad (4\text{-}30)$$

where $y' \equiv dy/dx$ and we choose the sign of the square root to give $ds > 0$. Then

$$t = \int_0^{s_1} \frac{ds}{v} = \int_0^{x_1} \sqrt{\frac{1+y'^2}{2gx}}\, dx \qquad (4\text{-}31)$$

Now we can identify the integral of Eq. (4-31) with the standard form of Eq. (4-19), where

$$f(y, y', x) = \sqrt{\frac{1 + y'^2}{2gx}} \qquad (4\text{-}32)$$

Since f is not an explicit function of y, the Euler-Lagrange equation of (4-28) becomes

$$\frac{d}{dx}\left(\frac{\partial f}{\partial y'}\right) = 0$$

or

$$\frac{\partial f}{\partial y'} = \frac{y'}{\sqrt{2gx(1 + y'^2)}} = c \qquad (4\text{-}33)$$

where c is a constant. This equation may be written in the form

$$\frac{dy}{dx} = \sqrt{\frac{2gc^2 x}{1 - 2gc^2 x}} \qquad (4\text{-}34)$$

Now let us introduce the parameter θ by the substitution

$$x = a(1 - \cos\theta) \qquad (4\text{-}35)$$

where $a = 1/4gc^2$. Then we obtain

$$y = \int a(1 - \cos\theta)\, d\theta$$

which is readily integrated to yield

$$y = a(\theta - \sin \theta) \tag{4-36}$$

Note that since the path starts from the origin, the constant of integration is zero.

Equations (4-35) and (4-36) are the parametric equations representing a cycloid. The parameter θ increases continuously as the particle proceeds along its path, even though x may actually decrease during the latter portion of certain trajectories. The constant a can always be chosen such that the path goes through the final point (x_1, y_1) for all $x_1 \geq 0$. Although we have shown only that the cycloidal path leads to a stationary value of t, a comparison with other neighboring paths will indicate that it is actually the path of minimum time.

Example 4-3. Another classical problem of the calculus of variations is the geodesic problem, that is, the problem of finding the shortest path between two points in a given space. As an example, let us consider the problem of finding the path of minimum length between two given points on the two-dimensional surface of a sphere of radius r.

Let us use the spherical coordinates (θ, ϕ) as variables. The differential element of length ds is given by

$$ds^2 = r^2\, d\theta^2 + r^2 \sin^2 \theta\, d\phi^2$$

or

$$ds = \pm r \sqrt{1 + \sin^2 \theta (d\phi/d\theta)^2}\, d\theta \tag{4-37}$$

where we choose the sign of the square root so that each increment ds is positive. Thus we obtain

$$s = r \int_{\theta_0}^{\theta_1} \sqrt{1 + \phi'^2 \sin^2 \theta}\, d\theta \tag{4-38}$$

where $\phi' \equiv d\phi/d\theta$. This integral has the standard form of Eq. (4-19) and we see that

$$f = \sqrt{1 + \phi'^2 \sin^2 \theta} \tag{4-39}$$

Noting that f is not an explicit function of ϕ, we can write the Euler-Lagrange equation in the form

$$\frac{\partial f}{\partial \phi'} = \frac{\phi' \sin^2 \theta}{\sqrt{1 + \phi'^2 \sin^2 \theta}} = c \tag{4-40}$$

Now we solve for ϕ' and obtain

$$\frac{d\phi}{d\theta} = \frac{c}{\sin \theta \sqrt{\sin^2 \theta - c^2}}$$

which results in

$$\phi = \int \frac{c\, d\theta}{\sin \theta \sqrt{\sin^2 \theta - c^2}} \tag{4-41}$$

Integrating, we obtain

$$\phi = \cos^{-1} \frac{c \cot \theta}{\sqrt{1 - c^2}} + \phi_0$$

or

$$\cos (\phi - \phi_0) = \frac{c}{\sqrt{1 - c^2}} \cot \theta \qquad (4\text{-}42)$$

where ϕ_0 is the constant of integration.

In order to obtain a better understanding of this result, let us transform Eq. (4-42) to Cartesian coordinates. If we use the trigonometric identity for $\cos (\phi - \phi_0)$ and the equations

$$x = r \sin \theta \cos \phi$$
$$y = r \sin \theta \sin \phi \qquad (4\text{-}43)$$
$$z = r \cos \theta$$

we obtain

$$x \cos \phi_0 + y \sin \phi_0 - \frac{c}{\sqrt{1 - c^2}} z = 0 \qquad (4\text{-}44)$$

which we recognize as the equation of a plane through the origin. This plane intersects the sphere in a *great circle* which is the geodesic in this example. The constants c and ϕ_0 are chosen so that the great circle goes through the required two points.

The same result can be obtained more easily if we choose a system of spherical coordinates which are oriented in such a manner that the two endpoints lie at the same value of ϕ, and θ is chosen in the interval $0 < \theta < \pi$. Then it is immediately apparent from Eq. (4-38) that the path length is minimized by taking $d\phi/d\theta = 0$ at all points. Once again the geodesic is a great circle.

It is interesting to note that the path giving a stationary value to the integral of Eq. (4-38) is always a great circle, but it is not necessarily of minimum length. As an example, if the length of the great circle is more than π radians, then there are shorter neighboring paths which are not great circles. On the other hand, if the great circle is less than π radians in length, and therefore does not contain two diametrically opposite or *conjugate* points, then the path is always the shortest one connecting the given points.

The Case of n Dependent Variables. Now let us extend our previous results to the problem of finding the functions $y_1(x), y_2(x), \ldots, y_n(x)$ which will lead to a stationary value of the integral

$$I = \int_{x_0}^{x_1} f(y_1, \ldots, y_n, y_1', \ldots, y_n', x) \, dx \qquad (4\text{-}45)$$

where the values of each function $y_i(x)$ are specified at the fixed end-points x_0 and x_1. Again we assume that the functions $y_i(x)$ and the variations $\delta y_i(x)$

have two continuous derivatives. Let the variations be of the form

$$\delta y_i = \alpha \eta_i(x) \tag{4-46}$$

where $\eta_i(x_0) = \eta_i(x_1) = 0$.

It can be seen that the integral I is a function of the parameter α for any given set of η's, and Eq. (4-24) remains valid. Hence, by analogy to Eq. (4-25), we obtain

$$\frac{dI}{d\alpha} = \int_{x_0}^{x_1} \sum_{i=1}^{n} \left[\frac{\partial f}{\partial y_i} \eta_i(x) + \frac{\partial f}{\partial y_i'} \eta_i'(x) \right] dx = 0 \tag{4-47}$$

If we integrate by parts and remember that the η's are zero at the end-points, we obtain

$$\int_{x_0}^{x_1} \sum_{i=1}^{n} \left[\frac{\partial f}{\partial y_i} - \frac{d}{dx} \left(\frac{\partial f}{\partial y_i'} \right) \right] \eta_i(x) \, dx = 0 \tag{4-48}$$

which is similar to Eq. (4-27).

We have chosen the δy's, and hence the η's, to be independent. Therefore each expression in brackets must be zero, and we obtain the *Euler-Lagrange equations* for this more general case, namely,

$$\frac{\partial f}{\partial y_i} - \frac{d}{dx} \left(\frac{\partial f}{\partial y_i'} \right) = 0 \qquad (i = 1, 2, \ldots, n) \tag{4-49}$$

These n equations represent the *necessary and sufficient* conditions that δI is zero.

Hamilton's Principle. Now let us turn to the important variational principle of dynamics known as Hamilton's principle. As a starting point, consider a system of N particles whose configuration relative to an inertial frame is given by the vectors $\mathbf{r}_1, \mathbf{r}_2, \ldots, \mathbf{r}_N$. Using the Lagrangian form of d'Alembert's principle, we see that

$$\sum_{i=1}^{N} (\mathbf{F}_i - m_i \ddot{\mathbf{r}}_i) \cdot \delta \mathbf{r}_i = 0 \tag{4-50}$$

where $\mathbf{F}_i$ is the *applied force* acting on the ith particle. We assume that the virtual displacements $\delta \mathbf{r}_i$ are reversible and consistent with the instantaneous constraints, which are considered to be workless.

Next let us obtain an expression for the variation in the kinetic energy. We can write

$$\delta T = \delta \left(\tfrac{1}{2} \sum_{i=1}^{N} m_i \dot{\mathbf{r}}_i^2 \right) = \sum_{i=1}^{N} m_i \dot{\mathbf{r}}_i \cdot \delta \dot{\mathbf{r}}_i \tag{4-51}$$

But

$$\frac{d}{dt} \left(\sum_{i=1}^{N} m_i \dot{\mathbf{r}}_i \cdot \delta \mathbf{r}_i \right) = \sum_{i=1}^{N} m_i \ddot{\mathbf{r}}_i \cdot \delta \mathbf{r}_i + \sum_{i=1}^{N} m_i \dot{\mathbf{r}}_i \cdot \delta \dot{\mathbf{r}}_i \tag{4-52}$$

where we note that the operations of variation and of differentiation with respect to time are interchangeable. In other words, we find that

$$\frac{d}{dt}(\delta \mathbf{r}_i) = \delta \dot{\mathbf{r}}_i \tag{4-53}$$

because each virtual displacement is assumed to take place without the passage of time. Hence we obtain from Eqs. (4-50), (4-51), and (4-52) that

$$\delta T + \sum_{i=1}^{N} \mathbf{F}_i \cdot \delta \mathbf{r}_i = \frac{d}{dt}\left(\sum_{i=1}^{N} m_i \dot{\mathbf{r}}_i \cdot \delta \mathbf{r}_i\right) \tag{4-54}$$

Now let us integrate this equation with respect to time between the limits t_0 and t_1. Using the notation that δW is the virtual work of the *applied forces*, as given in Eq. (1-39), we obtain

$$\int_{t_0}^{t_1}(\delta T + \delta W)\,dt = \left[\sum_{i=1}^{N} m_i \dot{\mathbf{r}}_i \cdot \delta \mathbf{r}_i\right]_{t_0}^{t_1} \tag{4-55}$$

Next we assume that the configuration of the system is specified at the times t_0 and t_1, implying that variations $\delta \mathbf{r}_i$ are zero at these times. Thus we conclude that

$$\int_{t_0}^{t_1}(\delta T + \delta W)\,dt = 0 \tag{4-56}$$

Up to this point we have given the configuration of the system in terms of the position vectors $\mathbf{r}_i$ of the particles. It is apparent, however, that for a given virtual displacement and time the values of δT and δW are independent of the coordinate system. So let us make a transformation to the generalized coordinates $q_1, q_2, \ldots, q_n$. Then the kinetic energy T is a function of the q's, $\dot{q}$'s, and t, while the virtual work is

$$\delta W = \sum_{i=1}^{n} Q_i\,\delta q_i \tag{4-57}$$

where the Q's are the *applied* generalized forces. Hence Eq. (4-56) can be written in the form

$$\int_{t_0}^{t_1}\left(\delta T + \sum_{i=1}^{n} Q_i\,\delta q_i\right)dt = 0 \tag{4-58}$$

where the δq's are equal to zero at times t_0 and t_1. For constrained systems, the δq's must also conform to the instantaneous constraints.

It is helpful to think of the actual and varied paths in an $(n + 1)$-dimensional space consisting of the n q's and t, as shown in Fig. 4-3. This is known as the *extended configuration space*. Note that the two end-points of the actual and varied paths are fixed in this space, and each $\delta \mathbf{q}$ vector is perpendicular to the t axis.

The result given by Eq. (4-56) or (4-58) is often considered as a generalized version of Hamilton's principle. It is essentially an integrated form of d'Alembert's principle, as given in Eq. (4-50), and is applicable to the same wide variety of mechanical systems. The usual form of Hamilton's principle, however, applies to a more restricted class of systems.

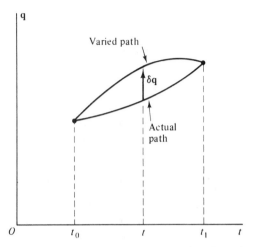

Fig. 4-3. The actual and varied paths in extended configuration space.

Let us proceed, then, in the derivation by assuming that *all the applied forces are derivable from a potential function* $V(q, t)$. If we recall from Eq. (1-91) that $\delta W = -\delta V$, Eq. (4-56) becomes

$$\int_{t_0}^{t_1} \delta(T - V)\, dt = 0 \qquad (4\text{-}59)$$

where we again assume fixed end-points. Now let us consider a *holonomic system*. For these systems the operations of integration and variation can be interchanged (see p. 159). Then, making the substitution $L = T - V$, we obtain

$$\delta \int_{t_0}^{t_1} L\, dt = 0 \qquad (4\text{-}60)$$

where both the actual and varied paths meet the conditions imposed by any holonomic constraints. Hence we obtain *Hamilton's principle*:

> *The actual path in configuration space followed by a holonomic dynamical system during the fixed interval* t_0 *to* t_1 *is such that the integral*
>
> $$I = \int_{t_0}^{t_1} L\, dt \qquad (4\text{-}61)$$
>
> *is stationary with respect to path variations which vanish at the end-points.*

If we compare the integral I of Eq. (4-61) with that of Eq. (4-45), we notice a similarity in form; that is, the independent variable t corresponds to x, the q's correspond to the y's, and $L(q, \dot{q}, t)$ corresponds to $f(y, y', x)$. Hence, assuming that the δq's are *independent*, the Euler-Lagrange equations of (4-49) take the following form:

$$\frac{d}{dt}\left(\frac{\partial L}{\partial \dot{q}_i}\right) - \frac{\partial L}{\partial q_i} = 0 \qquad (i = 1, 2, \ldots, n) \tag{4-62}$$

Here we obtain the standard form of Lagrange's equations as the necessary and sufficient conditions that $\delta I = 0$, provided that $L(q, \dot{q}, t)$ and $q(t)$ have the required smoothness.

Although Hamilton's principle and Lagrange's equations are equivalent for the assumed system, Hamilton's principle is broader in the sense that it applies to distributed systems which have an infinite number of degrees of freedom.

Nonholonomic Systems. In the development of Hamilton's principle we considered a *holonomic system*. For such a system having k degrees of freedom, it is always possible to express the configuration in terms of k independent q's. Any varied path in the $(k + 1)$-dimensional extended configuration space is a geometrically possible path and corresponds to geometrically possible δq's.

Now suppose we describe the *same system* in terms of $n = k + m$ generalized coordinates, where m is the number of independent equations of *holonomic* constraint. The same varied paths are possible, but now the configurations are represented by n q's which are not freely variable. Nevertheless, it is still true that any geometrically possible set of δq's corresponds to a geometrically possible varied path, and vice versa. This implies that the operations of variation and integration with respect to time are interchangeable for all holonomic systems. In other words, the variation of the integral is equal to the integral of the variation. Furthermore, since the value of the Lagrangian function L is invariant with respect to a coordinate transformation, Hamilton's principle applies to the given holonomic system, regardless of which set of generalized coordinates is used.

By contrast, let us next consider a *nonholonomic system*. Suppose there are n generalized coordinates and m nonholonomic constraint equations of the form

$$\sum_{i=1}^{n} a_{ji}(q, t)\dot{q}_i + a_{jt}(q, t) = 0 \qquad (j = 1, 2, \ldots, m) \tag{4-63}$$

Let us use the notation that $\mathbf{q}^*(t)$ is the actual path of the system and $\mathbf{q}(t)$ is a varied path. Then we can write

$$q_i = q_i^* + \delta q_i \tag{4-64}$$

and

$$\dot{q}_i = \dot{q}_i^* + \delta \dot{q}_i \tag{4-65}$$

Now assume that the varied path and the actual path both conform to the constraints of Eq. (4-63). We can represent the a's by a Taylor expansion about the reference value at each instant of time. Neglecting terms of order higher than the first in the δq's, we obtain

$$a_{ji}(q, t) = a_{ji}(q^*, t) + \sum_{k=1}^{n} \left(\frac{\partial a_{ji}}{\partial q_k}\right)_0 \delta q_k \tag{4-66}$$

$$a_{jt}(q, t) = a_{jt}(q^*, t) + \sum_{k=1}^{n} \left(\frac{\partial a_{jt}}{\partial q_k}\right)_0 \delta q_k \tag{4-67}$$

where a zero subscript indicates that the quantity is evaluated on the actual path. Then, substituting Eqs. (4-65)–(4-67) into Eq. (4-63) and noting that

$$\sum_{i=1}^{n} a_{ji}(q^*, t)\dot{q}_i^* + a_{jt}(q^*, t) = 0 \qquad (j = 1, 2, \ldots, m) \tag{4-68}$$

we find that

$$\sum_{i=1}^{n} a_{ji}(q^*, t)\delta\dot{q}_i + \sum_{i=1}^{n}\sum_{k=1}^{n} \left(\frac{\partial a_{ji}}{\partial q_k}\right)_0 \dot{q}_i^* \delta q_k + \sum_{k=1}^{n} \left(\frac{\partial a_{jt}}{\partial q_k}\right)_0 \delta q_k = 0 \tag{4-69}$$

where terms containing the products of small quantities are neglected. Now suppose we make the additional assumption that the δq's *meet the instantaneous constraint conditions,* namely,

$$\sum_{i=1}^{n} a_{ji}(q^*, t)\delta q_i = 0 \qquad (j = 1, 2, \ldots, m) \tag{4-70}$$

Differentiating this expression with respect to time and changing indices, we obtain

$$\sum_{k=1}^{n} \dot{a}_{jk}(q^*, t)\, \delta q_k + \sum_{i=1}^{n} a_{ji}(q^*, t)\delta\dot{q}_i = 0 \tag{4-71}$$

where

$$\dot{a}_{jk}(q^*, t) = \sum_{i=1}^{n} \left(\frac{\partial a_{jk}}{\partial q_i}\right)_0 \dot{q}_i^* + \left(\frac{\partial a_{jk}}{\partial t}\right)_0 \tag{4-72}$$

Hence, from Eqs. (4-71) and (4-72), we have

$$\sum_{i=1}^{n} a_{ji}(q^*, t)\delta\dot{q}_i + \sum_{k=1}^{n}\sum_{i=1}^{n} \left(\frac{\partial a_{jk}}{\partial q_i}\right)_0 \dot{q}_i^* \delta q_k + \sum_{k=1}^{n} \left(\frac{\partial a_{jk}}{\partial t}\right)_0 \delta q_k = 0 \tag{4-73}$$

Now let us subtract Eq. (4-73) from Eq. (4-69) and we see that

$$\sum_{i=1}^{n}\sum_{k=1}^{n} \left(\frac{\partial a_{ji}}{\partial q_k} - \frac{\partial a_{jk}}{\partial q_i}\right)_0 \dot{q}_i^* \delta q_k + \sum_{k=1}^{n} \left(\frac{\partial a_{jt}}{\partial q_k} - \frac{\partial a_{jk}}{\partial t}\right)_0 \delta q_k = 0 \tag{4-74}$$

Note that, in general, $\dot{q}_i^* \neq 0$. Hence, if Eq. (4-74) is to be valid continuously for any set of δq's which conform to the constraints of Eq. (4-70), we must have

$$\left(\frac{\partial a_{ji}}{\partial q_k} - \frac{\partial a_{jk}}{\partial q_i}\right)_0 = 0 \qquad \begin{matrix} (i, k = 1, 2, \ldots, n) \\ (j = 1, 2, \ldots, m) \end{matrix} \tag{4-75}$$

$$\left(\frac{\partial a_{jt}}{\partial q_k} - \frac{\partial a_{jk}}{\partial t}\right)_0 = 0 \qquad \begin{matrix} (k = 1, 2, \ldots, n) \\ (j = 1, 2, \ldots, m) \end{matrix} \tag{4-76}$$

Equations (4-75) and (4-76) represent the *exactness conditions* for the integrability of Eq. (4-63), as given previously in Eq. (1-23). In other words, if these conditions apply, the constraints are *holonomic*.

To summarize, we have shown that if the varied paths conform to the actual constraints, and if the δq's are consistent with the instantaneous constraints, then the system must be holonomic. For a nonholonomic system, on the other hand, a varied path in which the δq's are constrained by Eq. (4-70) will not be a geometrically possible path because it will not conform to Eq. (4-63). Hence we conclude once again that operations of variation and integration can be interchanged for the case of holonomic systems, but this is not possible for nonholonomic systems.

We also conclude that Hamilton's principle, as given by Eq. (4-60) is valid for *holonomic systems only*. Equations (4-58) and (4-59) apply to nonholonomic systems, but are not variational principles in the usual sense because the varied paths are not geometrically possible paths.

The Multiplier Rule. Now let us consider the problem of finding a stationary value of the integral I of Eq. (4-61), subject to m independent differential constraints (holonomic or nonholonomic) of the more general form

$$g_j(q, \dot{q}, t) = 0 \qquad (j = 1, 2, \ldots, m) \tag{4-77}$$

where we again assume that the end-points are fixed in extended configuration space. Using the terminology of the calculus of variations, this is an example of the problem of Lagrange. A standard method for the analysis of these problems is the *multiplier rule*. The multiplier rule states that the *constrained* stationary values of the integral of Eq. (4-61) are found by considering the *free variations* of

$$I = \int_{t_0}^{t_1} \Lambda \, dt \tag{4-78}$$

where $\Lambda(q, \dot{q}, \mu, t)$ is an *augmented Lagrangian function* given by

$$\Lambda(q, \dot{q}, \mu, t) \doteq L(q, \dot{q}, t) + \sum_{j=1}^{m} \mu_j g_j(q, \dot{q}, t) \tag{4-79}$$

The Lagrange multipliers $\mu_j(t)$ are treated as additional variables to be determined.

If we proceed with the free variation of the integral I of Eq. (4-78), we obtain the following Euler-Lagrange equations:†

$$\frac{d}{dt}\left(\frac{\partial \Lambda}{\partial \dot{q}_i}\right) - \frac{\partial \Lambda}{\partial q_i} = 0 \qquad (i = 1, 2, \ldots, n) \tag{4-80}$$

$$\frac{\partial \Lambda}{\partial \mu_j} = g_j(q, \dot{q}, t) = 0 \qquad (j = 1, 2, \ldots, m) \tag{4-81}$$

---

†The δq's are zero at the end-points, as usual. It can be seen, however, that the $\delta\mu$'s are not restricted in this fashion since Λ is not an explicit function of the $\dot{\mu}$'s; hence this assumption is not necessary in obtaining Eq. (4-81).

Here we have $(n + m)$ equations from which to solve for the n q's and m μ's as functions of time. Note that Eq. (4-81) is merely a reiteration of the requirement that the solution path conform to the constraints.

Now let us apply Eq. (4-80) to a system in which the constraint functions are *linear* in the $\dot{q}$'s. Using the familiar notation for differential constraints, we have

$$g_j(q, \dot{q}, t) = \sum_{i=1}^{n} a_{ji}(q, t)\dot{q}_i + a_{jt}(q, t) = 0 \qquad (j = 1, 2, \ldots, m) \qquad (4\text{-}82)$$

Then, from Eqs. (4-79) and (4-80), we obtain

$$\frac{d}{dt}\left(\frac{\partial L}{\partial \dot{q}_i}\right) - \frac{\partial L}{\partial q_i} = -\sum_{j=1}^{m}\frac{d}{dt}(\mu_j a_{ji}) + \sum_{j=1}^{m}\sum_{k=1}^{n}\mu_j\frac{\partial a_{jk}}{\partial q_i}\dot{q}_k + \sum_{j=1}^{m}\mu_j\frac{\partial a_{jt}}{\partial q_i}$$

or

$$\frac{d}{dt}\left(\frac{\partial L}{\partial \dot{q}_i}\right) - \frac{\partial L}{\partial q_i} = -\sum_{j=1}^{m}\dot{\mu}_j a_{ji} + \sum_{j=1}^{m}\sum_{k=1}^{n}\mu_j\left(\frac{\partial a_{jk}}{\partial q_i} - \frac{\partial a_{ji}}{\partial q_k}\right)\dot{q}_k$$

$$+ \sum_{j=1}^{m}\mu_j\left(\frac{\partial a_{jt}}{\partial q_i} - \frac{\partial a_{ji}}{\partial t}\right) \qquad (i = 1, 2, \ldots, n) \qquad (4\text{-}83)$$

These equations are the Euler-Lagrange equations for finding stationary values of the integral of Eq. (4-61), where both the actual and varied paths are required to meet the constraint conditions of Eq. (4-82). Comparing this result with the known form of Lagrange's equations for this system, namely,

$$\frac{d}{dt}\left(\frac{\partial L}{\partial \dot{q}_i}\right) - \frac{\partial L}{\partial q_i} = \sum_{j=1}^{m}\lambda_j a_{ji} \qquad (i = 1, 2, \ldots, n) \qquad (4\text{-}84)$$

we see that, in general, the equations are different. Hence we conclude that the use of the multiplier rule leads to *incorrect dynamical equations* for the general case of *nonholonomic constraints*. Conversely, the actual solution path of a nonholonomic system will not result, in general, in a stationary value of the integral of Eq. (4-61).

If, on the other hand, the constraints of Eq. (4-82) are actually *holonomic*, then we find that Eq. (4-83) reduces to the correct form given in Eq. (4-84). In this case, the integrability conditions of Eq. (4-75) and (4-76) apply, and the last two sums on the right-hand side of Eq. (4-83) vanish. Then, letting $\lambda_j = -\dot{\mu}_j$, we obtain the proper equations of motion.

4-2. HAMILTON'S EQUATIONS

Lagrange's equations, as we have seen, consist of n second-order differential equations in the n q's and time. Once the Lagrangian function $L(q, \dot{q}, t)$ is known for a given system, the equations of motion are obtained by a straightforward procedure. These n equations have a form which sometimes permits their complete integration, resulting in analytical solutions for the q's as functions of time.

From a theoretical viewpoint, however, it is often preferable to describe the system in terms of $2n$ first-order equations. There are many ways in which this can be accomplished, and all are not equally suited to our needs. For example, one can define n new variables v_i, where $v_i = \dot{q}_i$, and thereby convert the equations of motion to the desired $2n$ first-order equations. This is not particularly helpful, however. Instead, we shall define a new function $H(q, p, t)$ known as the Hamiltonian function, and use it to generate a set of first-order equations which are particularly symmetrical in form. These are the canonical equations of Hamilton.

Derivation of Hamilton's Equations. Let us consider a *holonomic* system which can be described by the standard form of Lagrange's equations, namely,

$$\frac{d}{dt}\left(\frac{\partial L}{\partial \dot{q}_i}\right) - \frac{\partial L}{\partial q_i} = 0 \qquad (i = 1, 2, \ldots, n) \tag{4-85}$$

Remembering that the generalized momentum conjugate to q_i is given by

$$p_i = \frac{\partial L}{\partial \dot{q}_i} \tag{4-86}$$

we can write Eq. (4-85) in the form

$$\dot{p}_i = \frac{\partial L}{\partial q_i} \qquad (i = 1, 2, \ldots, n) \tag{4-87}$$

Now let us define the *Hamiltonian function* $H(q, p, t)$ for the system as follows:

$$H(q, p, t) = \sum_{i=1}^{n} p_i \dot{q}_i - L(q, \dot{q}, t) \tag{4-88}$$

It is important to note that, in the general case, H is an explicit function of the q's, p's, and t. Since the right-hand side of Eq. (4-88) contains $\dot{q}$'s, we must eliminate these quantities by expressing them in terms of the p's. This is accomplished by recalling from Eq. (2-30) that

$$p_i = \sum_{j=1}^{n} m_{ij}(q, t)\dot{q}_j + a_i(q, t) \tag{4-89}$$

Then we solve for the $\dot{q}$'s and obtain

$$\dot{q}_i = \sum_{j=1}^{n} b_{ij}(p_j - a_j) \tag{4-90}$$

where $b_{ij}(q, t)$ is an element of the matrix $\mathbf{b} = \mathbf{m}^{-1}$. The inertia matrix $\mathbf{m}$ can always be inverted since it is positive definite.

Now consider an arbitrary variation in the Hamiltonian function $H(q, p, t)$. We have

$$\delta H = \sum_{i=1}^{n} \frac{\partial H}{\partial q_i} \delta q_i + \sum_{i=1}^{n} \frac{\partial H}{\partial p_i} \delta p_i + \frac{\partial H}{\partial t} \delta t \tag{4-91}$$

In a similar manner, we obtain from Eq. (4-88) that

$$\delta H = \sum_{i=1}^{n} p_i \, \delta \dot{q}_i + \sum_{i=1}^{n} \dot{q}_i \, \delta p_i - \sum_{i=1}^{n} \frac{\partial L}{\partial q_i} \delta q_i - \sum_{i=1}^{n} \frac{\partial L}{\partial \dot{q}_i} \delta \dot{q}_i - \frac{\partial L}{\partial t} \delta t$$

and, using the expression for p_i given in Eq. (4-86) to cancel the first and fourth terms on the right-hand side, we find that

$$\delta H = \sum_{i=1}^{n} \dot{q}_i \, \delta p_i - \sum_{i=1}^{n} \frac{\partial L}{\partial q_i} \delta q_i - \frac{\partial L}{\partial t} \delta t \qquad (4\text{-}92)$$

Thus far, the expressions given in Eqs. (4-91) and (4-92) follow directly from the definitions of H and p_i, but they do not contain any dynamical laws. Now let us inject some dynamics into the analysis by using Lagrange's equation in the form given by Eq. (4-87). Substituting this expression into Eq. (4-92), we obtain

$$\delta H = \sum_{i=1}^{n} \dot{q}_i \, \delta p_i - \sum_{i=1}^{n} \dot{p}_i \, \delta q_i - \frac{\partial L}{\partial t} \delta t \qquad (4\text{-}93)$$

The variations δq_i, δp_i, and δt are mutually independent, so their coefficients must be equal in Eqs. (4-91) and (4-93). Hence

$$\dot{q}_i = \frac{\partial H}{\partial p_i}$$

$$\dot{p}_i = -\frac{\partial H}{\partial q_i} \qquad\qquad (i = 1, 2, \ldots, n) \qquad (4\text{-}94)$$

and

$$\frac{\partial L}{\partial t} = -\frac{\partial H}{\partial t} \qquad (4\text{-}95)$$

The $2n$ first-order equations given in Eq. (4-94) are known as *Hamilton's canonical equations of motion*. The first n equations merely express the $\dot{q}$'s as linear functions of the p's and are equivalent to Eq. (4-90). On the other hand, the final n equations represent an application of the laws of motion to the system, and give the rates of change of the generalized momenta. Because of the symmetry of Hamilton's equations, however, one tends to accord the equations equal status. If we consider that the n q's and n p's together constitute a $2n$-vector $\mathbf{x}$, then Hamilton's equations can be written as a first-order nonlinear vector equation of the form

$$\dot{\mathbf{x}} = \mathbf{X}(\mathbf{x}, t) \qquad (4\text{-}96)$$

where we note that all the $\dot{q}$'s and $\dot{p}$'s occur linearly on the left-hand side of the equation.

One reason for the importance of the Hamiltonian form of the equations of motion is that it facilitates the use of transformations in obtaining solutions. In succeeding chapters, we shall discuss the application of *canonical transformations*, involving the paired quantities (q_i, p_i), in the solution of Hamilton's equations.

Comparing the Lagrangian and Hamiltonian formulations of the equations of motion, we see that either the Lagrangian function $L(q, \dot{q}, t)$ or the Hamiltonian function $H(q, p, t)$ can be regarded as a descriptive function for the system, from which a complete set of equations of motion can be derived. The standard holonomic forms of Lagrange's equations and Hamilton's equations are equivalent and apply to the same systems. Note that the q's must be independent in either case.

Now let us consider the necessary modifications to Hamilton's equations in order that they will apply to systems in which the generalized forces are not all derivable from a potential function. Suppose, for example, that we have a *holonomic system* obeying Lagrange's equations in the form of Eq. (2-29) or, equivalently,

$$\dot{p}_i = \frac{\partial L}{\partial q_i} + Q_i' \qquad (i = 1, 2, \ldots, n) \tag{4-97}$$

where Q_i' is that portion of the generalized applied force which is not derivable from a potential function. Hamilton's equations for this system are

$$\dot{q}_i = \frac{\partial H}{\partial p_i}, \qquad \dot{p}_i = -\frac{\partial H}{\partial q_i} + Q_i' \qquad (i = 1, 2, \ldots, n) \tag{4-98}$$

If each Q_i' can be expressed as a function of the q's, p's, and t, then it is clear that Eq. (4-98) is of the form $\dot{\mathbf{x}} = \mathbf{X}(\mathbf{x}, t)$ given in Eq. (4-96).

A *nonholonomic system* described by Lagrange's equations in the form

$$\dot{p}_i = \frac{\partial L}{\partial q_i} + \sum_{j=1}^{m} \lambda_j a_{ji} + Q_i' \qquad (i = 1, 2, \ldots, n) \tag{4-99}$$

has the following set of Hamilton's equations:

$$\dot{q}_i = \frac{\partial H}{\partial p_i}, \qquad \dot{p}_i = -\frac{\partial H}{\partial q_i} + \sum_{j=1}^{m} \lambda_j a_{ji} + Q_i' \qquad (i = 1, 2, \ldots, n) \tag{4-100}$$

where the m constraint equations are

$$\sum_{i=1}^{n} a_{ji}\dot{q}_i + a_{jt} = 0 \qquad (j = 1, 2, \ldots, m) \tag{4-101}$$

Here we have in Eqs. (4-100) and (4-101) a set of $(2n + m)$ first-order ordinary differential equations from which to solve for the n q's, n p's, and m λ's as functions of time.

The Form of the Hamiltonian Function. In order to consider the explicit form of the Hamiltonian function $H(q, p, t)$ in greater detail, let us first use Eq. (4-89) to obtain

$$\sum_{i=1}^{n} p_i\dot{q}_i = \sum_{i=1}^{n} \sum_{j=1}^{n} m_{ij}\dot{q}_i\dot{q}_j + \sum_{i=1}^{n} a_i\dot{q}_i = 2T_2 + T_1 \tag{4-102}$$

Then we find that the value of the Hamiltonian function is the same as the

Jacobi integral expression, namely,

$$H = \sum_{i=1}^{n} p_i \dot{q}_i - T + V = T_2 - T_0 + V \tag{4-103}$$

Using matrix notation, we recall from Eq. (2-6) that

$$T_2 = \tfrac{1}{2} \dot{\mathbf{q}}^T \mathbf{m} \dot{\mathbf{q}} \tag{4-104}$$

and, from Eq. (4-90),

$$\dot{\mathbf{q}} = \mathbf{b}(\mathbf{p} - \mathbf{a}) \tag{4-105}$$

Hence, noting that $\mathbf{b} = \mathbf{m}^{-1}$ and that both $\mathbf{b}$ and $\mathbf{m}$ are symmetric, we obtain

$$T_2 = \tfrac{1}{2}(\mathbf{p} - \mathbf{a})^T \mathbf{b}(\mathbf{p} - \mathbf{a}) \tag{4-106}$$

which may be expanded to yield

$$T_2 = \tfrac{1}{2} \sum_{i=1}^{n} \sum_{j=1}^{n} b_{ij} p_i p_j - \sum_{i=1}^{n} \sum_{j=1}^{n} b_{ij} a_i p_j + \tfrac{1}{2} \sum_{i=1}^{n} \sum_{j=1}^{n} b_{ij} a_i a_j \tag{4-107}$$

Since T_0 and V are functions of the q's and t, we obtain

$$H(q, p, t) = \tfrac{1}{2} \sum_{i=1}^{n} \sum_{j=1}^{n} b_{ij} p_i p_j - \sum_{i=1}^{n} \sum_{j=1}^{n} b_{ij} a_i p_j$$
$$+ \tfrac{1}{2} \sum_{i=1}^{n} \sum_{j=1}^{n} b_{ij} a_i a_j - T_0 + V \tag{4-108}$$

If we group the terms in the Hamiltonian according to their degree in p, we can write

$$H = H_2 + H_1 + H_0 \tag{4-109}$$

where

$$H_2 = \tfrac{1}{2} \sum_{i=1}^{n} \sum_{j=1}^{n} b_{ij} p_i p_j \tag{4-110}$$

$$H_1 = - \sum_{i=1}^{n} \sum_{j=1}^{n} b_{ij} a_i p_j \tag{4-111}$$

$$H_0 = \tfrac{1}{2} \sum_{i=1}^{n} \sum_{j=1}^{n} b_{ij} a_i a_j - T_0 + V \tag{4-112}$$

Hence the Hamiltonian function is, in general, quadratic in the p's.

Now let us consider a *scleronomic system*. Since the transformation equations from Cartesian coordinates to generalized coordinates do not contain t explicitly, it follows that the a's are zero and $T = T_2$. Consequently the Hamiltonian is of the form

$$H(q, p, t) = \tfrac{1}{2} \sum_{i=1}^{n} \sum_{j=1}^{n} b_{ij} p_i p_j + V(q, t) \tag{4-113}$$

where the b's are functions of the q's only. It can be seen that $H_2 = T_2$, $H_1 = 0$, and $H_0 = V$. Therefore the Hamiltonian function of a scleronomic system is equal to the total energy, that is,

$$H = T + V \tag{4-114}$$

It is interesting to consider how the Hamiltonian function varies with time. Its total time derivative is

$$\dot{H} = \sum_{i=1}^{n} \left(\frac{\partial H}{\partial q_i} \dot{q}_i + \frac{\partial H}{\partial p_i} \dot{p}_i \right) + \frac{\partial H}{\partial t} \tag{4-115}$$

For the common case of a standard holonomic system where the canonical equations of (4-94) apply, we see that

$$\dot{H} = \sum_{i=1}^{n} \left(\frac{\partial H}{\partial q_i} \frac{\partial H}{\partial p_i} - \frac{\partial H}{\partial p_i} \frac{\partial H}{\partial q_i} \right) + \frac{\partial H}{\partial t} = \frac{\partial H}{\partial t} \tag{4-116}$$

Thus the partial derivative of H with respect to time is equal to its total derivative. It is apparent from Eq. (4-95) that if L does not contain t explicitly, neither does H. Hence it follows that for a *conservative holonomic system*, the Hamiltonian function $H(q, p)$ has a constant value.

This result can be extended easily to *nonholonomic conservative systems*. Hamilton's equations in this case are

$$\dot{q}_i = \frac{\partial H}{\partial p_i}, \qquad \dot{p}_i = -\frac{\partial H}{\partial q_i} + \sum_{j=1}^{m} \lambda_j a_{ji} \tag{4-117}$$

We recall from Eq. (2-141) that the constraints for a conservative system can be expressed in the form

$$\sum_{i=1}^{n} a_{ji} \dot{q}_i = 0 \qquad (j = 1, 2, \ldots, m)$$

Hence we find once again that the Hamiltonian is constant, that is,

$$\dot{H} = \sum_{i=1}^{n} \left(\frac{\partial H}{\partial q_i} \frac{\partial H}{\partial p_i} - \frac{\partial H}{\partial p_i} \frac{\partial H}{\partial q_i} \right) + \sum_{i=1}^{n} \sum_{j=1}^{n} \lambda_j a_{ji} \dot{q}_i = 0 \tag{4-118}$$

We conclude that every conservative system, holonomic or nonholonomic, has a constant Hamiltonian function. Furthermore, we see from Eq. (4-103) that it is equal to the Jacobi integral, that is,

$$H = T_2 - T_0 + V = h \tag{4-119}$$

Finally, let us consider a *natural system*. In this case $T_0 = 0$ and Eq. (4-119) reduces to

$$H = T + V = h \tag{4-120}$$

Thus H is constant in value and is equal to the total energy.

Example 4-4. Given a mass-spring system consisting of a mass m and a linear spring of stiffness k, as shown in Fig. 4-4. Find the equations of motion using the Hamiltonian procedure. Assume that the displacement x is measured from the unstressed position of the spring.

First let us find the kinetic and potential energies in the usual form. We obtain

$$T = \tfrac{1}{2} m \dot{x}^2, \qquad V = \tfrac{1}{2} k x^2$$

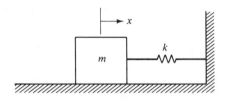

Fig. 4-4. A mass-spring system.

which results in

$$L = T - V = \tfrac{1}{2}m\dot{x}^2 - \tfrac{1}{2}kx^2 \tag{4-121}$$

The linear momentum is

$$p = \frac{\partial L}{\partial \dot{x}} = m\dot{x} \tag{4-122}$$

Hence we can write the kinetic energy in the form $T = p^2/2m$, and the Hamiltonian function is found to be

$$H(x, p) = p\dot{x} - L = \frac{p^2}{2m} + \frac{1}{2}kx^2 \tag{4-123}$$

Since this is a natural system, the Hamiltonian H is equal to the total energy $T + V$ and is constant.

To obtain the equations of motion, we apply Eq. (4-94) to the Hamiltonian H with the following result:

$$\dot{x} = \frac{\partial H}{\partial p} = \frac{p}{m}, \qquad \dot{p} = -\frac{\partial H}{\partial x} = -kx \tag{4-124}$$

These two first-order equations are equivalent to the single second-order equation

$$m\ddot{x} + kx = 0 \tag{4-125}$$

which is the familiar equation of motion that can be obtained by using Newton's law of motion or Lagrange's equation.

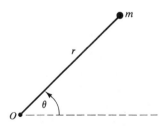

Fig. 4-5. The Kepler problem, using polar coordinates.

Example 4-5. A particle of mass m is attracted to a fixed point O by an inverse-square force, that is,

$$F_r = -\frac{\mu m}{r^2} \tag{4-126}$$

where μ is the gravitational coefficient. Using the plane polar coordinates (r, θ) to describe the position of the particle (Fig. 4-5), find the equations of motion. (See Example 2-7.)

First we obtain the Lagrangian function

$$L = T - V = \tfrac{1}{2}m(\dot{r}^2 + r^2\dot{\theta}^2) + \frac{\mu m}{r} \tag{4-127}$$

and find that the generalized momenta are

$$p_r = \frac{\partial L}{\partial \dot{r}} = m\dot{r}, \qquad p_\theta = \frac{\partial L}{\partial \dot{\theta}} = mr^2\dot{\theta} \tag{4-128}$$

The Hamiltonian function can be obtained most easily by noting that we are studying a natural system for which H is equal to the total energy. Hence we know immediately that

$$H = T + V = \frac{p_r^2}{2m} + \frac{p_\theta^2}{2mr^2} - \frac{\mu m}{r} \tag{4-129}$$

Using Hamilton's equations we now obtain

$$\dot{r} = \frac{\partial H}{\partial p_r} = \frac{p_r}{m}, \qquad \dot{p}_r = -\frac{\partial H}{\partial r} = \frac{p_\theta^2}{mr^3} - \frac{\mu m}{r^2} \tag{4-130}$$

and

$$\dot{\theta} = \frac{\partial H}{\partial p_\theta} = \frac{p_\theta}{mr^2}, \qquad \dot{p}_\theta = -\frac{\partial H}{\partial \theta} = 0 \tag{4-131}$$

These are the equations of motion. From the last equation we see that the angular momentum is

$$p_\theta = \beta \tag{4-132}$$

where β is a constant. Then Eq. (4-130) reduces to a single second-order equation

$$m\ddot{r} - \frac{\beta^2}{mr^3} + \frac{\mu m}{r^2} = 0 \tag{4-133}$$

An advantage of the Hamiltonian formulation can be illustrated by this example. Here θ is an *ignorable coordinate* and one can immediately consider the conjugate momentum p_θ as a constant β in writing the Hamiltonian. Thus we have

$$H = \frac{p_r^2}{2m} + \frac{\beta^2}{2mr^2} - \frac{\mu m}{r} \tag{4-134}$$

which is a function of r and p_r only, implying that only a single degree of freedom remains. By contrast, the Lagrangian approach to the same problem does not result in a reduction of the number of degrees of freedom because $\dot{\theta}$ is not eliminated as a variable in L.

If, on the other hand, we make the substitution $\dot{\theta} = \beta/mr^2$ in the Lagrangian function in an attempt to eliminate $\dot{\theta}$, incorrect equations of motion result. This can be seen by comparing the resulting incorrect Lagrangian with the correct function which is obtained by the standard Routhian method. (See Sec. 2-3).

Example 4-6. Another possible application of Hamilton's equations is in the description of a system having a velocity-dependent potential $U(q, \dot{q}, t)$. For example, let us find the equations of motion for a charged particle in an electromagnetic field.

We recall from Eq. (3-186) that the Lagrangian function is

$$L = T - U = \tfrac{1}{2}mv^2 - e(\phi - \mathbf{v \cdot A}) \tag{4-135}$$

and therefore the momentum is

$$\mathbf{p} = m\mathbf{v} + e\mathbf{A} \tag{4-136}$$

where e is the charge on the particle, $\mathbf{v}$ is its velocity, ϕ is the scalar potential, and $\mathbf{A}$ is the vector potential.

The Hamiltonian H is given in this case by

$$H = \mathbf{p \cdot v} - L = \tfrac{1}{2}mv^2 + e\phi \tag{4-137}$$

which we note is equal to the total energy. Expressing H as a function of the q's and p's, we have

$$H = \frac{1}{2m}(\mathbf{p} - e\mathbf{A})^2 + e\phi \tag{4-138}$$

where ϕ and $\mathbf{A}$ are functions of position and time. In terms of Cartesian coordinates, we find from the first of Hamilton's equations that

$$
\begin{aligned}
\dot{x} &= \frac{\partial H}{\partial p_x} = \frac{1}{m}(p_x - eA_x) \\
\dot{y} &= \frac{\partial H}{\partial p_y} = \frac{1}{m}(p_y - eA_y) \\
\dot{z} &= \frac{\partial H}{\partial p_z} = \frac{1}{m}(p_z - eA_z)
\end{aligned}
\tag{4-139}
$$

or

$$\mathbf{v} = \frac{1}{m}(\mathbf{p} - e\mathbf{A}) \tag{4-140}$$

Similarly, the second canonical equation yields

$$
\begin{aligned}
\dot{p}_x &= -e\frac{\partial \phi}{\partial x} + \frac{e}{m}\left[(p_x - eA_x)\frac{\partial A_x}{\partial x} + (p_y - eA_y)\frac{\partial A_y}{\partial x} + (p_z - eA_z)\frac{\partial A_z}{\partial x}\right] \\
\dot{p}_y &= -e\frac{\partial \phi}{\partial y} + \frac{e}{m}\left[(p_x - eA_x)\frac{\partial A_x}{\partial y} + (p_y - eA_y)\frac{\partial A_y}{\partial y} + (p_z - eA_z)\frac{\partial A_z}{\partial y}\right] \\
\dot{p}_z &= -e\frac{\partial \phi}{\partial z} + \frac{e}{m}\left[(p_x - eA_x)\frac{\partial A_x}{\partial z} + (p_y - eA_y)\frac{\partial A_y}{\partial z} + (p_z - eA_z)\frac{\partial A_z}{\partial z}\right]
\end{aligned}
\tag{4-141}
$$

which, with the aid of Eq. (4-140), can be written in the compact form

$$\dot{\mathbf{p}} = -e\nabla\phi + e\nabla(\mathbf{v \cdot A}) \tag{4-142}$$

In summary, Eqs. (4-139) and (4-141) are the scalar equations of motion of a charged particle in an electromagnetic field, and Eqs. (4-140) and (4-142) are the corresponding vector equations.

The Legendre Transformation. Another method of obtaining Hamilton's equations from Lagrange's equations is by means of a *Legendre transformation*. Let us consider this transformation, first in general terms, and then as it applies to the equations of motion.

Suppose a function $F(u_1, \ldots, u_n, w_1, \ldots, w_m, t)$ is associated with a given system. Let us call the u's *active variables*, while the w's and t are *passive variables*. We wish to describe the system in terms of a new set of active variables defined by

$$v_i = \frac{\partial F}{\partial u_i} \qquad (i = 1, 2, \ldots, n) \tag{4-143}$$

Let us assume that the $n \times n$ Hessian determinant for the transformation is nonzero, that is,

$$\left| \frac{\partial^2 F}{\partial u_i \, \partial u_j} \right| = \left| \frac{\partial v_i}{\partial u_j} \right| \neq 0 \tag{4-144}$$

This ensures that the n equations having the form of Eq. (4-143) can be solved for the u's as functions of the v's.

Now let us define a new function $G(v_1, \ldots, v_n, w_1, \ldots, w_m, t)$ according to the equation

$$G = \sum_{i=1}^{n} u_i v_i - F \tag{4-145}$$

It is important to note that the u's are expressed in terms of the v's, w's, and t whenever they appear in Eq. (4-145).

Consider a variation δG associated with arbitrary variations of the *active* variables. We see that

$$\delta G = \sum_{i=1}^{n} \frac{\partial G}{\partial v_i} \delta v_i = \sum_{i=1}^{n} \left(u_i \, \delta v_i + v_i \, \delta u_i - \frac{\partial F}{\partial u_i} \delta u_i \right)$$

or

$$\sum_{i=1}^{n} \frac{\partial G}{\partial v_i} \delta v_i = \sum_{i=1}^{n} \left[u_i \, \delta v_i + \left(v_i - \frac{\partial F}{\partial u_i} \right) \delta u_i \right] \tag{4-146}$$

Since the δu's and δv's are assumed to be arbitrary, the corresponding coefficients must be equal on each side of the equation. Hence the coefficient of δu_i is zero, in agreement with Eq. (4-143). Furthermore, comparing the coefficients of δv_i, we obtain

$$u_i = \frac{\partial G}{\partial v_i} \tag{4-147}$$

Note that Eqs. (4-143) and (4-147) are similar in form and, in fact, are duals in the sense that interchanging the u_i and v_i, and also F and G, results in transforming each equation into the other. Therefore, if we perform two Legendre transformations in sequence, using the u's and v's as active variables, the result is a return to the original function of the original set of variables.

Now let us consider a *holonomic system* which is described by the standard form of Lagrange's equation which we can write as

$$\dot{p}_i = \frac{\partial L}{\partial q_i} \qquad (i = 1, 2, \ldots, n) \tag{4-148}$$

where

$$p_i = \frac{\partial L}{\partial \dot{q}_i} \tag{4-149}$$

We wish to use a Legendre transformation to replace the $\dot{q}$'s by p's as active variables. Here we let the $\dot{q}$'s correspond to the u's, the q's correspond to w's, and the Lagrangian $L(\dot{q}, q, t)$ corresponds to $F(u, w, t)$. Furthermore the Hessian determinant is nonzero, that is,

$$\left| \frac{\partial^2 L}{\partial \dot{q}_i \, \partial \dot{q}_j} \right| = |m_{ij}| \neq 0 \tag{4-150}$$

since the inertia matrix is positive definite. Corresponding to the new function $G(v, w, t)$ we define the *Hamiltonian function*

$$H(p, q, t) = \sum_{i=1}^{n} p_i \dot{q}_i - L \tag{4-151}$$

in accordance with Eq. (4-145). Then, corresponding to Eq. (4-147), we obtain

$$\dot{q}_i = \frac{\partial H}{\partial p_i} \tag{4-152}$$

If we vary the q's (the passive variables) in Eq. (4-151) and equate corresponding coefficients, we obtain

$$\frac{\partial H(q, p, t)}{\partial q_i} = -\frac{\partial L(q, \dot{q}, t)}{\partial q_i} \tag{4-153}$$

in agreement with a similar comparison of Eqs. (4-91) and (4-92). Note that the partial derivatives imply that different quantities are held constant on the two sides of the equation. This result also confirms that if a certain coordinate is missing from L, it is also missing from H; that is, the coordinate is ignorable using either the Lagrangian or Hamiltonian approach.

Finally, we obtain from Eqs. (4-148) and (4-153) that

$$\dot{p}_i = -\frac{\partial H}{\partial q_i} \tag{4-154}$$

Thus the use of the Legendre transformation and Lagrange's equations yields the canonical equations of Hamilton, as given in Eqs. (4-152) and (4-154). In a similar fashion, one could use the Legendre transformation to obtain Lagrange's equations from Hamilton's equations.

4-3. OTHER VARIATIONAL PRINCIPLES

Modified Hamilton's Principle. Let us recall first that the usual form of Hamilton's principle, given by Eq. (4-60), applies to holonomic systems which can be described by the standard form of Lagrange's equations. The varied paths are taken in an n-dimensional configuration space and are restricted by the conditions that the δq's conform to the instantaneous constraints, if any, and vanish at the end-points. Furthermore, the $\delta \dot{q}$'s are related to the δq's by the equations

$$\delta \dot{q}_i = \frac{d}{dt} \delta q_i \qquad (i = 1, 2, \ldots, n) \tag{4-155}$$

The $\delta \dot{q}$'s are not zero, in general, at the end-points.

Now let us consider a *holonomic system* having n *independent* q's. In this case, however, we use Eq. (4-151) to write Hamilton's principle in the following *modified form*:

$$\delta \int_{t_0}^{t_1} \left(\sum_{i=1}^{n} p_i \dot{q}_i - H \right) dt = 0 \tag{4-156}$$

where, as usual, we regard H as a function of the q's, p's, and t. Next we carry out the indicated variation and obtain

$$\int_{t_0}^{t_1} \sum_{i=1}^{n} \left(p_i \, \delta \dot{q}_i + \dot{q}_i \, \delta p_i - \frac{\partial H}{\partial q_i} \delta q_i - \frac{\partial H}{\partial p_i} \delta p_i \right) dt = 0 \tag{4-157}$$

Note that the time is held constant during each variation, that is, $\delta t = 0$. If we integrate the term $p_i \, \delta \dot{q}_i \, dt$ by parts, using Eq. (4-155) and the end conditions on δq_i, we find that

$$\int_{t_0}^{t_1} p_i \, \delta \dot{q}_i \, dt = - \int_{t_0}^{t_1} \dot{p}_i \, \delta q_i \, dt \tag{4-158}$$

Hence, from Eqs. (4-157) and (4-158), we obtain

$$\int_{t_0}^{t_1} \sum_{i=1}^{n} \left[\left(\dot{q}_i - \frac{\partial H}{\partial p_i} \right) \delta p_i - \left(\dot{p}_i + \frac{\partial H}{\partial q_i} \right) \delta q_i \right] dt = 0 \tag{4-159}$$

At this point, let us remember that Hamilton's principle assumes varied paths in q-space. Therefore the $\delta \dot{q}$'s or δp's are directly related to the δq's by Eq. (4-155) and $p_i = \partial L / \partial \dot{q}_i$ (or its inverse, $\dot{q}_i = \partial H / \partial p_i$). This apparently implies that the δp's and δq's cannot be chosen independently in Eq. (4-159). We note, however, that the coefficient of δp_i is *identically zero* because of the canonical equation for $\dot{q}_i$, and therefore the value of the integral remains stationary even if the δp's are considered to be arbitrary. So now let us broaden our horizons and consider variations of the path in a *phase space* of $2n$ dimensions, namely, the n q's and n p's. The *modified Hamilton's principle* states that the actual path is such that the integral of Eq. (4-156) is stationary for arbitrary variations of the path in phase space, with the restriction that the δq's vanish at the fixed times t_0 and t_1. The δp's need not be zero at these end-points.

Because of the independence of the δq's in Eq. (4-159), it follows that the individual coefficients must be zero. Hence we obtain the n canonical equations $\dot{p}_i = -\partial H/\partial q_i$ as a consequence. Alternatively, if at the start we assume the validity of the modified form of Hamilton's principle, then Eq. (4-159) leads to the $2n$ canonical equations since both the δq's and δp's are arbitrary.

Principle of Least Action. The development of Hamilton's principle proceeds on the assumption that the variations δq_i of the generalized coordinates are *contemporaneous*; that is, a point $(\mathbf{q} + \delta\mathbf{q}, t)$ on the varied path corresponds to a point $(\mathbf{q}, t)$ on the actual path. Hence the variations are assumed to occur without the passage of time, as in a virtual displacement.

It is interesting, however, to consider a more general type of variation in which the point $(\mathbf{q} + \delta\mathbf{q}, t + \delta t)$ corresponds to $(\mathbf{q}, t)$, as shown in Fig. 4-6.

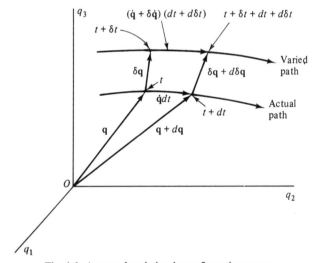

Fig. 4-6. A general variation in configuration space.

Here the varied point occurs, in general, at a *different time* than the corresponding point on the actual path. Note that we use the symbol d to indicate the differential changes in various quantities along an individual path; whereas the symbol δ refers to variations which occur in going from a point on the actual path to a corresponding point on a varied path.

If we consider the small quadrilateral in Fig. 4-6, we see that

$$\dot{\mathbf{q}}\, dt + \delta\mathbf{q} + d\delta\mathbf{q} = \delta\mathbf{q} + (\dot{\mathbf{q}} + \delta\dot{\mathbf{q}})(dt + d\delta t)$$

or, in terms of components,

$$\delta \dot{q}_i = \frac{d}{dt} \delta q_i - \dot{q}_i \frac{d}{dt} \delta t \tag{4-160}$$

where we neglect $\delta \dot{\mathbf{q}} \, (d/dt) \, \delta t$. Comparing this result with Eq. (4-155), we see that the sequence of operations involving variation and differentiation with respect to time can no longer be interchanged.

Now let us define an integral I as before, namely,

$$I = \int_{t_0}^{t_1} L(q, \dot{q}, t) \, dt \tag{4-161}$$

We can apply a general noncontemporaneous variation to this integral and obtain results which reduce to Hamilton's principle or to the principle of least action as particular cases. We obtain, using a procedure analogous to that of Eq. (4-160),

$$\delta I = \int_{t_0}^{t_1} \left(\delta L + L \frac{d}{dt} \delta t \right) dt \tag{4-162}$$

where we note that the interval t_0 to t_1 applies to the actual path and hence is fixed. The varied path, however, can have slightly different starting and stopping times because δt is not necessarily zero, even at the end points. Thus we regard the time t of the actual path as the variable of integration.

If we expand δL and use Eq. (4-160), we obtain

$$\delta I = \int_{t_0}^{t_1} \left[\sum_{i=1}^{n} \left(\frac{\partial L}{\partial q_i} \delta q_i + \frac{\partial L}{\partial \dot{q}_i} \frac{d}{dt} \delta q_i - \frac{\partial L}{\partial \dot{q}_i} \dot{q}_i \frac{d}{dt} \delta t \right) + \frac{\partial L}{\partial t} \delta t + L \frac{d}{dt} \delta t \right] dt \tag{4-163}$$

Then, noting that

$$\frac{d}{dt} \left(\sum_{i=1}^{n} \frac{\partial L}{\partial \dot{q}_i} \delta q_i \right) = \sum_{i=1}^{n} \frac{d}{dt} \left(\frac{\partial L}{\partial \dot{q}_i} \right) \delta q_i + \sum_{i=1}^{n} \frac{\partial L}{\partial \dot{q}_i} \frac{d}{dt} \delta q_i \tag{4-164}$$

we obtain

$$\delta I = \int_{t_0}^{t_1} \sum_{i=1}^{n} \frac{d}{dt} \left(\frac{\partial L}{\partial \dot{q}_i} \delta q_i \right) dt + \int_{t_0}^{t_1} \left[\frac{\partial L}{\partial t} \delta t - \left(\sum_{i=1}^{n} \frac{\partial L}{\partial \dot{q}_i} \dot{q}_i - L \right) \frac{d}{dt} \delta t \right] dt$$
$$- \int_{t_0}^{t_1} \sum_{i=1}^{n} \left[\frac{d}{dt} \left(\frac{\partial L}{\partial \dot{q}_i} \right) - \frac{\partial L}{\partial q_i} \right] \delta q_i \, dt \tag{4-165}$$

This is the general result which we shall now apply to particular systems. Let us limit the discussion to *holonomic systems* in order that the δq's corresponding to any virtual displacement will also represent a geometrically possible varied path.

Suppose we consider how Eq. (4-165) reduces to Hamilton's principle. First, we assume that the variations are *contemporaneous*, that is, δt is identically zero. This causes the second integral to vanish. Furthermore, let us assume that all the applied forces are derivable from a potential function

$V(q, t)$, and the δq's form a virtual displacement consistent with the instantaneous constraints. Then

$$\sum_{i=1}^{n} a_{ji}\, \delta q_i = 0 \qquad (j = 1, 2, \ldots, m) \tag{4-166}$$

and, in accordance with d'Alembert's principle as given in Eq. (2-22), we see that the third integral vanishes. Finally, let us assume fixed end-points in configuration space. Then Eq. (4-165) reduces to

$$\delta I = \left[\sum_{i=1}^{n} \frac{\partial L}{\partial \dot{q}_i}\, \delta q_i \right]_{t_0}^{t_1} = 0 \tag{4-167}$$

since the δq's are zero at t_0 and t_1. Hence we obtain *Hamilton's principle*, or $\delta I = 0$. Note once again that we do not require the system to be conservative.

Now let us derive the principle of least action with the aid of Eq. (4-165). We consider a *conservative holonomic system* in this case and assume that the δq's are consistent with the constraints. If all the applied forces can be obtained from a potential function $V(q)$ we find, as before, that the third integral vanishes. Furthermore, let us assume that all the varied paths have fixed end-points in q-space, with the result that the first integral vanishes in accordance with the last equality of Eq. (4-167).

The varied paths allowed by the principle of least action are restricted to those having an energy integral

$$\sum_{i=1}^{n} \frac{\partial L}{\partial \dot{q}_i} \dot{q}_i - L = h \tag{4-168}$$

where h is a constant for any varied path and is, in fact, equal to the total energy for the case of a natural system. In general, the time required to traverse a varied path is not the same as for the actual path since the $\dot{q}$'s are determined by energy considerations; hence we assume a *noncontemporaneous* variation in which $\delta t \neq 0$. Noting that $\partial L/\partial t = 0$ for a conservative system, we see that

$$\delta I = - \int_{t_0}^{t_1} h\, \frac{d}{dt}\,(\delta t)\, dt = -h(\delta t_1 - \delta t_0) \tag{4-169}$$

Let us define the *action* as the integral

$$A = \int_{t_0}^{t_1} \sum_{i=1}^{n} \frac{\partial L}{\partial \dot{q}_i} \dot{q}_i\, dt = \int_{t_0}^{t_1} \sum_{i=1}^{n} p_i \dot{q}_i\, dt \tag{4-170}$$

Then, for the assumed path variations, we obtain

$$\delta A = \delta \int_{t_0}^{t_1} (L + h)\, dt = \delta I + \delta h(t_1 - t_0) + h(\delta t_1 - \delta t_0) \tag{4-171}$$

Combining Eqs. (4-169) and (4-171), we have

$$\delta A = \delta h(t_1 - t_0) \tag{4-172}$$

Finally, let us restrict the varied paths to those for which h has the same value as the actual path. Then $\delta h = 0$ and it follows that

$$\delta A = \delta \int_{t_0}^{t_1} \sum_{i=1}^{n} p_i \dot{q}_i \, dt = 0 \tag{4-173}$$

This is the *principle of least action: the actual path of a conservative holonomic system is such that the action is stationary with respect to varied paths having the same energy integral h and the same end-points in q-space.*

In order to better understand the meaning of the action integral, notice that

$$\sum_{i=1}^{n} p_i \dot{q}_i = 2T_2 + T_1 \tag{4-174}$$

For a *natural system*, we see that $T_1 = 0$ and the principle of least action becomes

$$\delta \int_{t_0}^{t_1} 2T \, dt = 0 \tag{4-175}$$

In this case, the varied paths have the same total energy $T + V = h$. If we make the additional assumption that the kinetic energy is constant, then the actual path is that path which requires the least time.

Another form of the least action principle is that of Jacobi. Suppose we consider a *natural system* and write Eq. (4-175) in the form

$$\delta \int_{t_0}^{t_1} 2\sqrt{T(h - V)} \, dt = 0 \tag{4-176}$$

Now define a differential ds in accordance with the equation

$$ds^2 = \sum_{i=1}^{n} \sum_{j=1}^{n} m_{ij} \dot{q}_i \dot{q}_j \, dt^2 = \sum_{i=1}^{n} \sum_{j=1}^{n} m_{ij} \, dq_i \, dq_j \tag{4-177}$$

We see that

$$ds = \sqrt{2T} \, dt \tag{4-178}$$

and therefore we obtain

$$\delta A = \delta \int \sqrt{2(h - V)} \, ds = 0 \tag{4-179}$$

which is *Jacobi's form of the principle of least action.*

If we consider ds as an element of arc length, we note that the varied paths are not, in general, of the same total length, even though the end-points are fixed in q-space. Therefore, in order to obtain fixed limits on the integral, let us express ds in terms of another parameter such as one of the q's. Since V and the m_{ij}'s are functions of the q's only, we obtain an ordinary problem in the calculus of variations for which the Euler-Lagrange equations apply. The solution of these equations yields the path of the system in configuration space without expressing its motion as a function of time.

The result of Eq. (4-179) is similar to Fermat's principle of optics which states that the optical path between two given points is such that the time required for light to travel between these points is minimized, assuming that the speed of light is inversely proportional to the refractive index μ. Thus, noting that the time dt required for the light to travel a distance ds is given by $dt = \mu \, ds$, we see that $\sqrt{2(h - V)}$ corresponds to μ, and the action A corresponds to the total time required for the light to follow a given path.

Example 4-7. Let us use the Jacobi form of the principle of least action to obtain the orbit for the Kepler problem discussed in Example 4-5.

First we note that this is a natural system having a constant total energy

$$\frac{1}{2} m(\dot{r}^2 + r^2 \dot{\theta}^2) - \frac{\mu m}{r} = h \tag{4-180}$$

Applying the Jacobi form of the principle of least action, we obtain

$$\delta \int \sqrt{2\left(h + \frac{\mu m}{r}\right)} \, ds = 0 \tag{4-181}$$

where, in accordance with Eq. (4-177),

$$ds^2 = m(dr^2 + r^2 \, d\theta^2) \tag{4-182}$$

Now let us choose θ as the independent variable and use the notation $r' \equiv dr/d\theta$. Then Eq. (4-181) becomes

$$\delta \int_{\theta_0}^{\theta_1} \sqrt{2m\left(h + \frac{\mu m}{r}\right)(r^2 + r'^2)} \, d\theta = 0 \tag{4-183}$$

where the end-points in r and θ are fixed. This is of the form

$$\delta \int_{\theta_0}^{\theta_1} f(r, r') \, d\theta = 0 \tag{4-184}$$

where

$$f(r, r') = \sqrt{2m\left(h + \frac{\mu m}{r}\right)(r^2 + r'^2)} \tag{4-185}$$

The integral in Eq. (4-183) is therefore of the standard form of Eq. (4-19); hence the Euler-Lagrange equation applies, namely,

$$\frac{d}{d\theta}\left(\frac{\partial f}{\partial r'}\right) - \frac{\partial f}{\partial r} = 0 \tag{4-186}$$

Rather than write this differential equation in detail, let us look immediately for a first integral of the motion. We note that f is not an explicit function of the independent variable θ. Therefore, corresponding to the energy integral of Eq. (2-147), we obtain

$$\frac{\partial f}{\partial r'} r' - f = C \tag{4-187}$$

From Eq. (4-185) we see that

$$\frac{\partial f}{\partial r'} = \sqrt{\frac{2m(h + \mu m/r)}{r^2 + r'^2}} \, r' \tag{4-188}$$

which leads to

$$-\sqrt{\frac{2m(h+\mu m/r)}{r^2 + r'^2}}\, r^2 = C \qquad (4\text{-}189)$$

Using the substitution $\dot{r} = r'\dot{\theta}$, one can show that this constant of the motion is equal to the negative of the angular momentum, that is,

$$-mr^2\dot{\theta} = C \qquad (4\text{-}190)$$

In order to obtain the equation of the orbit, we first rearrange Eq. (4-189) in the form

$$\left(\frac{dr}{d\theta}\right)^2 = \frac{2mr^2}{C^2}\left(hr^2 + \mu mr - \frac{C^2}{2m}\right) \qquad (4\text{-}191)$$

Now choose the reference direction for measuring θ at the position of minimum r; that is, let $r = r_0 = r_{\min}$ when $\theta = \theta_0 = 0$. Then from Eq. (4-191) we obtain

$$\theta = \frac{C}{\sqrt{2m}} \int_{r_0}^{r} \frac{dr}{r\sqrt{hr^2 + \mu mr - C^2/2m}} \qquad (4\text{-}192)$$

This integral can be evaluated to yield

$$\theta = \sin^{-1}\left(\frac{\mu mr - C^2/m}{r\sqrt{\mu^2 m^2 + 2hC^2/m}}\right) - \frac{\pi}{2} \qquad (4\text{-}193)$$

In obtaining the value $\pi/2$ at the lower limit r_0, we note that

$$h = \frac{1}{2}\, mr_0^2\dot{\theta}_0^2 - \frac{\mu m}{r_0} \qquad (4\text{-}194)$$

and use the value of C from Eq. (4-190) to show that the argument of the inverse sine is equal to one.

Next we solve for r as a function of θ and obtain

$$r = \frac{C^2/\mu m^2}{1 + \sqrt{1 + 2hC^2/\mu^2 m^3}\,\cos\theta} \qquad (4\text{-}195)$$

This is the equation of a conic section and represents an orbit having a total energy h, an angular momentum $-C$, and an eccentricity

$$e = \sqrt{1 + 2hC^2/\mu^2 m^3} \qquad (4\text{-}196)$$

4-4. PHASE SPACE

Trajectories. In the previous discussion of the Lagrangian approach to dynamics we saw that a *holonomic system* can be described by n second-order ordinary differential equations, where n is the number of degrees of freedom. For specified q's and $\dot{q}$'s at some initial time t_0, the solution of the differential equations consists of n functions $q_i(t)$. It is convenient to think of this solution as a path or *trajectory* traced by a moving point in an n-dimensional *con-*

figuration space. We observe that a general point in the configuration space of a given mechanical system has many possible trajectories passing through it because the direction of the velocity $\dot{\mathbf{q}}$ is arbitrary at each point.

The canonical Hamiltonian formulation of the equations of motion, on the other hand, consists of a set of $2n$ first-order equations giving the $\dot{q}$'s and $\dot{p}$'s as functions of the q's, p's, and t. Even though each coordinate q_i is actually paired with the corresponding momentum p_i, it turns out that, because of the symmetry of the equations, the q's and p's are often considered as the components of a single vector $\mathbf{x}$. Thus the equations of motion for a standard holonomic system can be expressed in the form given previously in Eq. (4-96), namely,

$$\dot{\mathbf{x}} = \mathbf{X}(\mathbf{x}, t) \tag{4-197}$$

where $\mathbf{x}$ is a $2n$-dimensional vector consisting of the n q's and n p's. We might, for example, consider that the q's are represented by $x_1, x_2, \ldots, x_n$ and the p's by $x_{n+1}, x_{n+2}, \ldots, x_{2n}$. This $2n$-dimensional x-space is known as *phase space*.

A point in phase space specifies not only the configuration of the system, but also its state of motion as represented by the p's. If the n q's and n p's are known at some time t_0, these $2n$ parameters constitute the required $2n$ initial conditions for determining the further motion of the system, and therefore the trajectory in phase space.

Phase space is particularly convenient in representing the possible motions of a *conservative system*. In this case the Hamiltonian is not an explicit function of time and the equations of motion have the form

$$\dot{\mathbf{x}} = \mathbf{X}(\mathbf{x}) \tag{4-198}$$

A holonomic system described by equations of this form is called *autonomous* in contrast with the *nonautonomous* system of Eq. (4-197).

It can be seen from Eq. (4-198) that the direction of the tangent to a phase space trajectory corresponding to a conservative system is a function of the position $\mathbf{x}$ only. Hence there is only one trajectory through each point, and every trajectory is fixed. Furthermore, the whole of phase space with its trajectories represents the totality of all possible motions of the given system. Now let us recall that the Hamiltonian function of a conservative system is constant, that is,

$$H(p, q) = \sum_{i=1}^{n} p_i \dot{q}_i - L = h \tag{4-199}$$

where the constant h is usually evaluated from the initial conditions. This equation represents a surface in phase space and, of course, the corresponding trajectory must lie entirely on this surface. Other trajectories on the same surface are those with the same Jacobi integral and, frequently, the same total energy.

A point in phase space at which all the $\dot{q}$'s and $\dot{p}$'s are zero is known as an *equilibrium point* or *singular point*. Since $\dot{\mathbf{x}}$ remains zero at this point, the corresponding trajectory consists of the singular point only. Furthermore, if we make the usual assumption that the Hamiltonian function has at least two continuous partial derivatives in the q's and p's, it can be shown that there are no other trajectories which include the given singular point. This follows from the fact that, on the basis of Hamilton's equations, the magnitude of $\|\dot{\mathbf{x}}\|$ must be of order ϵ or smaller in the infinitesimal region of radius $\epsilon = \|\mathbf{x} - \mathbf{x}_0\|$ near a singular point at $\mathbf{x}_0$. As a consequence, an infinite time is required for an autonomous system to attain a condition of equilibrium from a nonequilibrium state, or vice versa. In other words, a trajectory starting at a singular point will never leave it in finite time; nor can a trajectory starting at an ordinary point arrive at a singular point in a finite time.

An alternate form for Eq. (4-197) or (4-198) is

$$\frac{dx_1}{X_1} = \frac{dx_2}{X_2} = \cdots = \frac{dx_{2n}}{X_{2n}} = dt \qquad (4\text{-}200)$$

For autonomous systems the X's are not explicit functions of time. Hence the trajectories in phase space can be obtained by omitting the last equality in Eq. (4-200). This will give their form in phase space without considering the time dependence. Under these conditions an ordinary point in phase space will have only one trajectory passing through it, but more than one trajectory may apparently pass through a singular point in some cases. As we have seen, however, a trajectory can approach arbitrarily close to a singular point only as t approaches infinity. Hence, with the given smoothness assumptions on the X's, we know that the trajectory cannot actually pass through, or even reach, a singular point in a finite time, although more than one trajectory may approach the given singular point.

Now let us consider a *nonconservative* holonomic system which is represented by Eq. (4-197). It is possible for the phase space trajectories of this nonautonomous system to intersect because the direction of $\dot{\mathbf{x}}$ at a given point $\mathbf{x}$ will, in general, change with time. If we consider, however, the trajectories in a $(2n + 1)$-dimensional space consisting of the q's, p's, and t, then the trajectories do not intersect. Once again they can be obtained from Eq. (4-200) for any set of initial conditions.

Extended Phase Space. Another approach, which applies particularly to nonconservative systems, is to consider the time t as an additional dependent variable; that is, let

$$q_{n+1} = t \qquad (4\text{-}201)$$

Now choose a parameter τ as the new independent variable. The solution of the corresponding differential equations for this system yields trajectories in a (q, p)-space of $(2n + 2)$ dimensions, consisting of the $(n + 1)$ q's and the corresponding $(n + 1)$ p's. Lanczos† calls this the *extended phase space*.

---

†C. Lanczos, *The Variational Principles of Mechanics*, (Toronto: University of Toronto Press, 1949), pp. 185–92.

An advantage of adding another degree of freedom to the analysis is that the system now resembles a *conservative* system because its Hamiltonian is not an explicit function of τ. Consequently, the trajectories are fixed in extended phase space and do not intersect.

Now let us consider the nature of the generalized momentum corresponding to t, that is, p_{n+1}. In order to obtain this momentum component in extended phase space we need to obtain a new Lagrangian function $L_1(q_1, \ldots, q_{n+1}, q'_1, \ldots, q'_{n+1})$, where $q'_i \equiv dq_i/d\tau$. Since either the old or the new Lagrangian function can be used to represent the given system, and since Hamilton's principle applies in either case, we can choose

$$ L \, dt = L_1 \, d\tau \tag{4-202} $$

Then, using Eq. (4-151) and noting that $dt = t' \, d\tau$, we find that

$$ L_1 \, d\tau = \left(\sum_{i=1}^{n} \frac{p_i q'_i}{t'} - H \right) t' \, d\tau $$

or

$$ L_1 = \sum_{i=1}^{n} p_i q'_i - Ht' \tag{4-203} $$

where H is the original Hamiltonian function. If we are careful to express L_1 as a function of the new variables, it can be shown that

$$ \frac{\partial}{\partial t'} \left(\sum_{i=1}^{n} p_i q'_i \right) - \frac{\partial H}{\partial t'} t' = 0 \tag{4-204} $$

and we obtain

$$ p_{n+1} = \frac{\partial L_1}{\partial t'} = -H \tag{4-205} $$

We conclude that $-H$, which has the units of energy, is the generalized momentum conjugate to the time t. For the particular case of a conservative system, we see that t is an ignorable coordinate since the Lagrangian function does not contain t explicitly. As a consequence, it follows that p_{n+1} (or H) is a constant, in agreement with our earlier results on the existence of an energy integral for conservative systems.

Liouville's Theorem. Let us consider a holonomic system described by n independent q's. Suppose we follow a group of phase points as they describe trajectories in a phase space of $2n$ dimensions. We can think of the points within a small volume element $dV = dq_1 \ldots dq_n \, dp_1 \ldots dp_n$ as constituting the moving particles of a fluid known as the *phase fluid*. The *phase velocity* **v** of a fluid particle is given in terms of its $2n$ components $(\dot{q}_i, \dot{p}_i)$ which can be expressed as a function of the q's, p's, and t by the canonical equations

$$ \dot{q}_i = \frac{\partial H}{\partial p_i}, \qquad \dot{p}_i = -\frac{\partial H}{\partial q_i} \tag{4-206} $$

As a given volume element of the phase fluid moves, it will, in general, change its shape, but neighboring particles will remain close to each other. Furthermore, it can be shown that the volume of each fluid element is constant during this motion. To see this, let us evaluate the divergence of the phase velocity $\mathbf{v}$ in $2n$-space. We obtain

$$\mathbf{V}\cdot\mathbf{v} = \sum_{i=1}^{n}\left(\frac{\partial \dot{q}_i}{\partial q_i} + \frac{\partial \dot{p}_i}{\partial p_i}\right) \tag{4-207}$$

which, upon substituting from Eq. (4-206), reduces to

$$\mathbf{V}\cdot\mathbf{v} = \sum_{i=1}^{n}\left(\frac{\partial^2 H}{\partial q_i\,\partial p_i} - \frac{\partial^2 H}{\partial p_i\,\partial q_i}\right) = 0 \tag{4-208}$$

A geometrical interpretation of this result is that the phase fluid is *incompressible*. This is *Liouville's theorem*.

Another statement of Liouville's theorem can be obtained by considering the equations

$$\begin{aligned} q_i &= q_i(q_{10},\ldots,q_{n0},p_{10},\ldots,p_{n0},t) \\ p_i &= p_i(q_{10},\ldots,q_{n0},p_{10},\ldots,p_{n0},t) \end{aligned} \qquad (i=1,2,\ldots,n) \tag{4-209}$$

which represent a trajectory in phase space for arbitrary initial conditions. We can also think of these equations as a coordinate transformation in $2n$-space for any given t. If the Jacobian of the transformation is nonzero, the transformation is one-to-one, implying the uniqueness of the solutions. For the case of an incompressible phase fluid, however, the Jacobian has the value $+1$, that is,

$$\frac{\partial(q_1,\ldots,q_n,p_1,\ldots,p_n)}{\partial(q_{10},\ldots,q_{n0},p_{10},\ldots,p_{n0})} = 1 \tag{4-210}$$

This alternate statement of Liouville's theorem clearly meets the limiting condition that as t approaches the initial time, the transformation equations must approach the identity transformation. It also implies that the volume of an infinitesimal element of phase space is conserved by the transformation.

The concept of considering the evolution of a dynamical system in terms of transformations in phase space is a useful one, particularly when one considers infinitesimal transformations for which the changes in the q's and p's, as well as the time interval are infinitesimal. We shall discuss this approach further in Chapter 6.

REFERENCES

1. LANCZOS, C., *The Variational Principles of Mechanics*. Toronto: University of Toronto Press, 1949. An enthusiastic introduction is presented to the variational approach to dynamics. This is a very readable treatment of Hamiltonian theory.

2. PARS, L. A., *A Treatise on Analytical Dynamics*. London: Heinemann, 1965. An excellent reference. The discussion of Hamilton's principle and the other variational principles is particularly well-done.

3. MARION, J. B., *Classical Dynamics of Particles and Systems*, second edition. New York: Academic Press, 1970. This text gives an excellent discussion of Hamilton's principle and its relation to Lagrange's and Hamilton's equations.

4. RUND, H., *The Hamilton-Jacobi Theory in the Calculus of Variations*. London: D. Van Nostrand Company, Ltd., 1966. Gives a good discussion of Hamilton's principle and nonholonomic systems from a geometric point of view.

PROBLEMS

4-1. Consider the problem of the top which is discussed in Example 3-5. Use the Eulerian angles as generalized coordinates and obtain the Hamiltonian function. Find Hamilton's canonical equations for this system.

4-2. Consider the vertical motion of a particle of mass m in a uniform gravitational field. Assuming a damping force of magnitude cv^2, where v is the velocity and c is a constant, obtain Hamilton's equations of motion.

4-3. A particle of mass m can slide without friction on the inside of a small tube bent in the form of a circle of radius r. The tube can rotate freely about a vertical axis

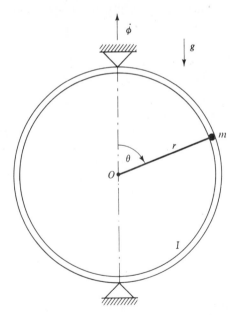

Fig. P4-3.

and has a moment of inertia I about this axis. Obtain Hamilton's equations of motion for this system. Now assume the initial conditions $\theta(0) = \pi/2$, $\dot{\theta}(0) = 0$, $\dot{\phi}(0) = 2\sqrt{g/r}$ and let $I = mr^2$. Solve for the maximum value of ϕ in the ensuing motion.

4-4. A particle of mass m moves without friction on a horizontal plane. It is subject to the rheonomic constraint

$$\sin t^2 \, dx - \cos t^2 \, dy = 0$$

where the Cartesian coordinates (x, y) specify its position. Obtain Hamilton's equations of motion. Assuming the initial conditions $p_x(0) = mv_0, p_y(0) = 0$, solve for p_x, p_y, and the Lagrange multiplier λ as functions of time.

4-5. Derive Hamilton's canonical equations by starting with the modified Hamilton's principle and applying the Euler-Lagrange equations in a phase space of $2n$ dimensions.

4-6. Consider the variational problem

$$\delta \int_{x_0}^{x_1} f(y_1, y_2, \ldots, y_n, y'_1, y'_2, \ldots, y'_n, x) \, dx = 0$$

with the usual smoothness assumptions on the $y_i(x)$ and f. Show that if any y'_i is missing from f, then the corresponding δy_i need not be zero at x_0 and x_1 in the derivation of the Euler-Lagrange equations.

4-7. Evaluate the canonical integral over the interval $0 \leq t \leq t_1$ for the problem of the free vertical motion of a particle under uniform gravity, with arbitrary initial conditions. Show that this integral yields a larger value if a constant average velocity is used in place of the actual velocity.

4-8. Given a holonomic system with a Lagrangian function

$$L = \tfrac{1}{2}m(\dot{x}_1^2 + \dot{x}_2^2 + \dot{x}_3^2) - mgx_3$$

and a constraint

$$\dot{x}_1 - \dot{x}_2 + \dot{x}_3 = 0$$

Use an augmented Lagrangian function to obtain the differential equations of motion. Solve for $\ddot{x}_1$.

4-9. Show that Eq. (4-165) can be expressed in the Hamiltonian form

$$\delta I = \left[\sum_{i=1}^{n} p_i \, \delta q_i - H \, \delta t \right]_{t_0}^{t_1} - \int_{t_0}^{t_1} \sum_{i=1}^{n} \left(\dot{p}_i + \frac{\partial H}{\partial q_i} \right) \delta q_i \, dt + \int_{t_0}^{t_1} \left(\dot{H} - \frac{\partial H}{\partial t} \right) \delta t \, dt$$

4-10. In the design of a two-stage rocket, it is desired to choose the sizes of the two stages such that the payload is given a specified final velocity with a minimum total weight. Mathematically, we wish to choose the mass ratios μ_1 and μ_2 to minimize

$$f = \ln \frac{\mu_1(1 - \beta_1)}{1 - \beta_1 \mu_1} + \ln \frac{\mu_2(1 - \beta_2)}{1 - \beta_2 \mu_2}$$

subject to the condition

$$\ln \mu_1 + \ln \mu_2 = C$$

where β_1, β_2, and C are known constants. $(1/\beta_1) > \mu_1 \geq 1$ and $(1/\beta_2) > \mu_2 \geq 1$. Solve for μ_1 and μ_2 as functions of the given constants. Evaluate $f_{\min}$ for the case $\beta_1 = 0.10$, $\beta_2 = 0.12$, $C = 2.5$. The function f represents the logarithm of the ratio of the gross weight to the payload weight.

4-11. Given a standard nonholonomic system having a Lagrangian function

$$L = \tfrac{1}{2}m(\dot{x}^2 + \dot{y}^2) + \tfrac{1}{2}I\dot{\theta}^2$$

where m and I are constants. The constraint equation is

$$\dot{x} \cos\theta + \dot{y} \sin\theta = 0$$

Assuming initial conditions $x(0) = y(0) = \theta(0) = 0$, $\dot{x}(0) = 0$, $\dot{y}(0) = v_0$, $\dot{\theta}(0) = \omega_0$, obtain the equations of motion and solve for x, y, and θ as functions of time. Show that Hamilton's principle does not apply to this system by calculating the first variation of the canonical integral and showing that it may have either sign for an assumed varied path with

$$y = \left[\frac{v_0}{\omega_0} + \eta(t)\right] \sin \omega_0 t$$

where $\eta(t)$ is zero at the end-points and is small elsewhere. Assume that the variations in x are consistent with the given constraint, but θ is not varied.

4-12. Consider a standard holonomic system with n degrees of freedom. Let $q_{n+1} = t$ and use a new independent variable $\tau = \tau(t)$. Show that the canonical integral can be written in the form

$$I = \int_{\tau_0}^{\tau_1} \sum_{i=1}^{n+1} p_i q_i' \, d\tau$$

where primes indicate differentiations with respect to τ. Also let $p_{n+1} = -H$. Next consider the variational problem with the auxiliary condition $p_{n+1} + H(q, p, t) = 0$. Obtain an augmented Hamiltonian function and show that it leads to the usual canonical equations.

4-13. Does the principle of least action require the q's to be independent? Explain.

4-14. Use the Jacobi form of the principle of least action to show that the path of a particle in a uniform gravitational field is given by

$$y = x \tan\theta_0 - \frac{gx^2}{2v_0^2 \cos^2\theta_0}$$

Assume that the particle leaves the origin with a velocity v_0 at an angle θ_0 above the horizontal.

5

HAMILTON-JACOBI THEORY

This chapter and the one following will be concerned primarily with various aspects of the transformation theory of dynamics. In the spirit of Hamiltonian dynamics we shall, in general, consider transformations in *phase space*; that is, we shall assume that the generalized momenta are included in the transformations, not just the generalized coordinates. Furthermore, we shall restrict ourselves to *canonical transformations* which preserve the Hamiltonian form of the equations of motion in the new variables.

In order to understand how transformation theory can be used in the solution of dynamical problems, let us recall first that in Chap. 2 we found that the complete solution for a holonomic system having n degrees of freedom is obtained by finding $2n$ independent functions known as *integrals of the motion*. These integrals of the motion can be written in the form

$$f_i(q, \dot{q}, t) = \gamma_i \quad (i = 1, 2, \ldots, 2n) \tag{5-1}$$

where the γ's are constants which are normally evaluated from the initial conditions. Another form is obtained by eliminating the $\dot{q}$'s in favor of the p's with the aid of Eq. (4-90), thereby producing integrals of the motion which are functions of the q's, p's, and t.

$$g_i(q, p, t) = \gamma_i \quad (i = 1, 2, \ldots, 2n) \tag{5-2}$$

Assuming that these functions are distinct; that is, none is algebraically derivable from the others, the corresponding Jacobian is nonzero.

$$\frac{\partial(g_1, \ldots, g_{2n})}{\partial(q_1, \ldots, p_n)} \neq 0 \tag{5-3}$$

Hence one can solve for the n q's and n p's as functions of the $2n$ γ's (in effect, the initial conditions) and time.

$$
\begin{aligned}
q_i &= q_i(\gamma_1, \ldots, \gamma_{2n}, t) \\
p_i &= p_i(\gamma_1, \ldots, \gamma_{2n}, t)
\end{aligned}
\quad (i = 1, 2, \ldots, n) \tag{5-4}
$$

Here we have the complete solution of Hamilton's canonical equations, commonly known as the solution of the *Hamilton problem*.

On the other hand, one might view Eq. (5-2) or (5-4) as representing a *transformation* in $2n$-space, namely, a transformation between a fixed point specified by the $2n$ γ's and the moving point (q, p). Because we are concerned with canonical transformations, we require that the γ's, as well as the q's and p's, obey Hamilton's equations. These canonical equations are particularly simple in the case of the γ's, namely,

$$\dot{\gamma}_i = 0 \qquad (i = 1, 2, \ldots, 2n) \tag{5-5}$$

since we have assumed that the γ's are constant. Here we consider that n γ's act as coordinates and the remaining n γ's act as momenta. A Hamiltonian function $K(\gamma, t)$ which yields these canonical equations is also very simple, that is,

$$K(\gamma, t) \equiv 0 \tag{5-6}$$

Although we have demonstrated that the solution for the motion of a holonomic system amounts to finding a canonical transformation in which the new variables are, in fact, constant, the question remains concerning how this transformation can be found. This is the fundamental problem. In this chapter we shall approach the problem by studying the *generating function* which is associated with the required canonical transformation. This generating function is the solution of a partial differential equation known as the *Hamilton-Jacobi equation*. The transformation equations, and hence the solution to the problem, are obtained from the generating function by a process of differentiations and algebraic manipulation.

5-1. HAMILTON'S PRINCIPAL FUNCTION

The Canonical Integral. Consider again the *canonical integral*

$$I = \int_{t_0}^{t_1} L \, dt \tag{5-7}$$

which is associated with Hamilton's principle. Suppose we evaluate this integral over the actual dynamical path of a *holonomic* system that obeys the standard form of Lagrange's or Hamilton's equations. If $2n$ independent initial conditions are specified at the time t_0; for example, if we know the q_0's and $\dot{q}_0$'s, then the further motion of the system is determined. Thus q and p can be found at any final time t_1. It turns out, however, that the trajectory is determined equally well by its end-points in q-space at times t_0 and t_1. In order to show this, let us consider the solution equations

$$q_{i1} = q_{i1}(q_0, \dot{q}_0, t_0, t_1) \qquad (i = 1, 2, \ldots, n) \tag{5-8}$$

and solve for the initial velocity components $\dot{q}_{i0}$. Here we assume that the Jacobian $\partial(q_{11}, \ldots, q_{n1})/\partial(\dot{q}_{10}, \ldots, \dot{q}_{n0})$ is nonzero. In other words, we assume that a trajectory connecting the given end-points $\mathbf{q}_0$ and $\mathbf{q}_1$, and having the given initial and final times, must have a unique initial velocity vector $\dot{\mathbf{q}}_0$. Hence we obtain

$$\dot{q}_{i0} = \eta_i(q_0, q_1, t_0, t_1) \qquad (i = 1, 2, \ldots, n) \tag{5-9}$$

Now consider t_1 as running time in Eq. (5-8) and evaluate the integral of Eq. (5-7) as a function of $(q_0, \dot{q}_0, t_0, t_1)$. Then, substituting for the $\dot{q}_{i0}$'s from Eq. (5-9), we obtain the canonical integral in the required form

$$S(q_0, q_1, t_0, t_1) = \int_{t_0}^{t_1} L \, dt \qquad (5\text{-}10)$$

The function $S(q_0, q_1, t_0, t_1)$ is assumed to be twice differentiable in all its arguments and is known as *Hamilton's principal function*. It will be shown that it is also the generating function of the desired canonical transformation.

In order to be able to associate changes in the principal function with the motion of a dynamical system, let us consider a general noncontemporaneous variation of the canonical integral, where the reference trajectory is an actual solution. Referring to the general result given in Eq. (4-165), we note that the last integral vanishes because the standard holonomic form of Lagrange's equation applies to the reference path. Now recall that the total derivative of the Hamiltonian with respect to time is

$$\dot{H} = \sum_{i=1}^{n} \left(\frac{\partial H}{\partial q_i} \dot{q}_i + \frac{\partial H}{\partial p_i} \dot{p}_i \right) + \frac{\partial H}{\partial t} \qquad (5\text{-}11)$$

which, with the aid of the canonical equations of (4-94) and Eq. (4-95), reduces to

$$\dot{H} = \frac{\partial H}{\partial t} = -\frac{\partial L}{\partial t} \qquad (5\text{-}12)$$

If we substitute $\partial L / \partial t = -\dot{H}$ in the second integral of Eq. (4-165) and use the defining equations for p_i and H, we obtain

$$\delta I = \int_{t_0}^{t_1} \sum_{i=1}^{n} \frac{d}{dt}(p_i \, \delta q_i) \, dt - \int_{t_0}^{t_1} \frac{d}{dt}(H \, \delta t) \, dt \qquad (5\text{-}13)$$

The principal function S is the canonical integral I expressed as a function of the end-points in extended configuration space, so we can identify δS with δI in Eq. (5-13) and perform the integration to obtain

$$\delta S = \left[\sum_{i=1}^{n} p_i \, \delta q_i - H \, \delta t \right]_{t_0}^{t_1} \qquad (5\text{-}14)$$

Now, if we write this equation in differential form, we obtain the important result that

$$dS = \sum_{i=1}^{n} p_{i1} \, dq_{i1} - \sum_{i=1}^{n} p_{i0} \, dq_{i0} - H_1 \, dt_1 + H_0 \, dt_0 \qquad (5\text{-}15)$$

On the other hand, a direct differentiation of $S(q_0, q_1, t_0, t_1)$ yields

$$dS = \sum_{i=1}^{n} \frac{\partial S}{\partial q_{i1}} \, dq_{i1} + \sum_{i=1}^{n} \frac{\partial S}{\partial q_{i0}} \, dq_{i0} + \frac{\partial S}{\partial t_1} \, dt_1 + \frac{\partial S}{\partial t_0} \, dt_0 \qquad (5\text{-}16)$$

Since we assume that the $(2n + 2)$ arguments of the principal function can be varied independently, the corresponding coefficients in Eqs. (5-15) and

(5-16) must be equal. Hence we obtain

$$p_{i1} = \frac{\partial S}{\partial q_{i1}}, \qquad p_{i0} = -\frac{\partial S}{\partial q_{i0}} \qquad (i = 1, 2, \ldots, n) \tag{5-17}$$

and

$$H_1 = -\frac{\partial S}{\partial t_1}, \qquad H_0 = \frac{\partial S}{\partial t_0} \tag{5-18}$$

Let us consider the second equation of (5-17) which gives p_{i0} as a function of $(q_{i0}, q_{i1}, t_0, t_1)$. Assuming that the following determinant is nonzero, that is,

$$\left| \frac{\partial^2 S}{\partial q_{i0}\, \partial q_{j1}} \right| \neq 0 \tag{5-19}$$

we can solve for each q_{i1} as a function of the initial conditions and time. Thus we have

$$q_{i1} = q_{i1}(q_0, p_0, t_0, t_1) \qquad (i = 1, 2, \ldots, n) \tag{5-20}$$

which is the solution of the *Lagrange problem* and gives the motion in *configuration space* as a function of time.

If we substitute this result into the first equation of (5-17) we obtain

$$p_{i1} = p_{i1}(q_0, p_0, t_0, t_1) \qquad (i = 1, 2, \ldots, n) \tag{5-21}$$

which, with Eq. (5-20), completes the solution of the *Hamilton problem*, giving the motion in *phase space* as a function of time.

Hence we see that, once the principal function is known, the complete solution for the motion of the system can be obtained by a process of differentiations and algebraic manipulation. What is needed, however, is a means of finding the principal function without first knowing the solution. This will be provided by the Hamilton-Jacobi equation.

Pfaffian Differential Forms. In the study of dynamics and, in particular, in the theory of canonical transformations, differential forms play an important part. Of special interest are the *Pfaffian differential forms*.

In general, a Pfaffian form Ω in the m variables $x_1, x_2, \ldots, x_m$ can be written as

$$\Omega = X_1(x)\, dx_1 + \cdots + X_m(x)\, dx_m \tag{5-22}$$

Here we notice the similarity to a virtual work expression in which the forces (X's) are functions of position. This Pfaffian form also leads naturally to a line integral over a path in x-space.

Now let us define

$$c_{ij} = \frac{\partial X_i}{\partial x_j} - \frac{\partial X_j}{\partial x_i} \tag{5-23}$$

If the Pfaffian form is an exact differential, then all the c's are zero. In the usual case, however, the differential form is not exact.

Let us consider the important differential form of Eq. (5-15). Here we see that the right-hand side consists of the difference of two Pfaffian expressions, each having the form

$$\sum_{i=1}^{n} p_i \, dq_i - H \, dt$$

where the $(2n + 1)$ variables are the p's, q's, and t. Hence $m = 2n + 1$ and each Pfaffian expression can be written more explicitly as

$$p_1 \, dq_1 + \cdots + p_n \, dq_n + 0 \cdot dp_1 + \cdots + 0 \cdot dp_n - H(q, p, t) \, dt$$

Another aspect of a Pfaffian differential form is that, if m is odd, there is an associated system of m differential equations known as the *first Pfaff's system*. These equations are of the form

$$\sum_{i=1}^{m} c_{ij} \, dx_i = 0 \qquad (j = 1, 2, \ldots, m) \tag{5-24}$$

and are obtained by setting equal to zero the coefficients of the δx's in the bilinear covariant expression for the given Pfaffian form (see Sec. 6-3). Applying Eq. (5-24) to the differential form, we obtain

$$dq_j - \frac{\partial H}{\partial p_j} \, dt = 0$$

$$-dp_j - \frac{\partial H}{\partial q_j} \, dt = 0 \tag{5-25}$$

and

$$\sum_{j=1}^{n} \left(\frac{\partial H}{\partial q_j} \, dq_j + \frac{\partial H}{\partial p_j} \, dp_j \right) = 0 \tag{5-26}$$

Eq. (5-25) can be written as the familiar

$$\dot{q}_j = \frac{\partial H}{\partial p_j}$$

$$\dot{p}_j = -\frac{\partial H}{\partial q_j} \qquad (j = 1, 2, \ldots, n) \tag{5-27}$$

which we recognize as Hamilton's canonical equations. If we take the total time derivative of $H(q, p, t)$, we observe that Eq. (5-26) reduces to

$$\dot{H} = \frac{\partial H}{\partial t} \tag{5-28}$$

It can be seen by referring to Eqs. (5-11) and (5-12) that Eq. (5-28) is not an independent result since it can be derived from Eq. (5-27).

Summarizing, then, we find that dS is equal to the difference between two Pfaffian differential forms, one involving the initial values and the other the final values of the p's, q's, and t. Each Pfaffian form, in turn, is associated with a set of canonical equations in the given variables. Hence the principal function S appears to be a connecting link between two sets of canonical variables. It is, in fact, the *generating function* for the canonical transformation between these variables.

Now let us generalize the differential form of Eq. (5-15) by using the $2n$ parameters $\gamma_1, \ldots, \gamma_{2n}$ to specify the initial conditions in place of the q_0's and p_0's. In other words, we assume a transformation

$$
\begin{aligned}
q_{i0} &= q_{i0}(\gamma_1, \ldots, \gamma_{2n}) \\
p_{i0} &= p_{i0}(\gamma_1, \ldots, \gamma_{2n})
\end{aligned}
\qquad (i = 1, 2, \ldots, n) \qquad (5\text{-}29)
$$

where the Jacobian of the transformation is nonzero, that is,

$$
\frac{\partial(q_{10}, \ldots, p_{n0})}{\partial(\gamma_1, \ldots, \gamma_{2n})} \neq 0 \qquad (5\text{-}30)
$$

Then, by direct substitution from Eq. (5-29), we obtain

$$
\sum_{i=1}^{n} p_{i0}\, dq_{i0} = \sum_{j=1}^{2n} \Gamma_j(\gamma)\, d\gamma_j \qquad (5\text{-}31)
$$

where

$$
\Gamma_j(\gamma) = \sum_{i=1}^{n} p_{i0} \frac{\partial q_{i0}}{\partial \gamma_j} \qquad (5\text{-}32)
$$

Now, it can be seen from Pfaff's theorem† that the $2n$ γ's can always be replaced by n α's and n β's, where the functions

$$
\begin{aligned}
\alpha_i &= \alpha_i(\gamma_1, \ldots, \gamma_{2n}) \\
\beta_i &= \beta_i(\gamma_1, \ldots, \gamma_{2n})
\end{aligned}
\qquad (i = 1, 2, \ldots, n) \qquad (5\text{-}33)
$$

are chosen so that

$$
\sum_{i=1}^{n} \beta_i\, d\alpha_i = \sum_{j=1}^{2n} \Gamma_j(\gamma)\, d\gamma_j \qquad (5\text{-}34)
$$

Then, in the same manner as for Eq. (5-32), we obtain

$$
\Gamma_j(\gamma) = \sum_{i=1}^{n} \beta_i \frac{\partial \alpha_i}{\partial \gamma_j} \qquad (5\text{-}35)
$$

It can be shown that the α's and β's are not unique.

A comparison of Eqs. (5-31) and (5-34) yields

$$
\sum_{i=1}^{n} p_{i0}\, dq_{i0} = \sum_{i=1}^{n} \beta_i\, d\alpha_i \qquad (5\text{-}36)
$$

where the α's and β's are another representation of the initial conditions. Eq. (5-36) implies that (q_0, p_0) and (α, β) are connected by a *homogeneous canonical transformation* at the given initial time t_0. This type of transformation will be discussed further in the next chapter.

---

†A. R. Forsyth, *Theory of Differential Equations*, Part I (New York: Dover Publications, Inc., 1959), pp. 112–14.

5-2. THE HAMILTON-JACOBI EQUATION

Let us consider once again the differential form

$$dS = \sum_{i=1}^{n} p_{i1} \, dq_{i1} - \sum_{i=1}^{n} p_{i0} \, dq_{i0} - H_1 \, dt_1 + H_0 \, dt_0 \qquad (5\text{-}37)$$

which is associated with a canonical transformation relating the initial and final points of a path in phase space. Now suppose that the initial conditions are specified by n α's and n β's, where

$$\begin{aligned} \alpha_i &= \alpha_i(q_{10}, \ldots, q_{n0}, p_{10}, \ldots, p_{n0}) \\ \beta_i &= \beta_i(q_{10}, \ldots, q_{n0}, p_{10}, \ldots, p_{n0}) \end{aligned} \qquad (i = 1, 2, \ldots, n) \qquad (5\text{-}38)$$

with the further stipulation that Eq. (5-36) applies. Note that the functions in Eq. (5-38) are not arbitrary, but represent a homogeneous canonical transformation.

From Eqs. (5-36) and (5-37) we obtain

$$dS = \sum_{i=1}^{n} p_{i1} \, dq_{i1} - \sum_{i=1}^{n} \beta_i \, d\alpha_i - H_1 \, dt_1 + H_0 \, dt_0 \qquad (5\text{-}39)$$

where we now consider S to be a function of $(q_{i1}, \alpha_i, t_1, t_0)$. Hence we can write

$$dS = \sum_{i=1}^{n} \frac{\partial S}{\partial q_{i1}} \, dq_{i1} + \sum_{i=1}^{n} \frac{\partial S}{\partial \alpha_i} \, d\alpha_i + \frac{\partial S}{\partial t_1} \, dt_1 + \frac{\partial S}{\partial t_0} \, dt_0 \qquad (5\text{-}40)$$

Now let us assume that the determinant

$$\left| \frac{\partial^2 S}{\partial q_{i1} \, \partial \alpha_j} \right| \neq 0 \qquad (5\text{-}41)$$

Then it is possible to solve for the α's in terms of $\partial S / \partial q_{i1}$, $(i = 1, 2, \ldots, n)$, which, in accordance with Eq. (5-43), will be identified with the p_1's. Thus the determinant in Eq. (5-41) is actually the Jacobian $\partial(p_{11}, \ldots, p_{n1}) / \partial(\alpha_1, \ldots, \alpha_n)$ where the p_1's are considered to be functions of (q_1, α, t_1, t_0). This implies that there are *no identical relations* of the form $\phi(q_1, \alpha, t_1, t_0) = 0$, that is, no relations exist which involve the q_1's and α's, but not the p_1's. Hence the q_1's and α's are independently variable, and therefore we can equate the corresponding coefficients in Eqs. (5-39) and (5-40) to obtain

$$-\beta_i = \frac{\partial S}{\partial \alpha_i} \qquad (i = 1, 2, \ldots, n) \qquad (5\text{-}42)$$

$$p_{i1} = \frac{\partial S}{\partial q_{i1}} \qquad (i = 1, 2, \ldots, n) \qquad (5\text{-}43)$$

As before, we have

$$H_1 = -\frac{\partial S}{\partial t_1}, \qquad H_0 = \frac{\partial S}{\partial t_0} \qquad (5\text{-}44)$$

Eq. (5-39) can be simplified by arbitrarily setting the initial time t_0 equal to zero with the result that $dt_0 = 0$. We measure the time t from this instant and drop the 1 subscripts for convenience. Then we obtain

$$dS = \sum_{i=1}^{n} p_i \, dq_i - \sum_{i=1}^{n} \beta_i \, d\alpha_i - H \, dt \qquad (5\text{-}45)$$

Another more general approach is to consider corresponding trajectories in the (α, β) and (q, p) phase spaces as functions of a common time t. These variables are related by a canonical transformation which can be specified by equating the total differential dS to the difference of two Pfaffian forms, namely,

$$\sum_{i=1}^{n} p_i \, dq_i - H \, dt$$

and

$$\sum_{i=1}^{n} \beta_i \, d\alpha_i - K \, dt$$

where $K(\alpha, \beta, t)$ is the Hamiltonian function which results from considering the α's and β's as variables. Hence we obtain

$$dS = \sum_{i=1}^{n} p_i \, dq_i - \sum_{i=1}^{n} \beta_i \, d\alpha_i - H \, dt + K \, dt \qquad (5\text{-}46)$$

Even though we consider the α's and β's as variables in the Hamiltonian formulation, we still would like to have them turn out to be the required $2n$ constants of the motion. In other words, we desire that the entire trajectory in the (α, β) phase space consist of a single fixed point. One way to ensure this result is to let K be identically zero. Then, in accordance with the canonical equations, we find that each $\dot{\alpha}_i$ and $\dot{\beta}_i$ is zero, and we obtain constant values for the α's and β's. So now let us set K equal to zero and we obtain once again the differential form of Eq. (5-45).

From Eq. (5-45) we see that the principal function is of the form $S(q, \alpha, t)$. Hence its total differential can be written as

$$dS = \sum_{i=1}^{n} \frac{\partial S}{\partial q_i} \, dq_i + \sum_{i=1}^{n} \frac{\partial S}{\partial \alpha_i} \, d\alpha_i + \frac{\partial S}{\partial t} \, dt \qquad (5\text{-}47)$$

Once again we assume a nonzero determinant, that is,

$$\left| \frac{\partial^2 S}{\partial q_i \, \partial \alpha_j} \right| \neq 0 \qquad (5\text{-}48)$$

and, equating coefficients in Eqs. (5-45) and (5-47), we obtain the following important results:

$$-\beta_i = \frac{\partial S}{\partial \alpha_i} \qquad (i = 1, 2, \ldots, n) \qquad (5\text{-}49)$$

$$p_i = \frac{\partial S}{\partial q_i} \qquad (i = 1, 2, \ldots, n) \qquad (5\text{-}50)$$

$$\frac{\partial S}{\partial t} = -H \qquad (5\text{-}51)$$

Eq. (5-49) can be solved for the q's as functions of (α, β, t), and this provides the solution of the Lagrange problem. We know that this is possible because Eq. (5-48) is essentially a statement that the Jacobian $\partial(\beta_1, \ldots, \beta_n)/$ $\partial(q_1, \ldots, q_n)$ is nonzero. Then, substituting these solutions for the q's into Eq. (5-50), we obtain expressions for the p's as functions of (α, β, t), thereby completing the solution of the Hamilton problem.

The Hamiltonian H is usually considered to be a function of (q, p, t). If we substitute for the p's from Eq. (5-50), however, we can write Eq. (5-51) in the form

$$\frac{\partial S}{\partial t} + H\left(q, \frac{\partial S}{\partial q}, t\right) = 0 \tag{5-52}$$

This first-order partial differential equation is known as the *Hamilton-Jacobi equation*. It has a single dependent variable S and $(n + 1)$ independent variables (q, t). Therefore, a complete solution of this equation contains $(n + 1)$ arbitrary constants. These constants are such that S and its first partial derivatives with respect to the q's may have arbitrary initial values, the initial value of $\partial S/\partial t$ then being determined by Eq. (5-52).

We note, however, that S itself does not appear explicitly in Eq. (5-52), but only its partial derivatives. Hence one of the α's, say α_{n+1}, is purely additive and may be disregarded since it disappears upon differentiation. The remaining nonadditive constants are evaluated by substituting the initial conditions (q_0, p_0) into Eq. (5-50) and solving for $\alpha_1, \alpha_2, \ldots, \alpha_n$. This solution is possible if the Jacobian $\partial(p_1, \ldots, p_n)/\partial(\alpha_1, \ldots, \alpha_n)$ is nonzero, and this was assumed previously in Eq. (5-48).

In summary, then, a complete solution of the Hamilton-Jacobi equation yields the principal function $S(q, \alpha, t)$ which provides the path of the system in phase space merely by differentiations and algebraic manipulation. This solution of the Hamilton problem is obtained by using Eqs. (5-49) and (5-50) and is equivalent to the complete integration of the canonical equations.

Jacobi's Theorem. In considering the solutions of the Hamilton-Jacobi partial differential equation, we have assumed that the principal function S can be expressed in terms of n q_0's or n α's, where we see from Eq. (5-38) that the α's may be any of a wide variety of functions of the q_0's and p_0's. The question arises whether a certain solution for S is required to obtain the complete integration of the canonical equations, or if *any complete solution* of the Hamilton-Jacobi equation will do.

The answer to this question is in the affirmative and is given by the Jacobi theorem:

If $S(q, \alpha, t)$ is any complete solution of the Hamilton-Jacobi equation

$$\frac{\partial S}{\partial t} + H\left(q, \frac{\partial S}{\partial q}, t\right) = 0 \tag{5-53}$$

and if the equations

$$-\beta_i = \frac{\partial S}{\partial \alpha_i} \qquad (i = 1, 2, \ldots, n) \tag{5-54}$$

$$p_i = \frac{\partial S}{\partial q_i} \qquad (i = 1, 2, \ldots, n) \tag{5-55}$$

where the β's are arbitrary constants, are used to solve for $q_i(\alpha, \beta, t)$ and $p_i(\alpha, \beta, t)$; then these expressions provide the general solution of the canonical equations associated with the Hamiltonian $H(q, p, t)$.

The proof of this theorem involves first a partial differentiation of the Hamilton-Jacobi equation with respect to α_i, yielding

$$\frac{\partial^2 S}{\partial \alpha_i \, \partial t} + \sum_{j=1}^{n} \frac{\partial H}{\partial p_j} \frac{\partial p_j}{\partial \alpha_i} = 0 \tag{5-56}$$

where p_j is considered as a function of (q, α, t), as in Eq. (5-55).

Now let us take the total time derivative of Eq. (5-54) for the case where we are following an actual solution path in phase space. We note that $\partial S/\partial \alpha_i$ is a function of (q, α, t), and the α's and β's are constants. Hence we obtain

$$\frac{\partial^2 S}{\partial t \, \partial \alpha_i} + \sum_{j=1}^{n} \frac{\partial^2 S}{\partial q_j \, \partial \alpha_i} \dot{q}_j = 0 \tag{5-57}$$

The order of the partial differentiations is immaterial because of the assumed smoothness of S. So, using Eqs. (5-55), (5-56), and (5-57), we find that

$$\sum_{j=1}^{n} \left(\dot{q}_j - \frac{\partial H}{\partial p_j} \right) \frac{\partial^2 S}{\partial q_j \, \partial \alpha_i} = 0 \qquad (i = 1, 2, \ldots, n) \tag{5-58}$$

The coefficients $\partial^2 S/\partial q_j \, \partial \alpha_i$ are the elements of a determinant which we assumed to be nonzero in Eq. (5-48). Therefore we see that

$$\dot{q}_j = \frac{\partial H}{\partial p_j} \qquad (j = 1, 2, \ldots, n) \tag{5-59}$$

which is the first of Hamilton's equations.

Now let us start again with the Hamilton-Jacobi equation, but differentiate partially with respect to q_j, assuming that p_i is a function of (q, α, t) in accordance with Eq. (5-55). We have

$$\frac{\partial^2 S}{\partial q_j \, \partial t} + \sum_{i=1}^{n} \frac{\partial H}{\partial p_i} \frac{\partial p_i}{\partial q_j} + \frac{\partial H}{\partial q_j} = 0 \tag{5-60}$$

Next take the total time derivative of Eq. (5-55), noting that each α_i is constant along a solution path, and obtain

$$\dot{p}_j - \frac{\partial^2 S}{\partial t \, \partial q_j} - \sum_{i=1}^{n} \frac{\partial^2 S}{\partial q_j \, \partial q_i} \dot{q}_i = 0 \tag{5-61}$$

Then, adding Eqs. (5-60) and (5-61), and using Eqs. (5-55) and (5-59), we find that

$$\dot{p}_j = -\frac{\partial H}{\partial q_j} \qquad (j = 1, 2, \ldots, n) \qquad (5\text{-}62)$$

which is the second canonical equation. Thus we see that *any complete solution* of the Hamilton-Jacobi equation leads to a solution of the Hamilton problem. This solution has the proper number of arbitrary constants and, of course, obeys the canonical equations.

Conservative Systems and Ignorable Coordinates. Now consider a *conservative holonomic system* whose configuration is described in terms of n independent q's. The Hamiltonian function for this system is not an explicit function of time and, in fact, is a constant of the motion. So we can write

$$H(q, p) = \alpha_n = h \qquad (5\text{-}63)$$

where h is the value of the familiar Jacobi integral or energy integral which we arbitrarily identify with α_n.

A suitable form for the principal function of this system is found by using Eqs.(5-53) and (5-63) to obtain

$$\frac{\partial S}{\partial t} = -H = -\alpha_n \qquad (5\text{-}64)$$

This suggests that S can be taken as a linear function of time, that is,

$$S(q, \alpha, t) = -\alpha_n t + W(q, \alpha) \qquad (5\text{-}65)$$

where we have omitted an arbitrary additive constant. The function $W(q_1, \ldots, q_n, \alpha_1, \ldots, \alpha_n)$ does not contain time explicitly and is known as the *characteristic function*. Note that

$$\frac{\partial S}{\partial \alpha_i} = \frac{\partial W}{\partial \alpha_i} \qquad (i = 1, 2, \ldots, n-1) \qquad (5\text{-}66)$$

$$\frac{\partial S}{\partial \alpha_n} = \frac{\partial W}{\partial \alpha_n} - t \qquad (5\text{-}67)$$

$$\frac{\partial S}{\partial q_i} = \frac{\partial W}{\partial q_i} \qquad (i = 1, 2, \ldots, n) \qquad (5\text{-}68)$$

From Eqs. (5-64) and (5-68) we see that the Hamilton-Jacobi equation reduces to

$$H\left(q, \frac{\partial W}{\partial q}\right) = \alpha_n \qquad (5\text{-}69)$$

Eq. (5-69) is the *modified Hamilton-Jacobi equation*. A complete solution of this equation involves $(n-1)$ nonadditive α's, $(\alpha_1, \alpha_2, \ldots, \alpha_{n-1})$, plus the energy constant α_n. The α's are arbitrary in the sense that their values are determined by the arbitrary initial values of the $\partial W / \partial q_i$, that is, the generalized momenta, at the given initial configuration.

A comparison of Eqs. (5-66)—(5-68) with Eqs. (5-54) and (5-55) shows that the solution of the Hamilton problem can be obtained from

$$-\beta_i = \frac{\partial W}{\partial \alpha_i} \qquad (i = 1, 2, \ldots, n-1) \tag{5-70}$$

$$t - \beta_n = \frac{\partial W}{\partial \alpha_n} \tag{5-71}$$

$$p_i = \frac{\partial W}{\partial q_i} \qquad (i = 1, 2, \ldots, n) \tag{5-72}$$

where β_n is the initial time t_0. Since W is not an explicit function of time, we see that Eq. (5-70) gives the path of the system in configuration space without reference to time. Eq. (5-71) then gives the relation of time to position along the path.

Now suppose we consider a system having *ignorable coordinates* q_1, $q_2, \ldots, q_k$. Initially we shall assume that the system is *not conservative*. We know that the p's associated with the ignorable q's are constant; hence we can take

$$p_i = \alpha_i \qquad (i = 1, 2, \ldots, k) \tag{5-73}$$

Then we see from Eq. (5-55) that we can assume a principal function of the form

$$S(q, \alpha, t) = \sum_{i=1}^{k} \alpha_i q_i + S'(q_{k+1}, \ldots, q_n, \alpha_1, \ldots, \alpha_n, t) \tag{5-74}$$

The Hamilton-Jacobi equation leads in this case to

$$\frac{\partial S'}{\partial t} + H\left(q_{k+1}, \ldots, q_n, \alpha_1, \ldots, \alpha_k, \frac{\partial S'}{\partial q_{k+1}}, \ldots, \frac{\partial S'}{\partial q_n}, t\right) = 0 \tag{5-75}$$

The complete solution of this partial differential equation involves $(n-k)$ nonadditive constants, exclusive of the constant momenta $\alpha_1, \alpha_2, \ldots, \alpha_k$. Once S' is known, the solution for the motion of the system is obtained from

$$-\beta_i = q_i + \frac{\partial S'}{\partial \alpha_i} \qquad (i = 1, 2, \ldots, k) \tag{5-76}$$

$$-\beta_i = \frac{\partial S'}{\partial \alpha_i} \qquad (i = k+1, \ldots, n) \tag{5-77}$$

$$p_i = \alpha_i \qquad (i = 1, 2, \ldots, k) \tag{5-78}$$

$$p_i = \frac{\partial S'}{\partial q_i} \qquad (i = k+1, \ldots, n) \tag{5-79}$$

where we note that

$$\beta_i = -q_{i0} \qquad (i = 1, 2, \ldots, k) \tag{5-80}$$

that is, each β corresponding to an ignorable coordinate is just the negative of the initial value of this coordinate.

Finally, let us consider a system which has *ignorable coordinates* q_1, $q_2, \ldots, q_k$ and is also *conservative*. Combining the previous results, we see that the principal function has the form

$$S(q, \alpha, t) = \sum_{i=1}^{k} \alpha_i q_i - \alpha_n t + W'(q_{k+1}, \ldots, q_n, \alpha_1, \ldots, \alpha_n) \qquad (5\text{-}81)$$

and the modified Hamilton-Jacobi equation becomes

$$H\left(q_{k+1}, \ldots, q_n, \alpha_1, \ldots, \alpha_k, \frac{\partial W'}{\partial q_{k+1}}, \ldots, \frac{\partial W'}{\partial q_n}\right) = \alpha_n \qquad (5\text{-}82)$$

The complete solution for W' in this case involves the $(n - k - 1)$ nonadditive constants $\alpha_{k+1}, \ldots, \alpha_{n-1}$ plus, of course, the energy constant α_n and the constant momenta $\alpha_1, \alpha_2, \ldots, \alpha_k$.

The motion of the system is given by

$$-\beta_i = q_i + \frac{\partial W'}{\partial \alpha_i} \qquad (i = 1, 2, \ldots, k) \qquad (5\text{-}83)$$

$$-\beta_i = \frac{\partial W'}{\partial \alpha_i} \qquad (i = k + 1, \ldots, n - 1) \qquad (5\text{-}84)$$

$$t - \beta_n = \frac{\partial W'}{\partial \alpha_n} \qquad (5\text{-}85)$$

$$p_i = \alpha_i \qquad (i = 1, 2, \ldots, k) \qquad (5\text{-}86)$$

$$p_i = \frac{\partial W'}{\partial q_i} \qquad (i = k + 1, \ldots, n) \qquad (5\text{-}87)$$

Example 5-1. As a first illustration of the Hamilton-Jacobi method, consider its application to a simple mass-spring system (Fig. 5-1). This is a natural system having kinetic and potential energies given by

$$T = \tfrac{1}{2}m\dot{x}^2, \qquad V = \tfrac{1}{2}kx^2 \qquad (5\text{-}88)$$

The momentum is the familiar

$$p = \frac{\partial T}{\partial \dot{x}} = m\dot{x} \qquad (5\text{-}89)$$

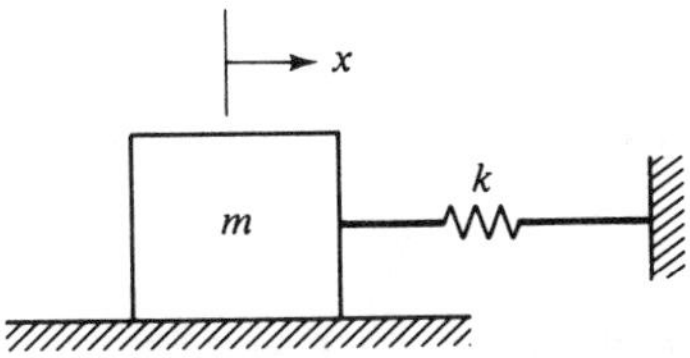

Fig. 5-1. A mass-spring system.

and we find that the Hamiltonian function is equal to the total energy, namely,

$$H = T + V = \frac{p^2}{2m} + \frac{1}{2}kx^2 \qquad (5\text{-}90)$$

Since we are considering a conservative system, we can use directly the modified Hamilton-Jacobi equation (5-69) which, in this case, is

$$\frac{1}{2m}\left(\frac{\partial W}{\partial x}\right)^2 + \frac{1}{2}kx^2 = \alpha \qquad (5\text{-}91)$$

where α is the energy constant. Hence we obtain

$$\frac{\partial W}{\partial x} = \sqrt{2m(\alpha - \tfrac{1}{2}kx^2)} \tag{5-92}$$

or

$$W(x, \alpha) = m\omega \int_{x_0}^{x} \sqrt{a^2 - \xi^2}\, d\xi \tag{5-93}$$

where

$$a = \sqrt{2\alpha/m\omega^2}, \qquad \omega = \sqrt{k/m} \tag{5-94}$$

In general, the lower limit x_0 of the integral is chosen to be either (1) a convenient absolute constant (not a function of the α's), or (2) a simple zero of $f(\xi)$, where $\sqrt{f(\xi)}$ is the integrand. This choice is made in order to simplify differentiation under the integral sign and will, of course, be reflected in the meaning attached to the various β's.

Let us continue, then, by applying Eq. (5-71) to the characteristic function given in Eq. (5-93). Differentiating with respect to α, we obtain

$$t - \beta = \frac{1}{\omega} \int_{x_0}^{x} \frac{d\xi}{\sqrt{a^2 - \xi^2}} = \frac{1}{\omega}\left[\cos^{-1}\frac{x_0}{a} - \cos^{-1}\frac{x}{a} \right] \tag{5-95}$$

which yields

$$x = \sqrt{\frac{2\alpha}{m\omega^2}} \cos\left[\omega(t - t_0) - \phi\right] \tag{5-96}$$

where

$$\cos\phi = \frac{x_0}{a} = \sqrt{\frac{m\omega^2}{2\alpha}}\, x_0 \tag{5-97}$$

and $\beta = t_0$. If we write the total energy α in terms of the initial conditions $x(t_0) = x_0$ and $\dot{x}(t_0) = v_0$, we have

$$\alpha = \frac{1}{2}mv_0^2 + \frac{1}{2}kx_0^2 = \frac{m\omega^2}{2}\left(x_0^2 + \frac{v_0^2}{\omega^2}\right) \tag{5-98}$$

Also

$$\sin\phi = \frac{1}{a}\sqrt{a^2 - x_0^2} = \frac{v_0}{a\omega} \tag{5-99}$$

and we can write Eq. (5-96) in the form

$$x = x_0 \cos\omega(t - t_0) + \frac{v_0}{\omega} \sin\omega(t - t_0) \tag{5-100}$$

which is identical with the result obtained by the direct solution of the ordinary differential equation describing the mass-spring system. The amplitude of the oscillation in x is

$$a = \sqrt{x_0^2 + \frac{v_0^2}{\omega^2}} \tag{5-101}$$

In evaluating an integral involving $\sqrt{f(\xi)}$, as in Eq. (5-95), a question arises concerning which sign is to be chosen for the square root. It frequently occurs that the variable of integration ξ oscillates between two zeros of $f(\xi)$, indicating a *librational* motion. Thus we see from Eq. (5-95), for example, that we must change the sign of $\sqrt{f(\xi)}$ at each *turning point* in the libration, that is, at the point where the direction of motion in ξ reverses. In this example, $\sqrt{f(\xi)}$ is positive for positive $d\xi$ and is negative for negative $d\xi$. Notice that the turning points occur at zeros of $f(\xi)$ and, since the integrand may become infinite at these points, it can result in an improper integral. For the usual case of simple zeros, however, this integral converges, indicating a finite period for the librational motion.

In this example we have found a solution for the motion of the mass-spring system without the necessity of evaluating the integral of Eq. (5-93) which leads to an explicit expression for $W(x, \alpha)$. Now let us perform this integration and substitute the result into Eq. (5-65) in order to obtain the principal function. We find that

$$S = -\alpha t + \frac{m\omega}{2}\left(x\sqrt{a^2 - x^2} - a^2\cos^{-1}\frac{x}{a} - x_0\sqrt{a^2 - x_0^2} + a^2\cos^{-1}\frac{x_0}{a}\right)$$

$$(5\text{-}102)$$

Here, for the sake of simplicity, we have expressed W as a function of a, which we recall is equal to $\sqrt{2\alpha/m\omega^2}$.

It is a straightforward process to check that this expression for the principal function obeys the Hamilton-Jacobi equation. Furthermore, it leads to the correct solution for the motion of the system upon the application o. Eq. (5-54). Thus we obtain

$$-\beta = \frac{\partial S}{\partial \alpha} = -t + \frac{\partial W}{\partial a}\frac{da}{d\alpha} = -t + \frac{1}{\omega}\left(\cos^{-1}\frac{x_0}{a} - \cos^{-1}\frac{x}{a}\right) \quad (5\text{-}103)$$

or

$$x = a\cos[\omega(t - t_0) - \phi] \quad (5\text{-}104)$$

where we have substituted t_0 for β. This result is identical with that found previously in Eq. (5-96).

Finally, let us calculate the principal function $S(x, x_0, t, t_0)$ by evaluating the canonical integral with the aid of the known solution given in Eq. (5-100). We have

$$S = \int_{t_0}^{t}(T - V)\,dt = \frac{m}{2}\int_{t_0}^{t}(\dot{x}^2 - \omega^2 x^2)\,dt \quad (5\text{-}105)$$

Performing the integration and simplifying, we obtain

$$S = -\frac{m\omega}{2}\sin\omega(t - t_0)\left[\left(x_0^2 - \frac{v_0^2}{\omega^2}\right)\cos\omega(t - t_0) + \frac{2x_0v_0}{\omega}\sin\omega(t - t_0)\right]$$

$$(5\text{-}106)$$

In order to express S in terms of the desired quantities, we note from Eq.

(5-100) that

$$\frac{v_0}{\omega} = \frac{x - x_0 \cos \omega(t - t_0)}{\sin \omega(t - t_0)} \tag{5-107}$$

Then, substituting Eq. (5-107) into Eq. (5-106), we obtain

$$S(x, x_0, t, t_0) = \tfrac{1}{2}m\omega(x^2 + x_0^2) \cot \omega(t - t_0) - m\omega x x_0 \csc \omega(t - t_0) \tag{5-108}$$

By comparing this result with the principal function obtained previously in Eq. (5-102), we see that there is a considerable difference in form. In particular, this S function has no linear term in t even though we are analyzing a conservative system. Nevertheless, it represents another complete solution of the Hamilton-Jacobi equation and, in accordance with Jacobi's theorem, it allows one to find the motion of the system by a process of differentiations and algebraic manipulations.

To illustrate this point, let us consider that x_0 assumes the role of α, that is, x_0 is an arbitrary constant which in this case describes the position of the system at a preassigned time t_0. Then we use Eq. (5-54) to obtain

$$-\beta = \frac{\partial S}{\partial x_0} = m\omega x_0 \cot \omega(t - t_0) - m\omega x \csc \omega(t - t_0) \tag{5-109}$$

Solving for x, we find that

$$x = x_0 \cos \omega(t - t_0) + \frac{\beta}{m\omega} \sin \omega(t - t_0) \tag{5-110}$$

In accordance with the theory, β is equal to the initial momentum mv_0. Hence Eq. (5-110) is identical to the previous result given in Eq. (5-100).

Example 5-2. Let us use the Hamilton-Jacobi method to analyze the Kepler problem (see Examples 2-7, 4-5, and 4-7). Suppose a particle of unit mass is attracted by an inverse-square gravitational force to a fixed point O (Fig. 5-2). The position of the particle is given in terms of the polar coordinates (r, θ) measured in the plane of the orbit.

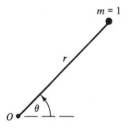

The kinetic and potential energies are

$$T = \frac{1}{2}(\dot{r}^2 + r^2\dot{\theta}^2), \qquad V = -\frac{\mu}{r} \tag{5-111}$$

where μ is the gravitational coefficient. The Lagrangian function is

$$L = \frac{1}{2}(\dot{r}^2 + r^2\dot{\theta}^2) + \frac{\mu}{r} \tag{5-112}$$

Fig. 5-2. The Kepler problem.

and we find that the generalized momenta are given by

$$p_r = \frac{\partial L}{\partial \dot{r}} = \dot{r}, \qquad p_\theta = \frac{\partial L}{\partial \dot{\theta}} = r^2\dot{\theta} \tag{5-113}$$

We are considering a natural system, so the Hamiltonian function is equal to the total energy, that is,

$$H = \frac{1}{2}\left(p_r^2 + \frac{p_\theta^2}{r^2}\right) - \frac{\mu}{r} = \alpha_t \qquad (5\text{-}114)$$

Here we use α_t to represent the constant value of the total energy.

The coordinate θ does not appear in H and therefore it is ignorable, implying that the conjugate momentum p_θ has a constant value which we shall designate by α_θ. Then, in accordance with Eq. (5-81), we see that the principal function can be written in the form

$$S = -\alpha_t t + \alpha_\theta \theta + W'(r, \alpha_t, \alpha_\theta) \qquad (5\text{-}115)$$

The modified Hamilton-Jacobi equation is

$$\frac{1}{2}\left(\frac{\partial W'}{\partial r}\right)^2 + \frac{\alpha_\theta^2}{2r^2} - \frac{\mu}{r} = \alpha_t \qquad (5\text{-}116)$$

and we obtain

$$\frac{\partial W'}{\partial r} = \sqrt{2\alpha_t + \frac{2\mu}{r} - \frac{\alpha_\theta^2}{r^2}} \qquad (5\text{-}117)$$

which results in

$$W' = \int_{r_0}^{r} \sqrt{2\alpha_t + \frac{2\mu}{r} - \frac{\alpha_\theta^2}{r^2}}\, dr \qquad (5\text{-}118)$$

where r_0 is the value of the radial distance r at the initial time t_0.

Now let us differentiate under the integral sign with respect to α_t. In accordance with Eq. (5-85) we obtain

$$t - t_0 = \frac{\partial W'}{\partial \alpha_t} = \int_{r_0}^{r} \frac{dr}{\sqrt{2\alpha_t + \dfrac{2\mu}{r} - \dfrac{\alpha_\theta^2}{r^2}}} \qquad (5\text{-}119)$$

Note that the square root is equal to $\dot{r}$, as may be seen from the energy equation (5-114).

In a similar manner, we use Eq. (5-83) to obtain

$$\theta - \theta_0 = -\frac{\partial W'}{\partial \alpha_\theta} = \int_{r_0}^{r} \frac{\alpha_\theta\, dr}{r\sqrt{2\alpha_t r^2 + 2\mu r - \alpha_\theta^2}} \qquad (5\text{-}120)$$

Comparing these last two integrals, we find that the first gives t as a function of r, whereas the second gives θ as a function of r, that is, it gives the *shape* of the orbit. It is convenient to measure θ from the position of minimum r. Then $\theta_0 = 0$ for the case where we choose $r_0 = r_{\min}$, and the integral can be evaluated in the same manner as in Example 4-7, yielding

$$\theta = \cos^{-1}\left(\frac{\alpha_\theta^2 - \mu r}{r\sqrt{\mu^2 + 2\alpha_t \alpha_\theta^2}}\right) \qquad (5\text{-}121)$$

Finally, solving for r, we obtain

$$r = \frac{\alpha_\theta^2/\mu}{1 + \sqrt{1 + 2\alpha_t\alpha_\theta^2/\mu^2} \cos \theta} \qquad (5\text{-}122)$$

which we recognize as the equation of a conic section having an eccentricity

$$e = \sqrt{1 + 2\alpha_t\alpha_\theta^2/\mu^2} \qquad (5\text{-}123)$$

5-3. SEPARABILITY

The idea of separability is associated with the solution of partial differential equations by a reduction to quadratures, that is, by expressing the solution in terms of integrals, each involving only one variable. In the context of the Hamilton-Jacobi partial differential equation, the possibility of obtaining a separation of variables depends partly upon the nature of the physical system and partly upon the coordinates used in its mathematical representation. Quite naturally, we would like to know the conditions under which such a separation is possible. Unfortunately, the answer to the basic question concerning what is the most general separable system is not known. Some progress can be made, however, if we restrict the investigation to a certain class of systems. In particular, we shall consider in the following discussion only *orthogonal systems*, that is, conservative holonomic systems whose kinetic energy function contains only squared terms in the $\dot{q}$'s (or p's), and no product terms in these variables. In other words, there are no inertial coupling terms.

We shall assume that the term *separability* implies that a characteristic function for the system can be found which has the form

$$W = \sum_{i=1}^{n} W_i(q_i) \qquad (5\text{-}124)$$

that is, it consists of the sum of n functions where each function W_i contains only one of the q's. Furthermore, we shall assume that W is a complete integral of the modified Hamilton-Jacobi equation and thus contains n nonadditive constants (including the energy constant), usually designated by α's. A particularly simple example of a separable system occurs in the case where all but one of the coordinates are ignorable.

Liouville's System. In accordance with the previous discussion of Sec. 2-3, let us define a *Liouville system* to be an orthogonal system which has kinetic and potential energies of the forms

$$T = \frac{1}{2}\left(\sum_{i=1}^{n} f_i(q_i)\right)\left(\sum_{i=1}^{n} \frac{\dot{q}_i^2}{R_i(q_i)}\right) = \frac{R_1 p_1^2 + \cdots + R_n p_n^2}{2(f_1 + \cdots + f_n)} \qquad (5\text{-}125)$$

$$V = \frac{v_1(q_1) + \cdots + v_n(q_n)}{f_1(q_1) + \cdots + f_n(q_n)} \qquad (5\text{-}126)$$

where f_i, R_i, and v_i are each functions of q_i, and we note that R_i is identical with M_i^{-1} of Eq. (2-201). Also, we assume that $\sum_i f_i(q_i) > 0$ and $R_i(q_i) > 0$.

These Liouville conditions are sufficient to ensure the separability of the given system, and therefore a reduction to quadratures is possible. As a proof, we can show that a complete solution $W(q)$ of the modified Hamilton-Jacobi equation exists, and this solution has the separable form given in Eq. (5-124).

The modified Hamilton-Jacobi equation for this system can be written in the form

$$\sum_{i=1}^{n} \left[\frac{1}{2} R_i \left(\frac{\partial W}{\partial q_i} \right)^2 + v_i \right] = h \sum_{i=1}^{n} f_i \tag{5-127}$$

Now let us substitute for W from Eq. (5-124). We shall find that each function $W_i(q_i)$ can be obtained in integral form and a complete solution results. First, let us group the terms in each coordinate q_i $(i = 1, 2, \ldots, n)$ and use $\alpha_1, \alpha_2, \ldots, \alpha_n$ as separation constants. Upon setting each group of terms equal to the corresponding α_i, we have

$$\frac{1}{2} R_i \left(\frac{\partial W_i}{\partial q_i} \right)^2 + v_i - hf_i = \alpha_i \qquad (i = 1, 2, \ldots, n) \tag{5-128}$$

where

$$\alpha_1 + \alpha_2 + \cdots + \alpha_n = 0 \tag{5-129}$$

Eq. (5-128) can be integrated and the resulting W_i's added in accordance with Eq. (5-124) to obtain

$$W = \sum_{i=1}^{n} \int \frac{1}{R_i} \sqrt{\phi_i(q_i)} \, dq_i \tag{5-130}$$

where

$$\phi_i(q_i) = 2R_i[hf_i(q_i) - v_i(q_i) + \alpha_i] \qquad (i = 1, 2, \ldots, n) \tag{5-131}$$

This solution actually contains the $(n + 1)$ constants $\alpha_1, \alpha_2, \ldots, \alpha_n, h$, but Eq. (5-129) represents one relation among the α's. Therefore, one α_i can be eliminated, leaving the required n independent constants. Hence we see that the expression for W given in Eq. (5-130) is a complete solution of the modified Hamilton-Jacobi equation, thereby confirming that the Liouville conditions are sufficient for the separability of an orthogonal system.

We turn next to the solution for the motion of the system, assuming that we have found the characteristic function $W(q)$. First, let us arbitrarily eliminate α_n by regarding it as a function of the other α's, that is,

$$\alpha_n = -(\alpha_1 + \alpha_2 + \cdots + \alpha_{n-1}) \tag{5-132}$$

Then we have

$$\frac{\partial W}{\partial \alpha_i} = \frac{\partial W_i}{\partial \alpha_i} + \frac{\partial W_n}{\partial \alpha_n} \frac{\partial \alpha_n}{\partial \alpha_i} = \frac{\partial W_i}{\partial \alpha_i} - \frac{\partial W_n}{\partial \alpha_n} \qquad (i = 1, 2, \ldots, n - 1) \tag{5-133}$$

and, with the aid of Eq. (5-70), we obtain

$$\frac{\partial W}{\partial \alpha_i} = \int \frac{dq_i}{\sqrt{\phi_i(q_i)}} - \int \frac{dq_n}{\sqrt{\phi_n(q_n)}} = -\beta_i \qquad (i = 1, 2, \ldots, n-1) \qquad (5\text{-}134)$$

Also, from Eq. (5-71) we find that

$$\frac{\partial W}{\partial h} = \sum_{i=1}^{n} \int \frac{f_i \, dq_i}{\sqrt{\phi_i(q_i)}} = t - \beta_n \qquad (5\text{-}135)$$

The solution to the Lagrange problem is given by Eqs. (5-134) and (5-135) and presents the path of the system in extended configuration space. The path in phase space is found by using the additional equations

$$p_i = \frac{\partial W}{\partial q_i} = \frac{1}{R_i} \sqrt{\phi_i(q_i)} \qquad (i = 1, 2, \ldots, n) \qquad (5\text{-}136)$$

Since β_i in Eq. (5-134) is a constant along any actual path of the system, we see that the increments in the values of any two of the given integrals must be equal for any interval of time. Hence we obtain

$$\frac{dq_1}{\sqrt{\phi_1(q_1)}} = \frac{dq_2}{\sqrt{\phi_2(q_2)}} = \cdots = \frac{dq_n}{\sqrt{\phi_n(q_n)}} = d\tau \qquad (5\text{-}137)$$

in agreement with Eq. (2-203).

Stäckel's Theorem. We have shown that the Liouville conditions are sufficient for the separability of an orthogonal system, and hence for its reduction to quadratures. The question arises whether these conditions are also necessary. It turns out that the answer is no, except for the special case in which $n = 2$. So now let us present a more general set of conditions.

Consider once again an orthogonal system and suppose its kinetic energy is given by

$$T = \frac{1}{2} \sum_{i=1}^{n} m_i \dot{q}_i^2 = \frac{1}{2} \sum_{i=1}^{n} c_i p_i^2 \qquad (5\text{-}138)$$

where $c_i(q_1, q_2, \ldots, q_n) > 0$.

Stäckel's theorem asserts that this is a separable system if and only if the following two Stäckel conditions are met, namely, that a nonsingular $n \times n$ matrix $[\Phi_{ij}(q_i)]$ and a column matrix $\{\psi_i(q_i)\}$ exist such that

$$\mathbf{c}^T \mathbf{\Phi} = (1, 0, \ldots, 0) \qquad (5\text{-}139)$$

and

$$\mathbf{c}^T \mathbf{\psi} = V \qquad (5\text{-}140)$$

where $V(q_1, q_2, \ldots, q_n)$ is the potential energy and $\mathbf{c}$ is a column matrix composed of the n c's.

To show the *necessity* of these conditions, assume that the given orthogonal system is separable, and therefore possesses a characteristic function $W(q, \alpha)$ which consists of the sum of terms of the form $W_i(q_i, \alpha_1, \alpha_2, \ldots, \alpha_n)$, as in Eq. (5-124). Of course, this characteristic function is a complete integral of the modified Hamilton-Jacobi equation

$$\frac{1}{2}\sum_{i=1}^{n} c_i \left(\frac{\partial W_i}{\partial q_i}\right)^2 + V = \alpha_1 \tag{5-141}$$

where we choose α_1 as the total energy. Because of the assumed separability, we know that $(\partial W_i/\partial q_i)^2$ is a function of $(q_i, \alpha_1, \alpha_2, \ldots, \alpha_n)$ and, furthermore, we can choose the separation constants such that the α's appear linearly. Hence we see that the most general form involving the single coordinate q_i is

$$\left(\frac{\partial W_i}{\partial q_i}\right)^2 = -2\psi_i(q_i) + 2\sum_{j=1}^{n} \Phi_{ij}(q_i)\alpha_j \tag{5-142}$$

where the numerical coefficients are chosen for convenience. Then, substituting this expression into Eq. (5-141) and using matrix notation, we obtain

$$-\mathbf{c}^T\boldsymbol{\psi} + \mathbf{c}^T\boldsymbol{\Phi}\boldsymbol{\alpha} + V = \alpha_1 \tag{5-143}$$

Comparing the terms containing α's, we find that

$$\mathbf{c}^T\boldsymbol{\Phi} = (1, 0, \ldots, 0) \tag{5-144}$$

which we recognize as the first Stäckel condition. Similarly, the terms in Eq. (5-143) which do not involve α's must sum to zero, leading to

$$\mathbf{c}^T\boldsymbol{\psi} = V \tag{5-145}$$

which is the second Stäckel condition. Thus, these two conditions are necessary for separability.

Next, let us prove the *sufficiency* of the Stäckel conditions. To simplify the notation, we define a column matrix **a** given by

$$a_i = \left(\frac{\partial W}{\partial q_i}\right)^2 \qquad (i = 1, 2, \ldots, n) \tag{5-146}$$

Then, upon writing the modified Hamilton-Jacobi equation in the form

$$\tfrac{1}{2}\mathbf{c}^T\mathbf{a} + V = \alpha_1 \tag{5-147}$$

and using the second Stäckel condition, we obtain

$$\tfrac{1}{2}\mathbf{c}^T\mathbf{a} + \mathbf{c}^T\boldsymbol{\psi} = (1, 0, \ldots, 0)\boldsymbol{\alpha} \tag{5-148}$$

But from the first Stäckel condition given in Eq. (5-139), we see that

$$\mathbf{c}^T = (1, 0, \ldots, 0)\boldsymbol{\Phi}^{-1} \tag{5-149}$$

Hence, from Eqs. (5-148) and (5-149), we have

$$(1, 0, \ldots, 0)(\tfrac{1}{2}\boldsymbol{\Phi}^{-1}\mathbf{a} + \boldsymbol{\Phi}^{-1}\boldsymbol{\psi}) = (1, 0, \ldots, 0)\boldsymbol{\alpha} \tag{5-150}$$

which has a solution

$$\mathbf{a} = -2\boldsymbol{\psi} + 2\boldsymbol{\Phi}\boldsymbol{\alpha} \tag{5-151}$$

This result is identical with Eq. (5-142), and indicates that the system is separable, thereby confirming the sufficiency of the Stäckel conditions.

The practical usefulness of Stäckel's theorem is limited by the fact that it presents no direct method for determining whether or not a given system is separable by looking at its kinetic and potential energies. But even though Φ and ψ cannot be obtained directly from T and V, these matrices can be constructed with the aid of the characteristic function, once the solution is known.†

As an example, we can take

$$\Phi_{ij} = \frac{\partial W_i}{\partial q_i} \frac{\partial^2 W_i}{\partial q_i\, \partial \alpha_j} \tag{5-152}$$

and, for the case of *linear* expressions in the α's, as in Eq. (5-142), we obtain

$$\psi_i = \sum_{j=1}^{n} \Phi_{ij}\alpha_j - \frac{1}{2}\left(\frac{\partial W_i}{\partial q_i}\right)^2 \tag{5-153}$$

The matrix Φ is nonsingular because the determinant $|\Phi|$ is nonzero, that is,

$$|\Phi| = \frac{\partial W_1}{\partial q_1} \frac{\partial W_2}{\partial q_2} \cdots \frac{\partial W_n}{\partial q_n} \left| \frac{\partial^2 W_i}{\partial q_i\, \partial \alpha_j} \right| \neq 0 \tag{5-154}$$

This follows from the fact that W is a complete integral of the modified Hamilton-Jacobi equation, and therefore in accordance with Eq. (5-48) we have

$$\left| \frac{\partial^2 W}{\partial q_i\, \partial \alpha_j} \right| \neq 0 \tag{5-155}$$

Furthermore, none of the factors $\partial W_i/\partial q_i$ are zero, in general, since they represent the generalized momenta.

Finally, let us consider the question of the uniqueness of the functions associated with the Liouville and Stäckel conditions. In general, these functions are not unique. It is apparent, for example, that $R_i(q_i), f_i(q_i)$, and $v_i(q_i)$ can be multiplied by the same positive constant without influencing the Liouville conditions. Furthermore, since the potential energy is determined only to within an additive constant, each $v_i(q_i)$ may be replaced by $v_i(q_i) + kf_i(q_i)$, where k is the same constant for all i.

When one considers the Stäckel theory, it is clear that the functions $c_i(q_i)$ are determined uniquely for a given kinetic energy function. On the other hand, the first Stäckel condition is unchanged if each column of Φ except the first is multiplied by an arbitrary nonzero constant. Also, we see from Eqs. (5-139) and (5-140) that each function $\psi_i(q_i)$ may have terms added of the form $\sum_{j=2}^{n} \Phi_{ij}k_j$, where each k_j ($j = 2, 3, \ldots, n$) is a constant and is associated with the jth column of Φ.

---

†See L. A. Pars, *A Treatise on Analytical Dynamics* (London: Heinemann, Ltd., 1965), pp. 322–23.

Example 5-3. Let us consider the Kepler problem once again, this time inquiring into its separability. Suppose we use spherical coordinates to specify the position of the particle of unit mass (Fig. 5-3) which is attracted to the fixed point O by an inverse-square gravitational force.

The kinetic and potential energies are

$$T = \frac{1}{2}(\dot{r}^2 + r^2\dot{\theta}^2 + r^2\dot{\phi}^2 \sin^2\theta)$$

$$= \frac{1}{2}p_r^2 + \frac{1}{2r^2}p_\theta^2 + \frac{1}{2r^2\sin^2\theta}p_\phi^2$$

$$(5\text{-}156)$$

and

$$V = -\frac{\mu}{r} \qquad (5\text{-}157)$$

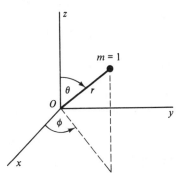

Fig. 5-3. The Kepler problem, using spherical coordinates.

where p_r, p_θ, p_ϕ are the generalized momenta. Then, since this is an orthogonal system, we know that the Hamiltonian $H = T + V$ is constant, and the modified Hamilton-Jacobi equation is

$$\frac{1}{2}\left(\frac{\partial W_r}{\partial r}\right)^2 + \frac{1}{2r^2}\left(\frac{\partial W_\theta}{\partial\theta}\right)^2 + \frac{1}{2r^2\sin^2\theta}\left(\frac{\partial W_\phi}{\partial\phi}\right)^2 - \frac{\mu}{r} = \alpha_t \qquad (5\text{-}158)$$

where

$$W = W_r(r) + W_\theta(\theta) + W_\phi(\phi) \qquad (5\text{-}159)$$

In other words, we seek a characteristic function which is a complete solution of the modified Hamilton-Jacobi equation and which has the separable form of Eq. (5-159).

At this point we note that ϕ is missing from T and V, and is therefore an ignorable coordinate. Hence

$$p_\phi = \frac{\partial W_\phi}{\partial\phi} = \alpha_\phi \qquad (5\text{-}160)$$

and it follows that

$$W_\phi(\phi) = \alpha_\phi\phi \qquad (5\text{-}161)$$

So far, we have obtained two of the required three α's. The third is found by obtaining a separation of variables through first multiplying Eq. (5-158) by $2r^2$, yielding

$$r^2\left(\frac{\partial W_r}{\partial r}\right)^2 - 2r^2\left(\frac{\mu}{r} + \alpha_t\right) + \left(\frac{\partial W_\theta}{\partial\theta}\right)^2 + \frac{\alpha_\phi^2}{\sin^2\theta} = 0 \qquad (5\text{-}162)$$

Here the first two terms are functions of r only, and the last two terms are functions of θ only. Hence, they are each equal to a separation constant, that is, we can take

$$\left(\frac{\partial W_\theta}{\partial \theta}\right)^2 + \frac{\alpha_\phi^2}{\sin^2 \theta} = \alpha_\theta^2 \tag{5-163}$$

and

$$r^2\left(\frac{\partial W_r}{\partial r}\right)^2 - 2r^2\left(\frac{\mu}{r} + \alpha_t\right) = -\alpha_\theta^2 \tag{5-164}$$

The separation constant is chosen to be α_θ^2 (rather than α_θ) as a matter of convenience. This choice permits α_θ to have the dimensions of angular momentum, but does not influence the validity of the Stäckel conditions.

We are now assured that the system is separable because we have

$$\frac{\partial W_r}{\partial r} = \sqrt{2\left(\frac{\mu}{r} + \alpha_t\right) - \frac{\alpha_\theta^2}{r^2}} \tag{5-165}$$

and

$$\frac{\partial W_\theta}{\partial \theta} = \sqrt{\alpha_\theta^2 - \frac{\alpha_\phi^2}{\sin^2 \theta}} \tag{5-166}$$

which are immediately integrable. W_ϕ was found previously.

Now let us check to see that the Stäckel conditions are met. From Eqs. (5-138) and (5-156) we see that

$$\mathbf{c} = \left\{1, \frac{1}{r^2}, \frac{1}{r^2 \sin^2 \theta}\right\} \tag{5-167}$$

Then, using Eq. (5-152), we obtain

$$\mathbf{\Phi} = \begin{bmatrix} 1 & -\dfrac{\alpha_\theta}{r^2} & 0 \\[2mm] 0 & \alpha_\theta & -\dfrac{\alpha_\phi}{\sin^2 \theta} \\[2mm] 0 & 0 & \alpha_\phi \end{bmatrix} \tag{5-168}$$

which confirms that the first condition of Eq. (5-139) is met. We immediately see that if we choose

$$\mathbf{\psi} = \left\{-\frac{\mu}{r}, 0, 0\right\} \tag{5-169}$$

the second condition, Eq. (5-140), is also satisfied.

It is interesting to note that this separable system does not meet the criteria of a Liouville system for $n = 3$, thereby illustrating that the Liouville conditions are not necessary conditions. By comparing Eq. (5-125) with the kinetic energy expression of Eq. (5-156), we see that the Liouville conditions require that

$$\frac{R_\phi(\phi)}{R_r(r)} = \frac{1}{r^2 \sin^2 \theta}$$

which is impossible.

Another approach to this problem is to take advantage immediately of the fact that ϕ is an ignorable coordinate, thereby reducing the number of degrees of freedom from three to two. We can write

$$H = \frac{1}{2}p_r^2 + \frac{1}{2r^2}p_\theta^2 + \frac{\alpha_\phi^2}{2r^2\sin^2\theta} - \frac{\mu}{r} = \alpha_t \qquad (5\text{-}170)$$

and then regard this as an orthogonal system with kinetic and potential energies given by

$$T' = \frac{1}{2}p_r^2 + \frac{1}{2r^2}p_\theta^2 \qquad (5\text{-}171)$$

and

$$V' = \frac{\alpha_\phi^2}{2r^2\sin^2\theta} - \frac{\mu}{r} \qquad (5\text{-}172)$$

The characteristic function is of the form

$$W = W_r(r) + W_\theta(\theta) \qquad (5\text{-}173)$$

and the modified Hamilton-Jacobi equation is

$$\frac{1}{2}\left(\frac{\partial W_r}{\partial r}\right)^2 + \frac{1}{2r^2}\left(\frac{\partial W_\theta}{\partial \theta}\right)^2 + \frac{\alpha_\phi^2}{2r^2\sin^2\theta} - \frac{\mu}{r} = \alpha_t \qquad (5\text{-}174)$$

Once again, the system proves to be separable and Eqs. (5-163) through (5-166) apply.

A check of the Stäckel conditions shows that

$$\mathbf{c} = \left\{1, \frac{1}{r^2}\right\} \qquad (5\text{-}175)$$

$$\mathbf{\Phi} = \begin{bmatrix} 1 & -\dfrac{\alpha_\theta}{r^2} \\ 0 & \alpha_\theta \end{bmatrix} \qquad (5\text{-}176)$$

and

$$\mathbf{\psi} = \left\{-\frac{\mu}{r}, \frac{\alpha_\phi^2}{2\sin^2\theta}\right\} \qquad (5\text{-}177)$$

which satisfy Eqs. (5-139) and (5-140).

For this case in which $n = 2$, we know that the Liouville and Stäckel criteria give identical results. A comparison of the expressions for T' and V' with Eqs. (5-125) and (5-126) reveals that

$$f_r(r) = r^2, \qquad f_\theta(\theta) = 0 \qquad (5\text{-}178)$$

$$R_r(r) = r^2, \qquad R_\theta(\theta) = 1 \qquad (5\text{-}179)$$

$$v_r(r) = -\mu r, \qquad v_\theta(\theta) = \frac{\alpha_\phi^2}{2\sin^2\theta} \qquad (5\text{-}180)$$

thereby confirming that the Liouville conditions apply. Notice that equivalent expressions for T' and V' would have been obtained if the Routhian procedure had been used with respect to the ignorable coordinate ϕ, rather than the Hamiltonian approach employed here.

REFERENCES

1. PARS, L. A., *A Treatise on Analytical Dynamics.* London: Heinemann, 1965. An excellent detailed treatment of the material of this chapter.

2. RUND, H., *The Hamilton-Jacobi Theory in the Calculus of Variations.* London: D. Van Nostrand Company, Ltd., 1966. An extensive and somewhat difficult treatment of Hamilton-Jacobi theory from a largely geometrical viewpoint.

3. CARATHÉODORY, C., *Calculus of Variations and Partial Differential Equations of the First Order, Part I.* (English translation by R. B. Dean and J. J. Brandstatter). San Francisco: Holden-Day, Inc., 1965. A good discussion is given on Pfaffian differential forms and also on the relationship of the Hamilton-Jacobi partial differential equation and the canonical equations.

4. GOLDSTEIN, H., *Classical Mechanics.* Reading, Mass: Addison-Wesley Press, Inc., 1950. A very readable account is given of principal features of the theory, with examples.

PROBLEMS

5-1. Consider a standard Hamiltonian system with $H = q + p$. Solve the Hamilton-Jacobi equation and show that $S = -\alpha t + \alpha q - \frac{1}{2}q^2$ is a solution. Show that the principal function leads to the correct solution of the canonical equations.

5-2. Analyze the top problem (Example 3-5) by using the Hamilton-Jacobi method. Choose the Eulerian angles as coordinates and obtain t and ψ as integrals of functions of θ.

5-3. Use the Hamilton-Jacobi method to solve for the motion of a uniform disk of mass m and radius r which rolls without slipping down a plane that is inclined at an angle γ with the horizontal. Assume initial conditions $\theta(0) = 0$, $\dot{\theta}(0) = \dot{\theta}_0$, where θ is the rotation angle of the disk. Evaluate the integral and solve for θ as a function of t.

5-4. A particle of mass m moves under the action of gravity in the vertical xy plane, where the positive y axis points vertically upward. Assuming that the particle starts at the origin with initial velocity components $(\dot{x}_0, \dot{y}_0)$, use the Hamilton-Jacobi method to solve for x and y as functions of time.

5-5. A particle of mass m moves in a plane orbit. It is attracted to a fixed point by a radial force $F_r = -kr$, where k is a constant. Use polar coordinates (r, θ) and Hamilton-Jacobi theory to obtain solutions in integral form for t and θ as functions of r. Evaluate the integrals and solve for $r(t)$ and $r(\theta)$. Assume that $\theta(0) = 0$, $\dot{\theta}(0) = \dot{\theta}_0$, $\dot{r}(0) = 0$, and $r(0) = r_{\max}$.

5-6. Given a conservative system having

$$T = \frac{1}{2}(q_1^2 + q_2^2)(\dot{q}_1^2 + \dot{q}_2^2), \qquad V = \frac{K}{q_1^2 + q_2^2}$$

where we assume that $K > 0$. Use the Hamilton-Jacobi method to solve for the path in configuration space. Show that

$$\int \frac{dq_1}{\sqrt{q_1^2 + a^2}} - \int \frac{dq_2}{\sqrt{q_2^2 - b^2}} = \text{constant}$$

where $a^2 = (2\alpha_1 - K)/2\alpha_t \geq 0$ and $b^2 = (2\alpha_1 + K)/2\alpha_t > 0$.

5-7. Show that the problem of the motion of a spherical pendulum described by the coordinates (θ, ϕ) is separable in accordance with the Stäckel and Liouville criteria. Use the Hamilton-Jacobi method to solve for the motion in integral form.

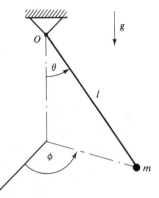

Fig. P5-7.

5-8. Consider a particle of mass m whose position is given by the spherical coordinates (r, θ, ϕ). Suppose the potential function is

$$V = \frac{v_1(r)}{m} + \frac{v_2(\theta)}{mr^2} + \frac{v_3(\phi)}{mr^2 \sin^2 \theta}$$

Let α_1 be the total energy and use α_θ, α_ϕ as separation constants. Obtain integral expressions for $W_1(r)$, $W_2(\theta)$, $W_3(\phi)$, and show that the system meets the Stäckel conditions.

5-9. Show that a general Liouville system meets the two Stäckel conditions for separability. Choose n independent α's, and let α_1 be the total energy. Let the remaining $(n - 1)$ α's be separation constants and, in order to avoid a confusion of notation, eliminate the explicit use of the first separation constant with the aid of Eq. (5-129). Choose $\psi_i = v_i(q_i)/R_i(q_i)$ and let $\Phi_{ij} = p_i(\partial p_i/\partial \alpha_j)$ where $p_i \equiv \partial W_i/\partial q_i$.

5-10. Sometimes it is convenient to use arbitrary differentiable functions $g_i(\alpha_i)$, $(i = 2, 3, \ldots, n)$, in place of the corresponding α's as independent separation constants. Consider the Liouville system of the previous problem and show that in applying the Stäckel conditions we can take

$$\psi_i = \sum_{j=1}^{n} \Phi_{ij} \frac{g_j}{g_j'} - \frac{1}{2} \left(\frac{\partial W_i}{\partial q_i} \right)^2$$

where $\Phi_{ij} = p_i(\partial p_i/\partial \alpha_j)$ and $p_i \equiv \partial W_i/\partial q_i$. Let $g_1(\alpha_1) \equiv \alpha_1$.

5-11. Prove the necessity of the Liouville conditions for the separability of an orthogonal system with $n = 2$. This can be accomplished by following the method for a similar proof of the Stäckel conditions and noting the following correspondences: $R_1 = 1/\Phi_{12}$, $R_2 = -1/\Phi_{22}$, $f_1 = \Phi_{11}/\Phi_{12}$, $f_2 = -\Phi_{21}/\Phi_{22}$, $v_1 = \psi_1/\Phi_{12}$, $v_2 = -\psi_2/\Phi_{22}$.

6

CANONICAL TRANSFORMATIONS

The previous chapter was concerned primarily with the use of the Hamilton-Jacobi method in obtaining the principal function $S(q, \alpha, t)$ and, for conservative systems, the characteristic function $W(q, \alpha)$. In either case, we found that the solution of the Hamilton problem (that is, the solution of the canonical equations) was obtained merely by a process of differentiations and algebraic manipulation.

Another view of the Hamilton problem, however, is that its solution represents a canonical transformation between two points in phase space, namely, a moving point (q, p) and a fixed point (α, β). The principal function is the generating function for this transformation.

With this as a background, we shall turn now to a more detailed exploration of the theory of canonical transformations and study its application to dynamical theory.

6-1. DIFFERENTIAL FORMS AND GENERATING FUNCTIONS

Canonical Transformations. In the discussion of canonical transformations, let us consider only *holonomic systems* which can be described by the standard form of either Hamilton's or Lagrange's equations. We recall from Sec. 4-1 that Hamilton's principle applies to these systems. Hence, if the configuration is specified by the generalized coordinates $q_1, q_2, \ldots, q_n$, we know that

$$\delta \int_{t_0}^{t_1} L(q, \dot{q}, t) \, dt = 0 \tag{6-1}$$

where the end-points of the varied paths are fixed in both configuration space and time.

Now, for a given inertial reference frame, the value of the Lagrangian function is equal to $T - V$, independent of the particular set of generalized coordinates used to specify the configuration of the system. It follows, then, that if a new set of coordinates $Q_1, Q_2, \ldots, Q_n$ is related to the former set by a *point transformation*

$$q_i = q_i(Q, t) \qquad (i = 1, 2, \ldots, n) \tag{6-2}$$

the resulting Lagrangian $L^*(Q, \dot{Q}, t)$ is given by

$$L^*(Q, \dot{Q}, t) = L(q, \dot{q}, t) = T - V \tag{6-3}$$

that is, L and L^* are *equal in value* even though they are, in general, different in form. Furthermore, Hamilton's principle applies to the new Lagrangian function. Thus,

$$\delta \int_{t_0}^{t_1} L^*(Q, \dot{Q}, t)\, dt = 0 \qquad (6\text{-}4)$$

So far we have chosen the Q's to be any set of variables which specify the configuration of the system. The question arises, however, whether it might not be possible for Hamilton's principle to apply following a more general transformation to an $L^*(Q, \dot{Q}, t)$ which differs from $L(q, \dot{q}, t)$ in *both form and value*. For example, we might try a new Lagrangian given by

$$L^*(Q, \dot{Q}, t) = L(q, \dot{q}, t) - \frac{d}{dt}\phi(q, Q, t) \qquad (6\text{-}5)$$

where $\phi(q, Q, t)$ is twice differentiable, but is otherwise arbitrary. Then

$$\delta \int_{t_0}^{t_1} L^*(Q, \dot{Q}, t)\, dt = \delta \int_{t_0}^{t_1} L(q, \dot{q}, t)\, dt - \delta \left[\phi(q, Q, t) \right]_{t_0}^{t_1} \qquad (6\text{-}6)$$

The last term vanishes because the end-points are fixed in extended configuration space, that is, the δq's and δQ's are zero at the fixed times t_0 and t_1. The integral on the right vanishes due to Hamilton's principle, Eq. (6-1). Hence we see that

$$\delta \int_{t_0}^{t_1} L^*(Q, \dot{Q}, t)\, dt = 0 \qquad (6\text{-}7)$$

for the more general case where Eq. (6-5) applies. This means that $L^*(Q, \dot{Q}, t)$ describes the given system just as effectively as $L(q, \dot{q}, t)$ and, furthermore, the standard form of Lagrange's equation is valid for the new coordinate system. We note, however, that $L^*(Q, \dot{Q}, t)$ is no longer equal in value to $L(q, \dot{q}, t)$ and therefore is not equal to $T - V$ in the original inertial frame.

Now suppose we consider the two corresponding Hamiltonian descriptions of the given holonomic system. These Hamiltonian functions are defined by the equations

$$H(q, p, t) = \sum_{i=1}^{n} p_i \dot{q}_i - L(q, \dot{q}, t) \qquad (6\text{-}8)$$

$$K(Q, P, t) = \sum_{i=1}^{n} P_i \dot{Q}_i - L^*(Q, \dot{Q}, t) \qquad (6\text{-}9)$$

where the generalized momenta are given by

$$p_i = \frac{\partial L}{\partial \dot{q}_i}, \qquad P_i = \frac{\partial L^*}{\partial \dot{Q}_i} \qquad (6\text{-}10)$$

Since Hamilton's principle applies to $L(q, \dot{q}, t)$ and $L^*(Q, \dot{Q}, t)$, it follows that the canonical equations are valid for both descriptions, that is,

$$\dot{q}_i = \frac{\partial H}{\partial p_i}, \qquad \dot{p}_i = -\frac{\partial H}{\partial q_i} \qquad (i = 1, 2, \ldots, n) \qquad (6\text{-}11)$$

and

$$\dot{Q}_i = \frac{\partial K}{\partial P_i} \qquad \dot{P}_i = -\frac{\partial K}{\partial Q_i} \qquad (i = 1, 2, \ldots, n) \qquad (6\text{-}12)$$

A transformation from (q, p) to (Q, P) which preserves the canonical form of the equations of motion is known as a *canonical transformation*, provided that the conditions apply to all Hamiltonian systems and not just to specially chosen ones. This suggests that it should be possible to test whether or not a given transformation is canonical without a specific knowledge of the system to which it is applied.

We shall discuss this point in some detail later, but first let us consider a system which has a specific Hamiltonian function $H(q, p, t)$. Suppose we are given transformation equations of the form

$$Q_i = Q_i(q, p, t), \qquad P_i = P_i(q, p, t) \qquad (i = 1, 2, \ldots, n) \qquad (6\text{-}13)$$

where each function is at least twice differentiable. If we solve for L and L^* from Eqs. (6-8) and (6-9) and substitute into Eq. (6-5), we obtain

$$\sum_{i=1}^{n} p_i \, dq_i - H \, dt - \sum_{i=1}^{n} P_i \, dQ_i + K \, dt = d\phi \qquad (6\text{-}14)$$

This important equation states that the difference of two Pfaffian differential forms (Sec. 5-1) is equal to an *exact differential* $d\phi$. A review of its derivation will show that it represents a *sufficient condition* for a canonical transformation from the old variables (q, p) and the associated Hamiltonian $H(q, p, t)$ to the new variables (Q, P) and the Hamiltonian $K(Q, P, t)$. The function $\phi(q, Q, t)$ is called the *generating function* for the transformation.

In testing the differential form of Eq. (6-14) for exactness, we need first to keep in mind the complete set of variables under consideration. Because the generating function is a function of $(2n + 1)$ variables, namely, the q's, Q's, and t, it is natural to express the differential form in terms of these variables. But there are other possibilities. Let us recall the previous discussions of exactness and integrability which occurred in the context of constraints. There we learned that these properties are essentially geometric in nature and are related to the dimensionality of the space. The implication here is that another set of $(2n + 1)$ variables such as (q, p, t) or (Q, P, t) may be used equally well in checking for exactness. It is frequently convenient to define a function $\psi(q, p, t)$ which is equal in value to the generating function, that is,

$$\psi(q, p, t) = \phi(q, Q, t) \qquad (6\text{-}15)$$

Then Eq. (6-14) becomes

$$\sum_{i=1}^{n} p_i \, dq_i - H \, dt - \sum_{i=1}^{n} P_i \, dQ_i + K \, dt = d\psi \qquad (6\text{-}16)$$

Here the differential form is expressed in terms of the variables (q, p, t) by using the transformation equations. Furthermore, we assume that either set of $(2n + 1)$ variables, namely, (q, Q, t) or (q, p, t), are independent. It is important to notice, however, that the function $\psi(q, p, t)$ does not, in general, qualify as a generating function because it contains none of the new variables.

Let us return now to a discussion of the generating function $\phi(q, Q, t)$ which we consider to be *arbitrary*. With this as a starting point, we wish to obtain the corresponding transformation equations. We see that its total differential is

$$d\phi = \sum_{i=1}^{n} \frac{\partial \phi}{\partial q_i} dq_i + \sum_{i=1}^{n} \frac{\partial \phi}{\partial Q_i} dQ_i + \frac{\partial \phi}{\partial t} dt \qquad (6\text{-}17)$$

Assuming that the q's, Q's, and t can be varied independently, we then equate the coefficients of the same differentials in Eqs. (6-14) and (6-17) and obtain

$$p_i = \frac{\partial \phi}{\partial q_i} \qquad (i = 1, 2, \ldots, n) \qquad (6\text{-}18)$$

$$P_i = -\frac{\partial \phi}{\partial Q_i} \qquad (i = 1, 2, \ldots, n) \qquad (6\text{-}19)$$

$$K = H + \frac{\partial \phi}{\partial t} \qquad (6\text{-}20)$$

The independence of the q's and Q's is assured if the following determinant is nonzero, namely,

$$\left| \frac{\partial^2 \phi}{\partial q_i \, \partial Q_j} \right| \neq 0 \qquad (6\text{-}21)$$

The actual transformation equations are obtained by solving Eqs. (6-18) and (6-19) for $q_i(Q, P, t)$ and $p_i(Q, P, t)$ or, conversely, for $Q_i(q, p, t)$ and $P_i(q, p, t)$. Furthermore, the new Hamiltonian function $K(Q, P, t)$ is found by using the transformation equations and Eq. (6-20).

Summarizing, if one is given a transformation in phase space, either of the differential forms given in Eqs. (6-14) and (6-16) can be used to show if it is canonical. On the other hand, if one is given an arbitrary generating function $\phi(q, Q, t)$, then Eqs. (6-18) and (6-19) yield the transformation equations.

Let us consider next some possible simplifications in the test for a canonical transformation. First we recall that a given canonical transformation must apply to *all* systems having the specified n degrees of freedom; therefore, it should be possible to state the criteria for a canonical transformation without reference to the Hamiltonian function. Furthermore, we note from Eq. (6-14) that the time t is unchanged by the transformation and may be regarded as an independent parameter. Since t is not directly involved, we may consider a contemporaneous variation with dt set equal to zero. Then Eqs. (6-14) and (6-16) reduce to

$$\sum_{i=1}^{n} p_i \, \delta q_i - \sum_{i=1}^{n} P_i \, \delta Q_i = \delta \phi \qquad (6\text{-}22)$$

and

$$\sum_{i=1}^{n} p_i \, \delta q_i - \sum_{i=1}^{n} P_i \, \delta Q_i = \delta \psi \qquad (6\text{-}23)$$

It is more convenient to use the latter equation (6-23) in testing whether or not a given transformation is canonical. The functions $Q_i(q, p, t)$ and $P_i(q, p, t)$ from Eq. (6-13) are used in expressing each $P_i \, \delta Q_i$ in terms of the old variables. If the differential form on the left side of Eq. (6-23) is exact, and if the functions $Q_i(q, p, t)$ and $P_i(q, p, t)$ are at least twice differentiable, then the given transformation is canonical, and a function $\psi(q, p, t)$ exists such that

$$p_i - \sum_{j=1}^{n} P_j \frac{\partial Q_j}{\partial q_i} = \frac{\partial \psi}{\partial q_i} \qquad (i = 1, 2, \dots, n) \tag{6-24}$$

$$- \sum_{j=1}^{n} P_j \frac{\partial Q_j}{\partial p_i} = \frac{\partial \psi}{\partial p_i} \qquad (i = 1, 2, \dots, n) \tag{6-25}$$

These equations are equivalent to Eq. (6-23) and may be verified by writing $\delta \psi$ in detail and comparing the coefficients of the terms in each δq_i and δp_i. Upon integrating to obtain $\psi(q, p, t)$, the new Hamiltonian $K(Q, P, t)$ is found by equating the coefficients of dt in Eq. (6-16). This leads to

$$K = H + \frac{\partial \psi}{\partial t} + \sum_{i=1}^{n} P_i \frac{\partial Q_i}{\partial t} \tag{6-26}$$

Notice that Eqs. (6-20) and (6-26) are somewhat different in form, the explanation being that different variables are held constant in the two cases when taking partial derivatives with respect to time.

In considering the Hamiltonian functions $H(q, p, t)$ and $K(Q, P, t)$, notice that they are determined only to within an additive constant for conservative systems. For nonconservative systems, the canonical equations of motion are unchanged even by the addition to the Hamiltonian of a term of the form $h(t)$. From Eqs. (6-20) and (6-26) we see that a similar comment applies to $\phi(q, Q, t)$ and $\psi(q, p, t)$. It is the usual practice, however, to omit terms of this sort.

We have considered that Eq. (6-14) is associated with a canonical transformation from the old variables (q, p) to the new variables (Q, P). But the symmetry of this equation, as well as Eqs. (6-18)–(6-20), shows that the inverse of a given canonical transformation is itself canonical, and is generated by the negative of $\phi(q, Q, t)$. Also, we note that the sum of two exact differentials, expressed in terms of (q, p, t), is exact. Hence it follows that two canonical transformations performed in sequence are equivalent to a single canonical transformation. Furthermore, the identity transformation is canonical. These properties of canonical transformations indicate that, for a given value of n, they form a *group*.

Finally, it is interesting to note from a comparison of Eqs. (5-45) and (6-14) that Hamilton's principal function $S(q, \alpha, t)$ is the generating function for a very special canonical transformation, namely, a transformation such that the new Hamiltonian function K for the given system is identically zero. In this case, the new variables (the α's and β's) are not variables at all, but are constants of the motion and can be considered as a representation of the initial conditions. Furthermore, the canonical transformation equations and the motion of the system are such that a trajectory in (q, p) space is transformed into a fixed point in (α, β) space. The transformation equations are, in effect, the solution of the canonical equations of motion. Hence we see that a given principal function is associated with a particular dynamical system. On the other hand, the test of a general canonical transformation can be made by using Eq. (6-23), without reference to the system under consideration.

Example 6-1. Let us consider the transformation

$$Q = \frac{1}{2}(q^2 + p^2), \qquad P = -\tan^{-1}\frac{q}{p} \qquad (6\text{-}27)$$

We wish to show that this transformation is canonical. Furthermore, the new Hamiltonian function is desired for the case where the old Hamiltonian is

$$H = \tfrac{1}{2}(q^2 + p^2) \qquad (6\text{-}28)$$

To show that the transformation is canonical, consider the differential form of Eq. (6-23) and express it in terms of the old variables. We have

$$p\,\delta q - P\,\delta Q = \left(p + q\tan^{-1}\frac{q}{p}\right)\delta q + p\tan^{-1}\frac{q}{p}\,\delta p \qquad (6\text{-}29)$$

This expression is exact, as may be demonstrated by noting that

$$\frac{\partial}{\partial p}\left(p + q\tan^{-1}\frac{q}{p}\right) = \frac{p^2}{q^2 + p^2} = \frac{\partial}{\partial q}\left(p\tan^{-1}\frac{q}{p}\right) \qquad (6\text{-}30)$$

Hence the transformation is canonical.

Next we observe that the right side of Eq. (6-29) is equal to $\delta\psi$, and integrate to obtain

$$\psi = \frac{1}{2}(q^2 + p^2)\tan^{-1}\frac{q}{p} + \frac{1}{2}qp \qquad (6\text{-}31)$$

The generating function $\phi(q, Q, t)$ is obtained by using Eqs. (6-15), (6-31) and the transformation equations, resulting in

$$\phi = Q\sin^{-1}\frac{q}{\sqrt{2Q}} + \frac{1}{2}q\sqrt{2Q - q^2} \qquad (6\text{-}32)$$

From Eq. (6-20) or (6-26) we see that $K = H$. Hence, using Eqs. (6-27) and (6-28), we find that the new Hamiltonian function is

$$K = Q \qquad (6\text{-}33)$$

The canonical equations of motion, expressed in terms of the new variables, are

$$\dot{Q} = \frac{\partial K}{\partial P} = 0, \qquad \dot{P} = -\frac{\partial K}{\partial Q} = -1 \qquad (6\text{-}34)$$

We see that Q is constant and P decreases linearly with time. Note that the new Hamiltonian K is constant.

Example 6-2. As an example of a rheonomic transformation, consider

$$Q = \sqrt{2q}\, e^t \cos p$$
$$P = \sqrt{2q}\, e^{-t} \sin p \tag{6-35}$$

Once again, we wish to show that the transformation is canonical.

First we obtain

$$p\, \delta q - P\, \delta Q = (p - \sin p \cos p)\, \delta q + 2q \sin^2 p\, \delta p \tag{6-36}$$

and the exactness check yields

$$\frac{\partial}{\partial p}(p - \sin p \cos p) = 2 \sin^2 p = \frac{\partial}{\partial q}(2q \sin^2 p) \tag{6-37}$$

which indicates that the transformation is canonical. To find ψ, we first note from Eq. (6-36) that

$$\frac{\partial \psi}{\partial q} = p - \sin p \cos p$$
$$\frac{\partial \psi}{\partial p} = 2q \sin^2 p = q(1 - \cos^2 p + \sin^2 p) \tag{6-38}$$

Integrating, we obtain

$$\psi = qp - q \sin p \cos p \tag{6-39}$$

In order to obtain the generating function $\phi(q, Q, t)$, we start with the function $\psi(q, p)$ and use the transformation equations to eliminate p in favor of Q. The result is

$$\phi = q \cos^{-1} \frac{Qe^{-t}}{\sqrt{2q}} - \frac{1}{2} Qe^{-t} \sqrt{2q - Q^2 e^{-2t}} \tag{6-40}$$

We note that, although $\phi(q, Q, t)$ contains t explicitly, ψ is not an explicit function of time for this rheonomic example.

Principal Forms of Generating Functions. The previous discussions of canonical transformations from the $2n$ old variables (q, p) to the $2n$ new variables (Q, P) have considered a generating function of the form $\phi(q, Q, t)$. This generating function contains half of the old variables and half of the new variables, namely, n q's and n Q's. Furthermore, we notice that only one variable from each pair of conjugate variables is included; that is, the p's and P's are missing from the generating function. They appear, however, in Eqs. (6-18) and (6-19) which are used to obtain the actual transformation equations.

Now let us recall that the derivation of Eqs. (6-18) and (6-19) required the independence of the q's and Q's. When this condition is not met, as in the case where one or more of the transformation equations has the form of Eq. (6-2), then it is convenient to use a generating function which involves different variables. We might try, for example, a generating function which is a function of the q's, P's, and t. This assumes that the q's and P's can be varied independently, and is in accordance with the Hamiltonian viewpoint that there is no *a priori* reason for favoring Q's over P's as variables.

In order to distinguish the various types of generating functions, let us designate $\phi(q, Q, t)$ as the first type $F_1(q, Q, t)$, that is,

$$F_1(q, Q, t) \equiv \phi(q, Q, t) \tag{6-41}$$

Three other types of generating functions come to mind, namely, $F_2(q, P, t)$, $F_3(p, Q, t)$, and $F_4(p, P, t)$. Generating functions involving other combinations of coordinates and momenta can also be used, provided that no coordinate and its conjugate momentum appear in the same generating function.

In order to show the relationship of $F_2(q, P, t)$ and $F_1(q, Q, t)$, let us start with the basic differential form

$$\sum_{i=1}^{n} p_i \, dq_i - \sum_{i=1}^{n} P_i \, dQ_i - H \, dt + K \, dt = dF_1(q, Q, t) \tag{6-42}$$

We desire to replace the Q's by P's as variables in the generating function and in the corresponding differential form. As a first step, let us add the equation

$$\sum_{i=1}^{n} Q_i \, dP_i + \sum_{i=1}^{n} P_i \, dQ_i = d\left(\sum_{i=1}^{n} Q_i \, P_i\right) \tag{6-43}$$

to Eq. (6-42). We obtain

$$\sum_{i=1}^{n} p_i \, dq_i + \sum_{i=1}^{n} Q_i \, dP_i - H \, dt + K \, dt = dF_2(q, P, t) \tag{6-44}$$

where

$$F_2(q, P, t) = F_1(q, Q, t) + \sum_{i=1}^{n} Q_i P_i \tag{6-45}$$

We can write the total differential of $F_2(q, P, t)$ in the form

$$dF_2 = \sum_{i=1}^{n} \frac{\partial F_2}{\partial q_i} \, dq_i + \sum_{i=1}^{n} \frac{\partial F_2}{\partial P_i} \, dP_i + \frac{\partial F_2}{\partial t} \, dt \tag{6-46}$$

Then, assuming that the variables (q, P, t) are independent, we can equate the coefficients in Eqs. (6-44) and (6-46) and obtain

$$p_i = \frac{\partial F_2}{\partial q_i} \qquad (i = 1, 2, \ldots, n) \tag{6-47}$$

from the dq_i terms. From the dP_i terms we have

$$Q_i = \frac{\partial F_2}{\partial P_i} \qquad (i = 1, 2, \ldots, n) \tag{6-48}$$

A comparison of the coefficients of dt yields

$$K = H + \frac{\partial F_2}{\partial t} \tag{6-49}$$

Here we have obtained the generating function $F_2(q, P, t)$ and the corresponding differential form given in Eq. (6-44). The canonical transformation equations are found from Eqs. (6-47) and (6-48), and the expression for the new Hamiltonian function has the standard form of Eq. (6-49).

Using a similar procedure, we can derive the following equations associated with $F_3(p, Q, t)$:

$$-\sum_{i=1}^{n} q_i \, dp_i - \sum_{i=1}^{n} P_i \, dQ_i - H \, dt + K \, dt = dF_3(p, Q, t) \tag{6-50}$$

$$F_3(p, Q, t) = F_1(q, Q, t) - \sum_{i=1}^{n} q_i p_i \tag{6-51}$$

$$q_i = -\frac{\partial F_3}{\partial p_i} \qquad (i = 1, 2, \ldots, n) \tag{6-52}$$

$$P_i = -\frac{\partial F_3}{\partial Q_i} \qquad (i = 1, 2, \ldots, n) \tag{6-53}$$

$$K = H + \frac{\partial F_3}{\partial t} \tag{6-54}$$

In a similar manner, the equations associated with $F_4(p, P, t)$ are:

$$-\sum_{i=1}^{n} q_i \, dp_i + \sum_{i=1}^{n} Q_i \, dP_i - H \, dt + K \, dt = dF_4(p, P, t) \tag{6-55}$$

$$F_4(p, P, t) = F_2(q, P, t) - \sum_{i=1}^{n} q_i p_i \tag{6-56}$$

$$q_i = -\frac{\partial F_4}{\partial p_i} \qquad (i = 1, 2, \ldots, n) \tag{6-57}$$

$$Q_i = \frac{\partial F_4}{\partial P_i} \qquad (i = 1, 2, \ldots, n) \tag{6-58}$$

$$K = H + \frac{\partial F_4}{\partial t} \tag{6-59}$$

Another approach to the problem of obtaining other generating functions is through the application of the Legendre transformation which was discussed in Chap. 4. If we look at Eqs. (6-45), (6-51), and (6-56), which give the relationships of the generating functions, we see that they have the form of Eq. (4-145) which is associated with the Legendre transformation. Furthermore, the active variables in each case are related by equations having the standard forms of Eqs. (4-143) and (4-147).

As a specific example, consider Eq. (6-45) which relates the generating functions $F_1(q, Q, t)$ and $F_2(q, P, t)$. Here the active variables in the Legendre transformation are the Q's (u's) and P's (v's). In order for the signs to be correct, we let $-F_1(Q)$ correspond to $F(u)$ and let $F_2(P)$ correspond to $G(v)$. We see that Eq. (4-147) leads to Eq. (6-48), and Eq. (4-143) corresponds to Eq. (6-19), or, replacing $\phi(q, Q, t)$ by $F_1(q, Q, t)$,

$$P_i = -\frac{\partial F_1}{\partial Q_i} \qquad (i = 1, 2, \ldots, n) \tag{6-60}$$

If we consider the partial derivatives with respect to the passive variables (q's), we see from Eqs. (6-18) and (6-47) that the form of the equation is the same for either generating function, that is,

$$p_i = \frac{\partial F_1}{\partial q_i} = \frac{\partial F_2}{\partial q_i} \qquad (i = 1, 2, \ldots, n) \tag{6-61}$$

Example 6-3. Consider the transformation

$$Q = \log \frac{\sin p}{q} \tag{6-62}$$

$$P = q \cot p$$

Let us obtain the four major types of generating functions associated with this transformation.

First let us check that the transformation is canonical. We have

$$p \, \delta q - P \, \delta Q = (p + \cot p) \, \delta q - q \cot^2 p \, \delta p \tag{6-63}$$

An exactness check shows that

$$\frac{\partial}{\partial p}(p + \cot p) = \frac{\partial}{\partial q}(-q \cot^2 p) \tag{6-64}$$

Therefore the transformation is canonical. An integration of the right side of Eq. (6-63) yields

$$\psi(q, p) = qp + q \cot p \tag{6-65}$$

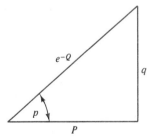

Figure 6-1. A geometrical representation of the transformation.

We know from Eqs. (6-15) and (6-41) that $F_1(q, Q)$ is equal to $\psi(q, p)$ but is expressed in terms of different variables. With the aid of Fig. 6-1, we can use Eq. (6-65) to obtain

$$F_1(q, Q) = q \cos^{-1}\sqrt{1 - q^2 e^{2Q}} + \sqrt{e^{-2Q} - q^2} \tag{6-66}$$

A substitution of $F_1(q, Q)$ into Eqs. (6-18) and (6-19) yields

$$\frac{\partial F_1}{\partial q} = \cos^{-1}\sqrt{1 - q^2 e^{2Q}} = p$$

$$\frac{\partial F_1}{\partial Q} = -\sqrt{e^{-2Q} - q^2} = -P \tag{6-67}$$

Next, let us find $F_2(q, P)$. Noting again that ψ and F_1 are equal, we can use Eqs. (6-45) and (6-65) to obtain

$$F_2 = qp + QP + q \cot p \tag{6-68}$$

Evaluating each of these terms, using (q, P) as variables, we see that

$$qp = q \tan^{-1} \frac{q}{P}$$

$$QP = -P \log\sqrt{q^2 + P^2}$$

$$q \cot p = P$$

which leads to

$$F_2(q, P) = q \tan^{-1} \frac{q}{P} + P(1 - \log\sqrt{q^2 + P^2}) \tag{6-69}$$

As a check, we see from Eqs. (6-47) and (6-48) that

$$\frac{\partial F_2}{\partial q} = \tan^{-1} \frac{q}{P} = p$$

$$\frac{\partial F_2}{\partial P} = -\log\sqrt{q^2 + P^2} = Q \tag{6-70}$$

A similar analysis using Eq. (6-51) yields the third generating function, namely,

$$F_3(p, Q) = F_1 - qp = e^{-Q} \cos p \tag{6-71}$$

We find that, in accordance with Eqs. (6-52) and (6-53),

$$\frac{\partial F_3}{\partial p} = -e^{-Q} \sin p = -q$$

$$\frac{\partial F_3}{\partial Q} = -e^{-Q} \cos p = -P \tag{6-72}$$

Finally, from Eqs. (6-56) and (6-68), we obtain

$$F_4(p, P) = F_2 - qp = QP + q \cot p$$

$$= P + P \log\left(\frac{\cos p}{P}\right) \tag{6-73}$$

A check of Eqs. (6-57) and (6-58) shows that

$$\frac{\partial F_4}{\partial p} = -P \tan p = -q$$

$$\frac{\partial F_4}{\partial P} = \log\left(\frac{\cos p}{P}\right) = Q \tag{6-74}$$

In all cases, the generating function is not an explicit function of time. Therefore we see from Eq. (6-49), for example, that

$$K(Q, P, t) = H(q, p, t) \tag{6-75}$$

regardless of which particular dynamical system is under consideration.

Further Comments on the Hamilton-Jacobi Method. With the perspective resulting from an increased knowledge of the theory of canonical transformations, let us again consider the Hamilton-Jacobi partial differential equation. We recall that the principal function $S(q, \alpha, t)$ is the generating function for a canonical transformation which leads to a new Hamiltonian K that is identically zero. It follows, then, from the canonical equations, that each of the new Q's and P's is constant. If we identify the Q_i's with α_i's and the P_i's with β_i's, we see that the principal function is of the form $F_1(q, \alpha, t)$. Furthermore, the Hamilton-Jacobi equation given in Eq. (5-53) applies and can be written in the form

$$\frac{\partial F_1}{\partial t} + H\left(q, \frac{\partial F_1}{\partial q}, t\right) = 0 \qquad (6\text{-}76)$$

where we note that

$$p_i = \frac{\partial F_1}{\partial q_i} \qquad (i = 1, 2, \ldots, n) \qquad (6\text{-}77)$$

Also, corresponding to Eq. (6-19), we have

$$-\beta_i = \frac{\partial F_1}{\partial \alpha_i} \qquad (i = 1, 2, \ldots, n) \qquad (6\text{-}78)$$

Here we assume that $F_1(q, \alpha, t)$ is at least twice differentiable, and the q's and α's are independently variable as indicated by the determinant

$$\left| \frac{\partial^2 F_1}{\partial q_i \, \partial \alpha_j} \right| \neq 0 \qquad (6\text{-}79)$$

If we now look at Eq. (6-49), it would seem possible to set K equal to zero and once again obtain the Hamilton-Jacobi equation, this time using the generating function $F_2(q, \alpha, t)$. Indeed it is possible, and we obtain

$$\frac{\partial F_2}{\partial t} + H\left(q, \frac{\partial F_2}{\partial q}, t\right) = 0 \qquad (6\text{-}80)$$

which is identical in form with Eq. (6-76). In this case, however, we identify the P_i's with α_i's and the Q_i's with β_i's. From Eqs. (6-47) and (6-48) we have

$$p_i = \frac{\partial F_2}{\partial q_i} \qquad (i = 1, 2, \ldots, n) \qquad (6\text{-}81)$$

and

$$\beta_i = \frac{\partial F_2}{\partial \alpha_i} \qquad (i = 1, 2, \ldots, n) \qquad (6\text{-}82)$$

Earlier in Eq. (6-45) we found that F_1 and F_2 differ by terms of the form $Q_i P_i$ which in the present case turn out to be the constant terms $\beta_i \alpha_i$. This is consistent with the fact that F_1 and F_2 are solutions of the same Hamilton-Jacobi partial differential equation in which F does not appear explicitly, but only in terms of its partial derivatives.

Another point to notice is that Eqs. (6-78) and (6-82) differ in sign, implying that each β_i obtained by using F_2 is equal to the corresponding $-\beta_i$ obtained from F_1. The convention of using the positive sign is widely used in celestial mechanics, but we shall, in general, continue with the negative sign which we have used previously.

In a similar fashion, we can write the Hamilton-Jacobi equation in terms of F_3 or F_4. Here we obtain

$$\frac{\partial F_3}{\partial t} + H\left(-\frac{\partial F_3}{\partial p}, p, t\right) = 0 \qquad (6\text{-}83)$$

or

$$\frac{\partial F_4}{\partial t} + H\left(-\frac{\partial F_4}{\partial p}, p, t\right) = 0 \qquad (6\text{-}84)$$

and the generating functions are of the form $F_3(p, \alpha, t)$ and $F_4(p, \alpha, t)$. In the former case, α_i corresponds to Q_i, while in the latter case, α_i corresponds to P_i. The generating functions F_3 and F_4 are not as widely used in the Hamilton-Jacobi equation because each q_i in the Hamiltonian is replaced by $-\partial F/\partial p_i$, and $H(q, p, t)$ is usually a more complicated function of the q's than the p's.

As a final comment on the Hamilton-Jacobi method, let us consider the use of the characteristic function $W(q, \alpha)$ as a generating function. Here we are concerned with *conservative systems* only, and we recall that $W(q, \alpha)$ is a complete solution of the modified Hamilton-Jacobi equation:

$$H\left(q, \frac{\partial W}{\partial q}\right) = \alpha_n \qquad (6\text{-}85)$$

In other words, the complete solution contains n independent nonadditive constants $\alpha_1, \alpha_2, \ldots, \alpha_n$. In order to have greater generality, let us define n P's which are related to the α's by n independent functions

$$P_i = P_i(\alpha) \qquad (i = 1, 2, \ldots, n) \qquad (6\text{-}86)$$

where each P_i is constant. Conversely, if we solve for the α's as functions of the P's, we have

$$\alpha_i = \alpha_i(P) \qquad (i = 1, 2, \ldots, n) \qquad (6\text{-}87)$$

Now let us eliminate the α's in favor of the P's as the required n independent constants in the characteristic function. The result is a function $W(q, P)$ which we shall use as a generating function of type F_2. From Eqs. (6-47) and (6-48) we obtain

$$p_i = \frac{\partial W}{\partial q_i} \qquad (i = 1, 2, \ldots, n) \qquad (6\text{-}88)$$

$$Q_i = \frac{\partial W}{\partial P_i} \qquad (i = 1, 2, \ldots, n) \qquad (6\text{-}89)$$

which can be solved to yield the canonical transformation equations relating (Q, P) and (q, p). The nature of the Q's, however, is more easily seen by finding the new Hamiltonian and obtaining the canonical equations of motion.

We are using a generating function $W(q, P)$ which is not an explicit function of time, and therefore we see from Eq. (6-49) that K and H are equal. Hence we obtain from Eqs. (6-85) and (6-87) that

$$K(P) = \alpha_n(P) \tag{6-90}$$

where we notice that all the Q's are ignorable and each corresponding P_i is constant. The first n canonical equations are

$$\dot{Q}_i = \frac{\partial K}{\partial P_i} = \nu_i \qquad (i = 1, 2, \ldots, n) \tag{6-91}$$

where the constants ν_i are functions of the P's. Upon integration, we obtain

$$Q_i = \nu_i t + \beta_i \qquad (i = 1, 2, \ldots, n) \tag{6-92}$$

The remaining canonical equations are

$$\dot{P}_i = -\frac{\partial K}{\partial Q_i} = 0 \qquad (i = 1, 2, \ldots, n) \tag{6-93}$$

confirming that all the P's are constant. Also, we see that the complete solution of the Hamilton problem contains the required $2n$ arbitrary constants, namely, the n P's and n β's. The ν's are not independent constants since they are functions of the P's.

In summary, we see that the use of $W(q, P)$ as a generating function has resulted in a new set of coordinates which, in general, vary linearly with time. We notice, however, that if any of the P's are missing from $\alpha_n(P)$, the corresponding ν_i will vanish and the Q_i will be constant. This theory finds application in the study of periodic motion. Specific examples are the use of action and angle variables in physics and the Delaunay variables of celestial mechanics.

6-2. SPECIAL TRANSFORMATIONS

After establishing some of the general criteria for canonical transformations, let us now consider a few simple examples. Then we shall turn to the theory of homogeneous canonical transformations and point transformations. This will, in turn, introduce new aspects of the general theory which apply to cases in which the differentials under consideration are not independent.

Some Simple Transformations. First let us consider the *identity transformation*, which is an obvious example of a canonical transformation. It is generated by a function of the form

$$F_2 = \sum_{i=1}^{n} q_i P_i \tag{6-94}$$

as can be confirmed by noting that

$$p_i = \frac{\partial F_2}{\partial q_i} = P_i, \qquad Q_i = \frac{\partial F_2}{\partial P_i} = q_i \qquad (i = 1, 2, \ldots, n) \qquad (6\text{-}95)$$

Also, it can be shown with the aid of Eqs. (6-52) and (6-53) that the identity transformation is generated by

$$F_3 = -\sum_{i=1}^{n} p_i Q_i \qquad (6\text{-}96)$$

The question arises whether functions of the form $F_1(q, Q, t)$ or $F_4(p, P, t)$ can be used to generate the identity transformation. The answer is no, for the reason that the variables in these generating functions are directly related by the transformation equations themselves, and therefore they cannot be independently varied as required by the theory. This is a special case of a point transformation, a topic to be discussed later in this section.

Now consider a related transformation which results in a translation in phase space. Let us begin with the F_2 generating function of Eq. (6-94). Then replace q_i by $(q_i + c_i)$ and P_i by $(P_i - d_i)$, omitting the product term $c_i d_i$ since it will not influence the transformation. Then we have

$$F_2 = \sum_{i=1}^{n} (q_i P_i + c_i P_i - d_i q_i) \qquad (6\text{-}97)$$

and we obtain

$$p_i = \frac{\partial F_2}{\partial q_i} = P_i - d_i, \qquad Q_i = \frac{\partial F_2}{\partial P_i} = q_i + c_i$$

or

$$Q_i = q_i + c_i, \qquad P_i = p_i + d_i \qquad (i = 1, 2, \ldots, n) \qquad (6\text{-}98)$$

which represents the required translation. Here we usually assume that the c's and d's are constant, although they could be functions of time.

The next simple transformation to be discussed is one which interchanges the roles of coordinates and momenta. Suppose we try

$$F_1 = \sum_{i=1}^{n} q_i Q_i \qquad (6\text{-}99)$$

Using Eqs. (6-18) and (6-19), we obtain

$$p_i = \frac{\partial F_1}{\partial q_i} = Q_i, \qquad P_i = -\frac{\partial F_1}{\partial Q_i} = -q_i \qquad (i = 1, 2, \ldots, n) \qquad (6\text{-}100)$$

The reason that a minus sign is required in one of these transformation equations is that the canonical equations themselves are not perfectly symmetrical with respect to an interchange of coordinates and momenta. A minus sign is involved in the equation for $\dot{p}_i$.

It is important to notice that this transformation emphasizes again the Hamiltonian viewpoint that coordinates and momenta are qualitatively indistinguishable but are necessarily paired as components of a vector in phase space. In other words, they do not always have the familiar attributes of position for coordinates and motion for momenta, but do occur together as paired quantities.

Another simple canonical transformation is the orthogonal transformation of the q's and p's generated by

$$F_2 = \sum_{i=1}^{n} \sum_{j=1}^{n} a_{ij} P_i q_j \qquad (6\text{-}101)$$

where the a's are constants meeting the orthogonality condition

$$\sum_{i=1}^{n} a_{ij} a_{ik} = \delta_{jk} \qquad (6\text{-}102)$$

and δ_{jk} is the Kronecker delta. The orthogonality condition can also be written with the aid of matrix notation as

$$\mathbf{a}^T \mathbf{a} = \mathbf{a} \mathbf{a}^T = \mathbf{1} \qquad (6\text{-}103)$$

implying that the transpose of $\mathbf{a}$ is equal to its inverse. The transformation equations are obtained in the usual manner, namely,

$$p_j = \frac{\partial F_2}{\partial q_j} = \sum_{i=1}^{n} a_{ij} P_i, \qquad Q_i = \frac{\partial F_2}{\partial P_i} = \sum_{j=1}^{n} a_{ij} q_j \qquad (6\text{-}104)$$

The first equation can be solved for the P's in terms of the p's. Then the transformation equations are

$$\begin{aligned} Q_i &= \sum_{j=1}^{n} a_{ij} q_j \\ P_i &= \sum_{j=1}^{n} a_{ij} p_j \end{aligned} \qquad (i = 1, 2, \ldots, n) \qquad (6\text{-}105)$$

For the usual case where the determinant $|a| = 1$, these equations can be considered to represent equal rotations in q-space and p-space. Notice that they reduce to the identity transformation when $a_{ij} = \delta_{ij}$, corresponding to a zero rotation.

Example 6-4. Consider a system having n degrees of freedom. Let us obtain a generating function for the resultant transformation equivalent to a sequence of two simple transformations, namely, a translation followed by a rotation. Suppose the first transformation involves going from (q, p) to (Q, P) and the second from (Q, P) to (Q^*, P^*). From Eq. (6-97), we see that the generating function for the translation is

$$F_2(q, P) = \sum_{i=1}^{n} [(q_i + c_i) P_i - d_i q_i] \qquad (6\text{-}106)$$

which yields the transformation equations

$$Q_i = q_i + c_i, \qquad P_i = p_i + d_i \qquad (i = 1, 2, \ldots, n) \qquad (6\text{-}107)$$

The generating function for the rotation is also of type 2, and, in accordance with Eq. (6-101), is of the form

$$F_2'(Q, P^*) = \sum_{i=1}^{n} \sum_{j=1}^{n} a_{ij} P_i^* Q_j \qquad (6\text{-}108)$$

This results in the second transformation

$$\begin{aligned} Q_i^* &= \sum_{j=1}^{n} a_{ij} Q_j \\ P_i^* &= \sum_{j=1}^{n} a_{ij} P_j \end{aligned} \qquad (i = 1, 2, \ldots, n) \qquad (6\text{-}109)$$

It follows from Eqs. (6-107) and (6-109) that the resultant transformation is

$$\begin{aligned} Q_i^* &= \sum_{j=1}^{n} a_{ij}(q_j + c_j) \\ P_i^* &= \sum_{j=1}^{n} a_{ij}(p_j + d_j) \end{aligned} \qquad (i = 1, 2, \ldots, n) \qquad (6\text{-}110)$$

The problem now facing us is to find the generating function for this resultant transformation. At this point, let us recall from Eq. (6-44) that the variational form (for fixed time) associated with F_2 is

$$\sum_{i=1}^{n} p_i \, \delta q_i + \sum_{i=1}^{n} Q_i \, \delta P_i = \delta F_2(q, P) \qquad (6\text{-}111)$$

In a similar fashion, for the second transformation we have

$$\sum_{i=1}^{n} P_i \, \delta Q_i + \sum_{i=1}^{n} Q_i^* \, \delta P_i^* = \delta F_2'(Q, P^*) \qquad (6\text{-}112)$$

Adding Eqs. (6-111) and (6-112), we obtain

$$\sum_{i=1}^{n} p_i \, \delta q_i + \sum_{i=1}^{n} Q_i^* \, \delta P_i^* = \delta F_2 + \delta F_2' - \sum_{i=1}^{n} Q_i \, \delta P_i - \sum_{i=1}^{n} P_i \, \delta Q_i \qquad (6\text{-}113)$$

Now, in order to have a single canonical transformation relating (q, p) and (Q^*, P^*), the right side of Eq. (6-113) must be equal to an exact differential which we shall call δF_2^*. Hence

$$\delta F_2^* = \delta F_2 + \delta F_2' - \delta\left(\sum_{i=1}^{n} Q_i P_i\right) \qquad (6\text{-}114)$$

In writing the expressions for δF_2 and $\delta F_2'$ let us use the (q, p) variables in order to facilitate the collection of terms and the checks of exactness. From Eqs. (6-106) and (6-107), we have

$$\delta F_2 = \sum_{j=1}^{n} [p_j \, \delta q_j + (q_j + c_j) \, \delta p_j] \qquad (6\text{-}115)$$

In the same manner, we obtain from Eqs. (6-108) and (6-110) that

$$\delta F_2' = \sum_{j=1}^{n} [(p_j + d_j) \, \delta q_j + (q_j + c_j) \, \delta p_j] \qquad (6\text{-}116)$$

where Eq. (6-102) is used to simplify the form. Finally, we have

$$\delta\left(\sum_{i=1}^{n} Q_i P_i\right) = \sum_{j=1}^{n} [(p_j + d_j)\,\delta q_j + (q_j + c_j)\,\delta p_j] \tag{6-117}$$

Hence we see from Eqs. (6-114)—(6-117) that

$$\delta F_2^* = \delta F_2 \tag{6-118}$$

for this particular sequence of transformations, that is, we can equate F_2^* and F_2. All that remains is to write F_2 in terms of the variables (q, P^*). From Eq. (6-109) we obtain

$$P_j = \sum_{i=1}^{n} a_{ij} P_i^* \qquad (j = 1, 2, \ldots, n) \tag{6-119}$$

and then, using Eq. (6-106), we find that

$$F_2^*(q, P^*) = \sum_{i=1}^{n} \sum_{j=1}^{n} [a_{ij}(q_j + c_j)P_i^* - d_j q_j] \tag{6-120}$$

This result can easily be checked to show that it yields the transformation given in Eq. (6-110). Of course, reversing the order of the transformations leads to a different F_2^*.

Homogeneous Canonical Transformations. Previously we showed that if the differential form

$$\sum_{i=1}^{n} p_i\,\delta q_i - \sum_{i=1}^{n} P_i\,\delta Q_i = \delta\psi \tag{6-121}$$

is an *exact differential* when expressed in terms of the variables (q, p), then the transformation given by the $2n$ functions $Q_i(q, p, t)$, $P_i(q, p, t)$ is *canonical*. In this derivation we observed that the time t is unchanged by the transformation, and therefore may be considered as a parameter. Also, we recall that ψ is equal in value to the generating function $\phi(q, Q, t)$ but is expressed in terms of (q, p, t).

Now let us consider the case where ϕ and ψ are *identically zero*. Then Eq. (6-121) becomes

$$\sum_{i=1}^{n} p_i\,\delta q_i - \sum_{i=1}^{n} P_i\,\delta Q_i = 0 \tag{6-122}$$

and the corresponding transformation is called a *homogeneous canonical transformation*. This transformation is also known as a *Mathieu transformation* or *contact transformation*.† Recall that we have previously encountered a transformation of this type in the process of introducing the (α, β) parameters in the development of the Hamilton-Jacobi theory.

What are some of the more important features of homogeneous canonical transformations? First, we notice from Eq. (6-5) that

$$L^*(Q, \dot{Q}, t) = L(q, \dot{q}, t) \tag{6-123}$$

---

†Many authors, however, use the term *contact transformation* in the same sense as our *canonical transformation*.

that is, the new and old Lagrangian functions are equal. It might seem plausible that the new and old Hamiltonian functions should be equal also. But from Eq. (6-26) we see that this is true only if the transformation equations are scleronomic and therefore do not contain time explicitly. In general, we have

$$K = H + \sum_{i=1}^{n} P_i \frac{\partial Q_i}{\partial t} \tag{6-124}$$

for a homogeneous canonical transformation.

Continuing with the case where $\psi \equiv 0$, Eqs. (6-24) and (6-25) reduce to

$$p_i = \sum_{j=1}^{n} P_j \frac{\partial Q_j}{\partial q_i} \qquad (i = 1, 2, \ldots, n) \tag{6-125}$$

$$\sum_{j=1}^{n} P_j \frac{\partial Q_j}{\partial p_i} = 0 \qquad (i = 1, 2, \ldots, n) \tag{6-126}$$

We assume that the P's are not all identically equal to zero, and therefore the determinant of the coefficients in Eq. (6-126) must equal zero. Hence we can write

$$\left| \frac{\partial Q_j}{\partial p_i} \right| = 0 \tag{6-127}$$

This we recognize as the Jacobian determinant for the Q's as functions of the p's. It implies that the n equations

$$Q_j = Q_j(q, p, t) \qquad (j = 1, 2, \ldots, n) \tag{6-128}$$

cannot be solved for the p's as functions of (q, Q, t). Therefore the matrix $(\partial Q_j/\partial p_i)$ has a rank less than n, namely, $(n - m)$ where $1 \leq m \leq n$. As a consequence, there exist m independent relations of the form

$$\Omega_j(q, Q, t) = 0 \qquad (j = 1, 2, \ldots, m) \tag{6-129}$$

that is, the Ω's are functions containing the q's and Q's, but not the p's.

Turning now to a consideration of generating functions associated with homogeneous canonical transformations, we note first that $\phi(q, Q, t)$, which we have identified with $F_1(q, Q, t)$, is identically zero. Referring to Eqs. (6-18) and (6-19), this might seem to imply that the p's and P's are zero. Upon more careful consideration, however, we realize that Eqs. (6-18) and (6-19) are not valid for this case because the q's and Q's cannot be varied independently. In fact, we see from Eq. (6-129) that the δq's and δQ's are constrained by

$$\sum_{i=1}^{n} \frac{\partial \Omega_j}{\partial q_i} \delta q_i + \sum_{i=1}^{n} \frac{\partial \Omega_j}{\partial Q_i} \delta Q_i = 0 \qquad (j = 1, 2, \ldots, m) \tag{6-130}$$

The question arises, then, concerning what can be done to avoid the difficulties caused by the nonindependence of the q's and Q's. One possibility is to use another type of generating function which has independent variables. In the usual case, $F_2(q, P, t)$, $F_3(p, Q, t)$, and $F_4(p, P, t)$ are all suitable. Another approach is to incorporate the constraints represented in Eq. (6-129) by using m Lagrange multipliers and the augmented generating function

$$F_1^*(\lambda, q, Q, t) = \sum_{j=1}^{m} \lambda_j \Omega_j(q, Q, t) \tag{6-131}$$

where we now consider the q's and Q's to be independently variable. Then we can replace Eqs. (6-18) and (6-19) by

$$p_i = \sum_{j=1}^{m} \lambda_j \frac{\partial \Omega_j}{\partial q_i} \qquad (i = 1, 2, \ldots, n) \tag{6-132}$$

$$P_i = -\sum_{j=1}^{m} \lambda_j \frac{\partial \Omega_j}{\partial Q_i} \qquad (i = 1, 2, \ldots, n) \tag{6-133}$$

These $2n$ equations plus the m equations of (6-129) can be solved for the λ's, Q's, and P's as functions of (q, p, t), thereby specifying the canonical transformation. Corresponding to Eq. (6-20) we obtain

$$K = H + \sum_{j=1}^{m} \lambda_j \frac{\partial \Omega_j}{\partial t} \tag{6-134}$$

Another important characteristic of homogeneous canonical transformations is obtained from Eqs. (6-125) and (6-126) and refers to certain homogeneity properties of the functions $Q_i(q, p, t)$ and $P_i(q, p, t)$. Let us take Eq. (6-125), multiply both sides by $\partial Q_k / \partial p_i$, and sum over i. We obtain

$$\sum_{i=1}^{n} p_i \frac{\partial Q_k}{\partial p_i} = \sum_{i=1}^{n} \sum_{j=1}^{n} P_j \frac{\partial Q_k}{\partial p_i} \frac{\partial Q_j}{\partial q_i} \tag{6-135}$$

Similarly, multiplying Eq. (6-126) by $\partial Q_k / \partial q_i$ and summing over i, we obtain

$$\sum_{i=1}^{n} \sum_{j=1}^{n} P_j \frac{\partial Q_k}{\partial q_i} \frac{\partial Q_j}{\partial p_i} = 0 \tag{6-136}$$

If we subtract Eq. (6-136) from Eq. (6-135), we find that

$$\sum_{i=1}^{n} p_i \frac{\partial Q_k}{\partial p_i} = \sum_{i=1}^{n} \sum_{j=1}^{n} P_j \left(\frac{\partial Q_k}{\partial p_i} \frac{\partial Q_j}{\partial q_i} - \frac{\partial Q_k}{\partial q_i} \frac{\partial Q_j}{\partial p_i} \right) \tag{6-137}$$

Next we anticipate a result of the discussion in Sec. 6-3 on Poisson brackets where it is shown that

$$\sum_{i=1}^{n} \left(\frac{\partial Q_k}{\partial p_i} \frac{\partial Q_j}{\partial q_i} - \frac{\partial Q_k}{\partial q_i} \frac{\partial Q_j}{\partial p_i} \right) = 0 \tag{6-138}$$

for all j, k if the transformation is canonical. Hence we conclude from Eqs. (6-137) and (6-138) that

$$\sum_{i=1}^{n} p_i \frac{\partial Q_k}{\partial p_i} = 0 \qquad (k = 1, 2, \ldots, n) \tag{6-139}$$

Using a similar procedure, whereby Eq. (6-125) is multiplied by $\partial P_k/\partial p_i$ and Eq. (6-126) by $\partial P_k/\partial q_i$, we obtain

$$\sum_{i=1}^{n} p_i \frac{\partial P_k}{\partial p_i} = \sum_{j=1}^{n} P_j \,\delta_{jk}$$

or

$$P_k = \sum_{i=1}^{n} p_i \frac{\partial P_k}{\partial p_i} \qquad (k = 1, 2, \ldots, n) \tag{6-140}$$

With the aid of Euler's relation concerning homogeneous functions, we conclude from Eq. (6-139) that $Q_k(q, p, t)$ is *homogeneous of degree zero* in the p's. From Eq. (6-140) we see that $P_k(q, p, t)$ is *homogeneous* and of the *first degree* in the p's. These results apply to all homogeneous canonical transformations.

Point Transformations. Suppose we consider next the class of *homogeneous canonical transformations* for which a full set of n independent functions $\Omega_j(q, Q, t)$ exist and are equal to zero. This implies that the two $n \times n$ coefficient matrices of Eq. (6-130) are nonsingular and the corresponding determinants are nonzero, that is,

$$\left| \frac{\partial \Omega_j}{\partial q_i} \right| \neq 0, \qquad \left| \frac{\partial \Omega_j}{\partial Q_i} \right| \neq 0 \tag{6-141}$$

Hence we can solve for the Q's in terms of (q, t), or for the q's as functions of (Q, t). Assume that this has been accomplished and we have the n equations

$$Q_i = f_i(q, t) \qquad (i = 1, 2, \ldots, n) \tag{6-142}$$

where the f's are twice differentiable. These equations represent a *point transformation*, that is, they represent a mapping of points in configuration space. If one also includes the equations for the P's, the entire set represents a transformation in phase space and is known as an *extended point transformation*.

The momentum transformation equations are readily obtained from Eq. (6-125) which in this case takes the form

$$p_i = \sum_{j=1}^{n} P_j \frac{\partial f_j}{\partial q_i} \qquad (i = 1, 2, \ldots, n) \tag{6-143}$$

Thus the p's are homogeneous linear functions of the P's, and vice versa. If we define n Ω's of the form

$$\Omega_j = Q_j - f_j(q, t) \qquad (j = 1, 2, \ldots, n) \tag{6-144}$$

then Eq. (6-133) reduces to

$$P_i = -\lambda_i \qquad (i = 1, 2, \ldots, n) \tag{6-145}$$

and Eq. (6-132) leads directly to Eq. (6-143). Corresponding to Eqs. (6-124) and (6-134) we have

$$K = H + \sum_{i=1}^{n} P_i \frac{\partial f_i}{\partial t} \tag{6-146}$$

indicating once again that K and H are equal only for the scleronomic case.

The question arises whether all point transformations are necessarily homogeneous canonical transformations. If so, as we can see from Eq. (6-123), this would imply that $L(q, \dot{q}, t)$ and $L^*(Q, \dot{Q}, t)$ are equal. But it is quite possible for L and L^* to be unequal, and this will be the case if two different inertial frames are used, one translating at a constant velocity relative to the other (see Example 6-5). In this instance the function $\phi(q, Q, t) \equiv F_1(q, Q, t)$ does not vanish and we can choose an augmented generating function of the form

$$F_1^*(\lambda, q, Q, t) = F_1(q, Q, t) + \sum_{j=1}^{m} \lambda_j \Omega_j(q, Q, t) \tag{6-147}$$

where we note that $m = n$ for a point transformation. Then, corresponding to Eqs. (6-132) and (6-133), we obtain

$$p_i = \frac{\partial F_1^*}{\partial q_i} = \frac{\partial F_1}{\partial q_i} + \sum_{j=1}^{m} \lambda_j \frac{\partial \Omega_j}{\partial q_i} \qquad (i = 1, 2, \ldots, n) \tag{6-148}$$

$$P_i = -\frac{\partial F_1^*}{\partial Q_i} = -\frac{\partial F_1}{\partial Q_i} - \sum_{j=1}^{m} \lambda_j \frac{\partial \Omega_j}{\partial Q_i} \qquad (i = 1, 2, \ldots, n) \tag{6-149}$$

The Hamiltonian functions are related by

$$K = H + \frac{\partial F_1^*}{\partial t} = H + \frac{\partial F_1}{\partial t} + \sum_{j=1}^{m} \lambda_j \frac{\partial \Omega_j}{\partial t} \tag{6-150}$$

Note that Eqs. (6-147) through (6-150) are valid for $m < n$ also.

Now let us apply the basic differential form of Eq. (6-23) to this case of a *nonhomogeneous point transformation*. Using Eq. (6-142), we obtain

$$\sum_{i=1}^{n} p_i \, \delta q_i - \sum_{i=1}^{n} P_i \, \delta Q_i = \delta \psi = \sum_{i=1}^{n} \left[p_i - \sum_{j=1}^{n} P_j \frac{\partial f_j}{\partial q_i} \right] \delta q_i \tag{6-151}$$

where, in general,

$$\delta \psi = \sum_{i=1}^{n} \frac{\partial \psi}{\partial q_i} \, \partial q_i + \sum_{i=1}^{n} \frac{\partial \psi}{\partial p_i} \, \delta p_i \tag{6-152}$$

Assuming that the δq's and δp's are independent, we can equate the corresponding coefficients and obtain

$$p_i = \sum_{j=1}^{n} P_j \frac{\partial f_j}{\partial q_i} + \frac{\partial \psi}{\partial q_i} \qquad (i = 1, 2, \ldots, n) \tag{6-153}$$

$$\frac{\partial \psi}{\partial p_i} = 0 \qquad (i = 1, 2, \ldots, n) \tag{6-154}$$

We see from Eq. (6-154) that $\psi(q, t)$ is not an explicit function of the p's. Furthermore, it has the proper form for a generating function of type F_1.

Hence we can choose F_1 to be *identically equal* to ψ for a point transformation. Eq. (6-153) shows that, once again, the p's and P's are linearly related, but here the equations are not homogeneous as they were in Eq. (6-143). Also, we see from Eq. (6-149) that P_i and $-\lambda_i$ are equal, as in the homogeneous case. Notice that these results are valid for *arbitrary* $f_j(q, t)$ and $\psi(q, t)$, provided that the assumptions of independence and differentiability apply.

Example 6-5. In order to illustrate the theory for the nonhomogeneous case, suppose we are given the scleronomic extended point transformation

$$Q = \tan q, \qquad P = (p - mv_0)\cos^2 q \qquad (6\text{-}155)$$

where m and v_0 are constants. Suppose we show first that the transformation is canonical, and then let us apply it to a specific mass-spring system.

Substituting the expressions of Eq. (6-155) into the basic differential form of Eq. (6-121), we obtain

$$p\,\delta q - P\,\delta Q = \delta\psi = mv_0\,\delta q \qquad (6\text{-}156)$$

which is exact. Integrating, we find that

$$\psi = mv_0 q \qquad (6\text{-}157)$$

Hence this is a nonhomogeneous canonical transformation.

A further check of the results thus far is obtained by substituting

$$f(q) = \tan q \qquad (6\text{-}158)$$

into Eq. (6-153) which yields the momentum transformation equation of Eq. (6-155). Also, we see that ψ is not an explicit function of p, in accordance with Eq. (6-154).

For this example, we have

$$F_1 \equiv \psi = mv_0 q \qquad (6\text{-}159)$$

$$\Omega = Q - \tan q \qquad (6\text{-}160)$$

and therefore Eq. (6-150) yields

$$K = H \qquad (6\text{-}161)$$

Thus the two Hamiltonian functions are equal, even though L and L^* are unequal due to the fact that the transformation is not homogeneous.

It is interesting to note that we can also use a generating function of the form

$$F_2(q, P) = F_1 + QP = mv_0 q + P\tan q \qquad (6\text{-}162)$$

from which we obtain

$$p = \frac{\partial F_2}{\partial q} = mv_0 + P\sec^2 q$$
$$\qquad\qquad\qquad\qquad\qquad (6\text{-}163)$$
$$Q = \frac{\partial F_2}{\partial P} = \tan q$$

in agreement with our earlier results. Similarly, we can use $F_3(p, Q)$ or $F_4(p, P)$ as generating functions.

Now let us apply this transformation to a specific physical situation. Consider a mass-spring system which is attached to a frame that is translating with a constant velocity v_0, as shown in Fig. 6-2. The unstressed length of the spring is l_0. Here we have a conservative holonomic system (see Example 2-8). Therefore the standard form of either Lagrange's or Hamilton's equations apply, provided that we use an inertial reference frame. In this instance there are two obvious inertial frames, namely, the fixed frame and the frame that is translating with a constant velocity v_0. By associating the Lagrangian L with the fixed frame, and L^* with the moving frame, it turns out that ϕ (or ψ) is given by Eq. (6-159).

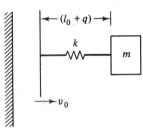

Figure 6-2. A translating mass-spring system.

To see how this comes about, let us obtain the Lagrangian and Hamiltonian functions in terms of the old variables. We have

$$L = T - V = \tfrac{1}{2}m(\dot{q} + v_0)^2 - \tfrac{1}{2}kq^2 \tag{6-164}$$

and therefore

$$p = \frac{\partial L}{\partial \dot{q}} = m(\dot{q} + v_0) \tag{6-165}$$

The Hamiltonian function is

$$H = T_2 - T_0 + V = \frac{p^2}{2m} - v_0 p + \frac{1}{2}kq^2 \tag{6-166}$$

Hence the canonical equations of motion for the (q, p) variables are

$$\dot{q} = \frac{\partial H}{\partial p} = \frac{p}{m} - v_0, \qquad \dot{p} = -\frac{\partial H}{\partial q} = -kq \tag{6-167}$$

Now let us obtain $L^*(Q, \dot{Q})$ relative to the moving frame. In terms of the old coordinates we have

$$T = \tfrac{1}{2}m\dot{q}^2, \qquad V = \tfrac{1}{2}kq^2 \tag{6-168}$$

But, from Eq. (6-163) we find that

$$q = \tan^{-1} Q, \qquad \dot{q} = \frac{\dot{Q}}{1 + Q^2} \tag{6-169}$$

Hence

$$L^* = T - V = \frac{m\dot{Q}^2}{2(1 + Q^2)^2} - \frac{1}{2}k(\tan^{-1} Q)^2 \tag{6-170}$$

and

$$P = \frac{\partial L^*}{\partial \dot{Q}} = \frac{m\dot{Q}}{(1 + Q^2)^2} \tag{6-171}$$

Now, from the first equation of (6-169) we see that

$$\cos q = \frac{1}{\sqrt{1 + Q^2}} \tag{6-172}$$

Hence, we can use Eqs. (6-167), (6-169), and (6-171) to verify the momentum transformation equation given in Eq. (6-155).

In this case the kinetic energy consists of a T_2 term only; hence we have a natural system and therefore the new Hamiltonian function is

$$K = T + V = \frac{P^2}{2m}(1 + Q^2)^2 + \frac{1}{2}k(\tan^{-1} Q)^2 \tag{6-173}$$

The canonical equations are

$$\dot{Q} = \frac{\partial K}{\partial P} = \frac{P}{m}(1 + Q^2)^2$$

$$\dot{P} = -\frac{\partial K}{\partial Q} = -\frac{2P^2Q}{m}(1 + Q^2) - \frac{k \tan^{-1} Q}{1 + Q^2} \tag{6-174}$$

If we now use the transformation equations to transform these equations of motion back to the old variables, we find that indeed they are identical with those of Eq. (6-167).

What has been illustrated by this example? First, we see that a given point transformation represented by $Q(q)$ may not result in a homogeneous canonical transformation. Whether or not ψ vanishes will depend upon the momentum transformation $P(q, p)$ as well. This, in turn, depends upon the choice of reference frames. In this example, L and L^* were found with respect to different inertial frames. But if we had chosen the same inertial frame for both L and L^*, then we would have obtained $L = L^*$, $H = K$, and $\psi \equiv 0$. The canonical transformation would have been homogeneous, and the momentum transformation would have been of the form

$$P = p \cos^2 q \tag{6-175}$$

in accordance with Eq. (6-143).

Momentum Transformations. In the discussion of point transformations, we assumed that the functions $f_i(q, t)$ are given for $i = 1, 2, \ldots, n$. These functions represent a point transformation in configuration space. In order to specify the complete transformation, however, we also need to write the momentum transformation equations. This requires further information which is often given in the form of the function $\psi(q, t)$. Then the momentum transformation equations are found with the aid of Eq. (6-153).

Now suppose that we begin with momentum transformation equations which are given explicitly and have the form

$$P_i = h_i(p, t) \qquad (i = 1, 2, \ldots, n) \tag{6-176}$$

This represents a point transformation in momentum space, and we shall call it a *momentum transformation*. From the Hamiltonian viewpoint, it is analogous to the point transformation in configuration space which was discussed previously. Hence, if care is taken with respect to signs, we can interchange the q's and p's in the previous results. More specifically, let us substitute q_i for p_i and $-p_i$ for q_i. Also replace $F_1(q, Q, t)$ by $F_4(p, P, t)$ and define the functions

$$\omega_j(p, P, t) = 0 \qquad (j = 1, 2, \ldots, m) \tag{6-177}$$

Then, in place of Eqs. (6-148)–(6-150), we have

$$q_i = -\frac{\partial F_4}{\partial p_i} - \sum_{j=1}^{m} \lambda_j \frac{\partial \omega_j}{\partial p_i} \qquad (i = 1, 2, \ldots, n) \tag{6-178}$$

$$Q_i = \frac{\partial F_4}{\partial P_i} + \sum_{j=1}^{m} \lambda_j \frac{\partial \omega_j}{\partial P_i} \qquad (i = 1, 2, \ldots, n) \tag{6-179}$$

$$K = H + \frac{\partial F_4}{\partial t} + \sum_{j=1}^{m} \lambda_j \frac{\partial \omega_j}{\partial t} \tag{6-180}$$

where $m \leq n$.

For the case of a momentum transformation in which we have a full set of n ω's, it is convenient to take

$$\omega_j = P_j - h_j(p, t) = 0 \qquad (j = 1, 2, \ldots, n) \tag{6-181}$$

and then we obtain

$$q_i = -\frac{\partial F_4}{\partial p_i} + \sum_{j=1}^{n} Q_j \frac{\partial h_j}{\partial p_i} \tag{6-182}$$

$$K = H + \frac{\partial F_4}{\partial t} - \sum_{j=1}^{n} Q_j \frac{\partial h_j}{\partial t} \tag{6-183}$$

where we note that F_4 is a function of (p, t) only and the multiplier λ_j is equal to Q_j.

Example 6-6. Consider the transformation

$$\begin{aligned} Q &= q - tp + \tfrac{1}{2}gt^2 \\ P &= p - gt \end{aligned} \tag{6-184}$$

Find $K - H$ and the generating functions.

These equations represent the solution for the one-dimensional motion of a particle of unit mass in a uniform gravitational field, where $Q = q(0)$ and $P = p(0)$. Notice that this is the solution that would be obtained from the Hamilton-Jacobi method, and corresponds to a canonical transformation in which the new Hamiltonian is identically zero.

A check of the canonical differential form yields

$$p\,\delta q - P\,\delta Q = gt\,\delta q + (tp - gt^2)\,\delta p = \delta \psi \tag{6-185}$$

Upon integrating, we obtain

$$\psi = gtq + \tfrac{1}{2}tp^2 - gt^2p \tag{6-186}$$

which confirms that the transformation is canonical. Note, however, that it is not homogeneous.

Now let us consider the Hamiltonian functions for the assumed physical system. Since this is a natural system, we have

$$H(q, p) = T + V = \tfrac{1}{2}p^2 - gq \tag{6-187}$$

We recall from Eq. (6-26) that

$$K - H = \frac{\partial \psi}{\partial t} + P \frac{\partial Q}{\partial t} \tag{6-188}$$

where, in this case,

$$\frac{\partial \psi}{\partial t} = gq + \frac{1}{2}p^2 - 2gtp$$

$$P \frac{\partial Q}{\partial t} = -p^2 + 2gtp - g^2t^2$$

Hence we find that

$$K - H = gq - \tfrac{1}{2}p^2 - g^2t^2 \tag{6-189}$$

From Eqs. (6-187) and (6-189), it would seem that K is equal to $-g^2t^2$. But since this is a function of t only, it will not contribute to the equations of motion and can be omitted.† Therefore we can take

$$K \equiv 0 \tag{6-190}$$

in accordance with the constant values of Q and P.

F_1 is found by expressing ψ in terms of (q, Q, t), using a substitution for p obtained from the first transformation equation. This results in

$$F_1 = \frac{1}{2t}(q - Q)^2 + \frac{1}{2}gt(q + Q) \tag{6-191}$$

where we have again omitted any terms in t only. As a check of this generating function, we obtain

$$p = \frac{\partial F_1}{\partial q} = \frac{1}{t}(q - Q) + \frac{1}{2}gt$$

$$P = -\frac{\partial F_1}{\partial Q} = \frac{1}{t}(q - Q) - \frac{1}{2}gt \tag{6-192}$$

which is in agreement with Eq. (6-184). In a similar fashion, we can obtain expressions for $F_2(q, P, t)$ and $F_3(p, Q, t)$. They are

$$F_2 = (q - \tfrac{1}{2}tP)(P + gt) \tag{6-193}$$

and

$$F_3 = (Q + \tfrac{1}{2}tp)(gt - p) \tag{6-194}$$

Finally, let us obtain $F_4(p, P, t)$. We find that

$$F_4 = F_2 - qp = (q - \tfrac{1}{2}tP)p - qp$$

$$= \tfrac{1}{2}gt^2p - \tfrac{1}{2}tp^2 \tag{6-195}$$

---

†Terms of this sort can also be carried along, of course, and will provide an additional check on the generating functions through the expressions for $\partial F/\partial t$.

Referring again to Eq. (6-184) we see that we have a momentum transformation with the form of Eq. (6-176). In other words, the variables p and P are not independent. In these cases, we can always use a generating function of the form $F_4(p, t)$ since any P's can be expressed in terms of p's. Then, if we let

$$\omega = P - p + gt = 0 \tag{6-196}$$

we can use Eqs. (6-178) and (6-179) to obtain

$$q = -\frac{\partial F_4}{\partial p} - \lambda \frac{\partial \omega}{\partial p} = \lambda - \frac{1}{2}gt^2 + tp$$

$$Q = \frac{\partial F_4}{\partial P} + \lambda \frac{\partial \omega}{\partial P} = \lambda \tag{6-197}$$

These equations are equivalent to the given coordinate transformation equation, thereby checking the result for F_4. A further check on K is obtained by a substitution into Eq. (6-180).

6-3. LAGRANGE AND POISSON BRACKETS

Now let us consider several additional methods which can be used to test a given transformation to determine if it is canonical. These methods are characterized by the fact that they are independent of the associated physical system.

Lagrange Brackets. Suppose we are given the transformation equations

$$Q_i = Q_i(q, p, t), \qquad P_i = P_i(q, p, t) \qquad (i = 1, 2, \ldots, n) \tag{6-198}$$

Once again let us consider the canonical differential form and test it for exactness. We have

$$\sum_{i=1}^{n} p_i \, \delta q_i - \sum_{i=1}^{n} P_i \, \delta Q_i = \sum_{j=1}^{n} \left(p_j - \sum_{i=1}^{n} P_i \frac{\partial Q_i}{\partial q_j} \right) \delta q_j - \sum_{i=1}^{n} \sum_{j=1}^{n} P_i \frac{\partial Q_i}{\partial p_j} \delta p_j \tag{6-199}$$

It may be seen directly, or with the aid of Eqs. (6-24) and (6-25), that the right side of this equation is equal to $\delta\psi$ if all of the following conditions are met, namely,

$$\frac{\partial}{\partial q_k}\left(p_j - \sum_{i=1}^{n} P_i \frac{\partial Q_i}{\partial q_j} \right) = \frac{\partial}{\partial q_j}\left(p_k - \sum_{i=1}^{n} P_i \frac{\partial Q_i}{\partial q_k} \right) \tag{6-200}$$

$$\frac{\partial}{\partial p_k}\left(\sum_{i=1}^{n} P_i \frac{\partial Q_i}{\partial p_j} \right) = \frac{\partial}{\partial p_j}\left(\sum_{i=1}^{n} P_i \frac{\partial Q_i}{\partial p_k} \right) \tag{6-201}$$

$$\frac{\partial}{\partial p_k}\left(p_j - \sum_{i=1}^{n} P_i \frac{\partial Q_i}{\partial q_j} \right) = \frac{\partial}{\partial q_j}\left(-\sum_{i=1}^{n} P_i \frac{\partial Q_i}{\partial p_k} \right) \tag{6-202}$$

where $j, k = 1, 2, \ldots, n$. Upon performing these differentiations, we see that these equations can be written in the following form:

$$\sum_{i=1}^{n} \left(\frac{\partial Q_i}{\partial q_j} \frac{\partial P_i}{\partial q_k} - \frac{\partial P_i}{\partial q_j} \frac{\partial Q_i}{\partial q_k} \right) = 0 \tag{6-203}$$

$$\sum_{i=1}^{n} \left(\frac{\partial Q_i}{\partial p_j} \frac{\partial P_i}{\partial p_k} - \frac{\partial P_i}{\partial p_j} \frac{\partial Q_i}{\partial p_k} \right) = 0 \tag{6-204}$$

$$\sum_{i=1}^{n} \left(\frac{\partial Q_i}{\partial q_j} \frac{\partial P_i}{\partial p_k} - \frac{\partial P_i}{\partial q_j} \frac{\partial Q_i}{\partial p_k} \right) = \delta_{jk} \tag{6-205}$$

where δ_{jk} is the Kronecker delta.

Now let us introduce the *Lagrange bracket* expression for the two variables (u, v) by using the notation

$$[u, v] \equiv \sum_{i=1}^{n} \left(\frac{\partial Q_i}{\partial u} \frac{\partial P_i}{\partial v} - \frac{\partial P_i}{\partial u} \frac{\partial Q_i}{\partial v} \right) \tag{6-206}$$

where u and v are any two of the variables $q_1, q_2, \ldots, q_n, p_1, p_2, \ldots, p_n$. With this notation, we can express the sufficient conditions for a canonical transformation, as given by Eqs. (6-203)—(6-205), in the form

$$[q_j, q_k] = 0, \qquad [p_j, p_k] = 0, \qquad [q_j, p_k] = \delta_{jk} \tag{6-207}$$

for all (j, k). Here we assume that the q's and p's are independently variable.

An alternate method of defining the Lagrange bracket involves the sum of n Jacobian determinants, that is,

$$[u, v] \equiv \sum_{i=1}^{n} \frac{\partial(Q_i, P_i)}{\partial(u, v)} \tag{6-208}$$

which follows directly from Eq. (6-206).

As a consequence of the skew symmetry of the Lagrange brackets, it is clear that

$$[u, v] = -[v, u] \tag{6-209}$$

and

$$[u, u] = [v, v] = 0 \tag{6-210}$$

A general characteristic of the Lagrange bracket is that its value is invariant under a canonical transformation. For example, if the $2n$ variables (q, p) and the $2n$ variables (Q, P) are related by a canonical transformation, then *either* set may be used in evaluating a given Lagrange bracket expression. Thus we have

$$[u, v] = \sum_{i=1}^{n} \left(\frac{\partial q_i}{\partial u} \frac{\partial p_i}{\partial v} - \frac{\partial p_i}{\partial u} \frac{\partial q_i}{\partial v} \right) = \sum_{i=1}^{n} \left(\frac{\partial Q_i}{\partial u} \frac{\partial P_i}{\partial v} - \frac{\partial P_i}{\partial u} \frac{\partial Q_i}{\partial v} \right) \tag{6-211}$$

where each set of dynamical variables is assumed to be a function of (u, v), as well as other variables.

As an illustration, if the functions $Q_i(q, p, t)$ and $P_i(q, p, t)$ are such that the Lagrange bracket expression is met for $u = q_j$ and $v = p_k$, then the first Lagrange bracket expression of Eq. (6-211) is also obviously satisfied. Similar reasoning applies to other functions obtained by sequences of canonical transformations, and therefore canonically related. Hence any set of

$2n$ dynamical variables can be used in the Lagrange bracket if they are related by a canonical transformation. It is for this reason that these variables are not explicitly stated in the bracket notation.

Poisson Brackets. Suppose we are given two functions of the dynamical variables and time, namely, $u(q, p, t)$ and $v(q, p, t)$. The *Poisson bracket* expression for these two functions is

$$(u, v) \equiv \sum_{i=1}^{n} \left(\frac{\partial u}{\partial q_i} \frac{\partial v}{\partial p_i} - \frac{\partial u}{\partial p_i} \frac{\partial v}{\partial q_i} \right) \tag{6-212}$$

We observe immediately that, as for the case of Lagrange brackets, we have

$$(u, v) = -(v, u) \tag{6-213}$$

$$(u, u) = (v, v) = 0 \tag{6-214}$$

The Poisson bracket is useful in testing whether a given transformation is canonical and, in fact, the sufficient conditions for a canonical transformation are that

$$(Q_j, Q_k) = 0, \qquad (P_j, P_k) = 0, \qquad (Q_j, P_k) = \delta_{jk} \tag{6-215}$$

for all (j, k).

In order to show how these equations follow from similar Lagrange bracket expressions, let us consider the $2n$ independent functions $Q_i(q, p, t)$ and $P_i(q, p, t)$ and, conversely, the $2n$ independent functions $q_i(Q, P, t)$ and $p_i(Q, P, t)$. We know that, in general, for any given value of t,

$$\delta q_j = \sum_{i=1}^{n} \left(\frac{\partial q_j}{\partial Q_i} \delta Q_i + \frac{\partial q_j}{\partial P_i} \delta P_i \right) \tag{6-216}$$

$$\delta p_j = \sum_{i=1}^{n} \left(\frac{\partial p_j}{\partial Q_i} \delta Q_i + \frac{\partial p_j}{\partial P_i} \delta P_i \right) \tag{6-217}$$

where

$$\delta Q_i = \sum_{k=1}^{n} \left(\frac{\partial Q_i}{\partial q_k} \delta q_k + \frac{\partial Q_i}{\partial p_k} \delta p_k \right) \tag{6-218}$$

$$\delta P_i = \sum_{k=1}^{n} \left(\frac{\partial P_i}{\partial q_k} \delta q_k + \frac{\partial P_i}{\partial p_k} \delta p_k \right) \tag{6-219}$$

After substituting from Eqs. (6-218) and (6-219) into Eqs. (6-216) and (6-217) and comparing coefficients of the δq's and δp's, we obtain

$$\sum_{i=1}^{n} \left(\frac{\partial q_j}{\partial Q_i} \frac{\partial Q_i}{\partial p_k} + \frac{\partial q_j}{\partial P_i} \frac{\partial P_i}{\partial p_k} \right) = 0 \tag{6-220}$$

$$\sum_{i=1}^{n} \left(\frac{\partial p_j}{\partial Q_i} \frac{\partial Q_i}{\partial q_k} + \frac{\partial p_j}{\partial P_i} \frac{\partial P_i}{\partial q_k} \right) = 0 \tag{6-221}$$

$$\sum_{i=1}^{n} \left(\frac{\partial q_j}{\partial Q_i} \frac{\partial Q_i}{\partial q_k} + \frac{\partial q_j}{\partial P_i} \frac{\partial P_i}{\partial q_k} \right) = \delta_{jk} \tag{6-222}$$

$$\sum_{i=1}^{n} \left(\frac{\partial p_j}{\partial Q_i} \frac{\partial Q_i}{\partial p_k} + \frac{\partial p_j}{\partial P_i} \frac{\partial P_i}{\partial p_k} \right) = \delta_{jk} \tag{6-223}$$

for $j, k = 1, 2, \ldots, n$.

Equations (6-220)–(6-223) are generally valid. At this point, however, we introduce the additional assumption that the transformation equations are such that the Lagrange bracket expressions of Eq. (6-207) apply. Now suppose we compare the expression for $[q_j, p_k]$ given in Eq. (6-205) with Eq. (6-223). In order for these equations to be valid for all the possible Q's and P's, the coefficients of $\partial P_i / \partial p_k$ must be equal, and a similar comment applies to $\partial Q_i / \partial p_k$. Hence we see that

$$\frac{\partial p_j}{\partial P_i} = \frac{\partial Q_i}{\partial q_j}, \qquad \frac{\partial p_j}{\partial Q_i} = -\frac{\partial P_i}{\partial q_j} \tag{6-224}$$

Also, comparing $[q_k, p_j]$ and Eq. (6-222), we obtain

$$\frac{\partial q_j}{\partial Q_i} = \frac{\partial P_i}{\partial p_j}, \qquad \frac{\partial q_j}{\partial P_i} = -\frac{\partial Q_i}{\partial p_j} \tag{6-225}$$

Further comparisons between $[q_j, q_k]$ and Eq. (6-221) and also between $[p_k, p_j]$ and Eq. (6-220) confirm these results. Finally, if we substitute from Eqs. (6-224) and (6-225) into the Lagrange bracket expressions given in Eqs. (6-203)–(6-205), we obtain the corresponding Poisson bracket expressions, namely,

$$\sum_{i=1}^{n} \left(\frac{\partial p_j}{\partial Q_i} \frac{\partial p_k}{\partial P_i} - \frac{\partial p_j}{\partial P_i} \frac{\partial p_k}{\partial Q_i} \right) = 0 \tag{6-226}$$

$$\sum_{i=1}^{n} \left(\frac{\partial q_j}{\partial Q_i} \frac{\partial q_k}{\partial P_i} - \frac{\partial q_j}{\partial P_i} \frac{\partial q_k}{\partial Q_i} \right) = 0 \tag{6-227}$$

$$\sum_{i=1}^{n} \left(\frac{\partial q_j}{\partial Q_i} \frac{\partial p_k}{\partial P_i} - \frac{\partial q_j}{\partial P_i} \frac{\partial p_k}{\partial Q_i} \right) = \delta_{jk} \tag{6-228}$$

which can be written in the form

$$(q_j, q_k) = 0, \qquad (p_j, p_k) = 0, \qquad (q_j, p_k) = \delta_{jk} \tag{6-229}$$

These Poisson bracket expressions, or those given in Eq. (6-215), can be used to confirm that a given transformation is canonical.

The Poisson bracket, in a manner similar to that found previously for the Lagrange bracket, is invariant with respect to a canonical transformation. In other words, the value of (u, v) is independent of which set of canonical variables is used to express the functions u and v, provided that all these (q, p) sets are related by canonical transformations.

Another important property of Poisson brackets is *Jacobi's identity*, namely,

$$(u, (v, w)) + (v, (w, u)) + (w, (u, v)) = 0 \tag{6-230}$$

where u, v, and w are functions of some set of q's and p's.

It is interesting to note that Hamilton's canonical equations can be written using the Poisson bracket notation. We have

$$(q_i, H) = \sum_{k=1}^{n} \left(\frac{\partial q_i}{\partial q_k} \frac{\partial H}{\partial p_k} - \frac{\partial q_i}{\partial p_k} \frac{\partial H}{\partial q_k} \right) = \frac{\partial H}{\partial p_i} \tag{6-231}$$

$$(p_i, H) = \sum_{k=1}^{n} \left(\frac{\partial p_i}{\partial q_k} \frac{\partial H}{\partial p_k} - \frac{\partial p_i}{\partial p_k} \frac{\partial H}{\partial q_k} \right) = -\frac{\partial H}{\partial q_i} \tag{6-232}$$

and therefore the canonical equations can be written in the form

$$\dot{q}_i = (q_i, H), \qquad \dot{p}_i = (p_i, H) \tag{6-233}$$

More generally, if some function $f(q, p, t)$ arises in connection with a dynamical system described by Hamilton's equations, then

$$\frac{df}{dt} = \sum_{i=1}^{n} \left(\frac{\partial f}{\partial q_i} \dot{q}_i + \frac{\partial f}{\partial p_i} \dot{p}_i \right) + \frac{\partial f}{\partial t} \tag{6-234}$$

and, substituting for $\dot{q}_i$ and $\dot{p}_i$ from the canonical equations, we have

$$\frac{df}{dt} = \sum_{i=1}^{n} \left(\frac{\partial f}{\partial q_i} \frac{\partial H}{\partial p_i} - \frac{\partial f}{\partial p_i} \frac{\partial H}{\partial q_i} \right) + \frac{\partial f}{\partial t}$$

$$= (f, H) + \frac{\partial f}{\partial t} \tag{6-235}$$

Hence, for the case where f is not an explicit function of time, we see that $f(q, p)$ is a constant of the motion if

$$(f, H) = 0 \tag{6-236}$$

Thus we can use Poisson brackets to identify constants of the motion.

Now let us state *Poisson's theorem*: If $u(q, p)$ and $v(q, p)$ are integrals of a Hamiltonian system, then the Poisson bracket (u, v) is also an integral, that is, (u, v) is a constant of the motion.

The proof of this theorem begins by noting first that, since u and v are integrals of the motion, Eq. (6-235) implies that

$$(u, H) + \frac{\partial u}{\partial t} = 0 \tag{6-237}$$

$$(v, H) + \frac{\partial v}{\partial t} = 0 \tag{6-238}$$

Also,

$$\frac{d}{dt}(u, v) = ((u, v), H) + \frac{\partial}{\partial t}(u, v)$$

$$= ((u, v), H) + \left(\frac{\partial u}{\partial t}, v \right) + \left(u, \frac{\partial v}{\partial t} \right) \tag{6-239}$$

which, with the aid of Eqs. (6-237) and (6-238), becomes

$$\frac{d}{dt}(u, v) = ((u, v), H) - ((u, H), v) - (u, (v, H))$$

$$= ((u, v), H) + ((H, u), v) + ((v, H), u) = 0 \tag{6-240}$$

the latter expression being zero because of Jacobi's identity. Hence we have the desired result that (u, v) is constant.

Finally, an observation should be made concerning the relationship between the Lagrange and Poisson brackets. Suppose we consider $2n$ independent functions $u_1(q, p), \ldots, u_{2n}(q, p)$ and, conversely, the q's and p's are considered to be functions of the u's. Then, from the defining equations (6-206) and (6-212), we have

$$\sum_{k=1}^{2n} [u_i, u_k](u_j, u_k)$$

$$= \sum_{k=1}^{2n} \sum_{r=1}^{n} \sum_{s=1}^{n} \left(\frac{\partial q_r}{\partial u_i} \frac{\partial p_r}{\partial u_k} - \frac{\partial p_r}{\partial u_i} \frac{\partial q_r}{\partial u_k} \right) \left(\frac{\partial u_j}{\partial q_s} \frac{\partial u_k}{\partial p_s} - \frac{\partial u_j}{\partial p_s} \frac{\partial u_k}{\partial q_s} \right) \quad (6\text{-}241)$$

If we note that

$$\sum_{k=1}^{2n} \frac{\partial q_r}{\partial u_k} \frac{\partial u_k}{\partial q_s} = \delta_{rs}, \qquad \sum_{k=1}^{2n} \frac{\partial p_r}{\partial u_k} \frac{\partial u_k}{\partial p_s} = \delta_{rs} \quad (6\text{-}242)$$

and

$$\sum_{k=1}^{2n} \frac{\partial q_r}{\partial u_k} \frac{\partial u_k}{\partial p_s} = 0, \qquad \sum_{k=1}^{2n} \frac{\partial p_r}{\partial u_k} \frac{\partial u_k}{\partial q_s} = 0 \quad (6\text{-}243)$$

we obtain

$$\sum_{k=1}^{2n} [u_i, u_k](u_j, u_k) = \sum_{r=1}^{n} \left(\frac{\partial u_j}{\partial q_r} \frac{\partial q_r}{\partial u_i} + \frac{\partial u_j}{\partial p_r} \frac{\partial p_r}{\partial u_i} \right) = \frac{\partial u_j}{\partial u_i} = \delta_{ij} \quad (6\text{-}244)$$

This result indicates that the Lagrange and Poisson brackets are, in some sense, reciprocal quantities. To clarify this point, suppose we define the matrix elements

$$L_{ik} = [u_i, u_k], \qquad P_{kj} = (u_j, u_k) \quad (6\text{-}245)$$

Then Eq. (6-244) can be written in the form

$$\mathbf{LP} = \mathbf{1} \quad (6\text{-}246)$$

where $\mathbf{L}$ and $\mathbf{P}$ are both $2n \times 2n$ matrices. Hence we find that $\mathbf{L} = \mathbf{P}^{-1}$.

Example 6-7. Consider the transformation

$$Q = \sqrt{e^{-2q} - p^2}, \qquad P = \cos^{-1}(pe^q) \quad (6\text{-}247)$$

Use the Poisson bracket to show that it is canonical.

This is a simple example for which $n = 1$. Let us make the substitutions $u \equiv Q$ and $v \equiv P$ in Eq. (6-212), and we obtain

$$\frac{\partial Q}{\partial q} = \frac{-e^{-2q}}{\sqrt{e^{-2q} - p^2}}, \qquad \frac{\partial Q}{\partial p} = \frac{-p}{\sqrt{e^{-2q} - p^2}}$$

$$\frac{\partial P}{\partial q} = \frac{-p}{\sqrt{e^{-2q} - p^2}}, \qquad \frac{\partial P}{\partial p} = \frac{-1}{\sqrt{e^{-2q} - p^2}}$$

Therefore

$$(Q, P) = \frac{\partial Q}{\partial q} \frac{\partial P}{\partial p} - \frac{\partial Q}{\partial p} \frac{\partial P}{\partial q} = 1 \quad (6\text{-}248)$$

Furthermore, from Eq. (6-214) we have

$$(Q, Q) = 0, \qquad (P, P) = 0 \quad (6\text{-}249)$$

Hence we see from Eq. (6-215) that the transformation is canonical.

For this case where $n = 1$, we see that the Lagrange bracket $[q, p]$ is identical with (Q, P). Of course, $[q, q]$ and $[p, p]$ are both zero. Hence the Lagrange bracket criteria for a canonical transformation are also fulfilled.

The Bilinear Covariant. Another method which may be used in testing whether a given transformation is canonical involves the bilinear covariant. Suppose we consider the Pfaffian differential form

$$\Omega = \sum_{i=1}^{n} X_i(x)\, dx_i \qquad (6\text{-}250)$$

where the dx's represent an infinitesimal displacement from some reference position in n-space. Similarly, let

$$\theta = \sum_{j=1}^{n} X_j(x)\, \delta x_j \qquad (6\text{-}251)$$

where the δx's are an independent set of infinitesimal displacements from the same reference point. Now we can write

$$\delta\Omega = \sum_{i=1}^{n} (\delta X_i\, dx_i + X_i\, \delta dx_i) \qquad (6\text{-}252)$$

$$d\theta = \sum_{j=1}^{n} (dX_j\, \delta x_j + X_j\, d\delta x_j) \qquad (6\text{-}253)$$

where

$$\delta X_i = \sum_{j=1}^{n} \frac{\partial X_i}{\partial x_j}\, \delta x_j \qquad (6\text{-}254)$$

$$dX_j = \sum_{i=1}^{n} \frac{\partial X_j}{\partial x_i}\, dx_i \qquad (6\text{-}255)$$

Then we obtain

$$\delta\Omega - d\theta = \sum_{i=1}^{n}\sum_{j=1}^{n} \left(\frac{\partial X_i}{\partial x_j} - \frac{\partial X_j}{\partial x_i}\right) dx_i\, \delta x_j \qquad (6\text{-}256)$$

where we note that

$$\delta dx_i = d\delta x_i \qquad (6\text{-}257)$$

since we are now considering contemporaneous variations. If we next recall the notation of Eq. (5-23) and define

$$c_{ij} = \frac{\partial X_i}{\partial x_j} - \frac{\partial X_j}{\partial x_i} \qquad (6\text{-}258)$$

we find that

$$\delta\Omega - d\theta = \sum_{i=1}^{n}\sum_{j=1}^{n} c_{ij}\, dx_i\, \delta x_j \qquad (6\text{-}259)$$

This is the *bilinear covariant* associated with the differential form of Eq. (6-250).

So far we have been concerned with a general Pfaffian differential form. Now let us assume that the variables (q, p) and (Q, P) are related by a canonical transformation and consider the differential form

$$\sum_{i=1}^{n} p_i \, dq_i - \sum_{i=1}^{n} P_i \, dQ_i = d\psi(q, p) \tag{6-260}$$

where $d\psi$ is an exact differential and we consider time as a parameter. Then we have

$$\delta \left(\sum_{i=1}^{n} p_i \, dq_i - \sum_{i=1}^{n} P_i \, dQ_i \right) = \delta d\psi \tag{6-261}$$

In a similar fashion, we can write

$$d \left(\sum_{i=1}^{n} p_i \, \delta q_i - \sum_{i=1}^{n} P_i \, \delta Q_i \right) = d\delta\psi \tag{6-262}$$

Now subtract Eq. (6-262) from Eq. (6-261). With the aid of Eq. (6-257) we obtain

$$\sum_{i=1}^{n} (\delta p_i \, dq_i - dp_i \, \delta q_i) = \sum_{i=1}^{n} (\delta P_i \, dQ_i - dP_i \, \delta Q_i) \tag{6-263}$$

Hence we find that the bilinear covariant of

$$\sum_{i=1}^{n} p_i \, dq_i$$

is invariant with respect to a canonical transformation. On the other hand, if the differential form on the left side of Eq. (6-260) is not exact, when expressed in terms of (q, p) variables, then its bilinear covariant is not zero and Eq. (6-263) does not apply. Hence, we can use Eq. (6-263) as a criterion for canonical transformations in the same manner as the Lagrange bracket or Poisson bracket.

In order to illustrate this method, suppose we consider the transformation of Example 6-7, namely,

$$Q = \sqrt{e^{-2q} - p^2}, \qquad P = \cos^{-1}(pe^q) \tag{6-264}$$

In this instance we have

$$\delta P \, dQ = \frac{(\delta p + p \, \delta q)(e^{-2q} \, dq + p \, dp)}{e^{-2q} - p^2}$$

$$dP \, \delta Q = \frac{(dp + p \, dq)(e^{-2q} \, \delta q + p \, \delta p)}{e^{-2q} - p^2}$$

and therefore the bilinear covariant is

$$\delta P \, dQ - dP \, \delta Q = \frac{(e^{-2q} - p^2) \, \delta p \, dq + (p^2 - e^{-2q}) \, dp \, \delta q}{e^{-2q} - p^2}$$

$$= \delta p \, dq - dp \, \delta q \tag{6-265}$$

Thus Eq. (6-263) applies, confirming that the transformation is canonical.

6-4. MORE GENERAL TRANSFORMATIONS

Necessary Conditions. Let us recall that in this chapter we have been considering holonomic systems which can be described by the standard form of Hamilton's canonical equations. We have defined a transformation between (q, p) and (Q, P) to be a *canonical transformation* if it preserves the canonical form of the equations of motion, irrespective of the particular system to which it is applied. The previous development, however, has resulted in criteria such as the exactness test and the Lagrange and Poisson bracket expressions which represent *sufficient* conditions for a canonical transformation. The question arises, then, whether these conditions are also necessary. It turns out that they are not. More general criteria can be obtained.

Suppose we begin again with the fact that Hamilton's principle applies to a system if and only if it can be described by the standard holonomic forms of Lagrange's and Hamilton's equations. We ask whether Hamilton's principle applies to a Lagrangian function $L^*(Q, \dot{Q}, t)$ which is more general than that given in Eq. (6-5). In answering this question we note first that the Lagrangian equations of motion are unchanged if the Lagrangian function is multiplied by a nonzero constant. So let us consider a new Lagrangian function

$$L^*(Q, \dot{Q}, t) = \mu L(q, \dot{q}, t) - \frac{d}{dt}\phi(q, Q, t) \qquad (6\text{-}266)$$

where μ is a nonzero constant. Applying the same line of reasoning as we used in obtaining Eq. (6-7) from Eq. (6-5), we conclude that if Hamilton's principle applies to $L(q, \dot{q}, t)$, it also applies to $L^*(Q, \dot{Q}, t)$ since all the varied paths have fixed end-points in both configuration space and time. Hence, with the application of the Lagrange equation, we see that L^* describes the system just as effectively as L.

Now recall that

$$L = \sum_{i=1}^{n} p_i \dot{q}_i - H \qquad (6\text{-}267)$$

$$L^* = \sum_{i=1}^{n} P_i \dot{Q}_i - K \qquad (6\text{-}268)$$

and substitute these expressions for L and L^* into Eq. (6-266). We obtain

$$\mu \sum_{i=1}^{n} p_i \, dq_i - \sum_{i=1}^{n} P_i \, dQ_i + (K - \mu H) \, dt = d\phi \qquad (6\text{-}269)$$

This result is a generalization of Eq. (6-14). For a fixed time, we can write the variational equation

$$\mu \sum_{i=1}^{n} p_i \, \delta q_i - \sum_{i=1}^{n} P_i \, \delta Q_i = \delta\psi \qquad (6\text{-}270)$$

where $\psi(q, p, t)$ is equal to $\phi(q, Q, t)$, as in Eq. (6-15).

At this time the necessary and sufficient conditions for a canonical transformation can be stated, namely, that a nonzero constant μ can be found such that the differential form of Eq. (6-270) is exact when expressed in terms of one set of canonical variables. This exactness condition is equivalent to the statement that a function $\psi(q, p, t)$ and a nonzero constant μ exist such that

$$\mu p_i - \sum_{j=1}^{n} P_j \frac{\partial Q_j}{\partial q_i} = \frac{\partial \psi}{\partial q_i} \tag{6-271}$$

$$- \sum_{j=1}^{n} P_j \frac{\partial Q_j}{\partial p_i} = \frac{\partial \psi}{\partial p_i} \tag{6-272}$$

Notice that these equations are identical to Eqs. (6-24) and (6-25) except that p_i is replaced by μp_i.

An equation for the new Hamiltonian function $K(Q, P, t)$ can be found by equating the coefficients of dt on the two sides of Eq. (6-269) after expanding $d\phi$. We obtain

$$K = \mu H + \frac{\partial \phi}{\partial t} \tag{6-273}$$

or, in terms of $\psi(q, p, t)$,

$$K = \mu H + \frac{\partial \psi}{\partial t} + \sum_{i=1}^{n} P_i \frac{\partial Q_i}{\partial t} \tag{6-274}$$

In a similar manner, the bracket and bilinear covariant criteria for a canonical transformation can be generalized. For example, the necessary and sufficient conditions in terms of Poisson brackets are

$$(Q_j, Q_k) = 0, \qquad (P_j, P_k) = 0, \qquad (Q_j, P_k) = \mu \delta_{jk} \tag{6-275}$$

for all j, k. The corresponding Lagrange bracket conditions are

$$[q_j, q_k] = 0, \qquad [p_j, p_k] = 0, \qquad [q_j, p_k] = \mu \delta_{jk} \tag{6-276}$$

The bilinear covariant condition becomes

$$\mu \sum_{i=1}^{n} (\delta p_i \, dq_i - dp_i \, \delta q_i) = \sum_{i=1}^{n} (\delta P_i \, dQ_i - dP_i \, \delta Q_i) \tag{6-277}$$

A few comments are in order. First we notice that the values of the Lagrange and Poisson brackets are no longer invariant with respect to a canonical transformation, but may change by a constant factor. Similarly, the bilinear covariant is not invariant with respect to a canonical transformation. Furthermore, care must be taken in the application of Jacobi's identity to be sure that a consistent set of canonical variables is used. On the other hand, if we should restrict the definition of canonical transformations to the case $\mu = 1$, then the sufficient conditions of the previous development become necessary as well. Most authors follow this approach of adopting a somewhat restricted definition of a canonical transformation which leads to a more clean-cut theory. We also shall assume that μ is restricted to a value of unity unless we specifically state otherwise.

Example 6-8. Consider the transformation

$$Q = \sqrt{q} \, \cos p, \qquad P = \sqrt{q} \, \sin p \qquad (6\text{-}278)$$

Show that it represents a canonical transformation with $\mu \neq 1$. Solve for the new Hamiltonian function for a system having

$$T = \tfrac{1}{2}m\dot{q}^2, \qquad V = \tfrac{1}{2}kq^2 \qquad (6\text{-}279)$$

For this example we shall use the exactness test on the differential form of Eq. (6-270). First we see that

$$P \, \delta Q = \tfrac{1}{2} \sin p \cos p \, \delta q - q \sin^2 p \, \delta p$$

and therefore

$$\mu p \, \delta q - P \, \delta Q = (\mu p - \tfrac{1}{2} \sin p \cos p) \, \delta q + q \sin^2 p \, \delta p \qquad (6\text{-}280)$$

Then we ask if a constant nonzero value of μ exists such that

$$\frac{\partial}{\partial p}(\mu p - \frac{1}{2} \sin p \cos p) = \frac{\partial}{\partial q}(q \sin^2 p) \qquad (6\text{-}281)$$

or

$$\mu - \tfrac{1}{2}\cos^2 p + \tfrac{1}{2}\sin^2 p = \sin^2 p$$

We see that $\mu = \tfrac{1}{2}$ and the transformation is canonical.

To find the new Hamiltonian function K we first obtain the old Hamiltonian. Since this is a natural system, we have

$$H = T + V = \frac{p^2}{2m} + \frac{1}{2}kq^2 \qquad (6\text{-}282)$$

It can be shown that the inverse transformation equations are

$$q = Q^2 + P^2, \qquad p = \tan^{-1}\frac{P}{Q} \qquad (6\text{-}283)$$

and therefore

$$K = \mu H = \frac{1}{4m}\left(\tan^{-1}\frac{P}{Q}\right)^2 + \frac{1}{4}k(Q^2 + P^2)^2 \qquad (6\text{-}284)$$

As a check on whether $K(Q, P)$ correctly represents the system, let us obtain the equations of motion. We find that

$$\dot{Q} = \frac{\partial K}{\partial P} = \frac{Q \tan^{-1}\dfrac{P}{Q}}{2m(Q^2 + P^2)} + kP(Q^2 + P^2) \qquad (6\text{-}285)$$

$$\dot{P} = -\frac{\partial K}{\partial Q} = \frac{P \tan^{-1}\dfrac{P}{Q}}{2m(Q^2 + P^2)} - kQ(Q^2 + P^2) \qquad (6\text{-}286)$$

On the other hand, a differentiation of Eq. (6-283) yields

$$\dot{q} = 2(Q\dot{Q} + P\dot{P}), \qquad \dot{p} = \frac{Q\dot{P} - \dot{Q}P}{Q^2 + P^2} \tag{6-287}$$

Then, substituting from Eqs. (6-285) and (6-286) into Eq. (6-287), we obtain

$$\dot{q} = \frac{p}{m}, \qquad \dot{p} = -kq \tag{6-288}$$

These are the correct equations of motion in the original (q, p) variables.

Time Transformations. The previous discussions of canonical equations considered time as a parameter which does not enter directly into the transformations. In other words, we have assumed that the time variable t is the same before and after the transformation.

Now, however, let us broaden the discussion to include canonical transformations which map a point (q, p, t) into (Q, P, T) where T is the new time. Furthermore, let us introduce τ as the new independent variable. In general, its relation to t or T is unspecified, but we shall assume that it increases continuously as the trajectory is traced out in configuration space. We might, for example, take τ as the distance along the path from a given reference point.

Suppose we consider a holonomic system described by the canonical equations. We can write the modified Hamilton's principle for this system in the form

$$\delta \int_{\tau_0}^{\tau_1} \left(\sum_{i=1}^{n} p_i q_i' - H t' \right) d\tau = 0 \tag{6-289}$$

where the primes indicate differentiations with respect to τ. Here the endpoints of the varied paths are fixed in the q's and t, implying fixed end-points in τ also. Now suppose we consider t as an additional coordinate and take $-H$ as its conjugate momentum. Then we can write the variational equation in the form

$$\delta \int_{\tau_0}^{\tau_1} \sum_{i=1}^{n+1} p_i q_i' \, d\tau = 0 \tag{6-290}$$

In other words, we now have an extended phase space of $(2n + 2)$ dimensions in which we take

$$q_{n+1} = t, \qquad p_{n+1} = -H \tag{6-291}$$

It is important to notice that the δq's and δt associated with the varied path of Eq. (6-290) are not completely independent, but are constrained by the auxiliary condition

$$p_{n+1} + H(q, p, t) = 0 \tag{6-292}$$

In a similar manner, we can describe the same system in terms of the new canonical variables and we obtain

$$\delta \int_{\tau_0}^{\tau_1} \left(\sum_{i=1}^{n} P_i Q_i' - K T' \right) d\tau = 0 \tag{6-293}$$

or

$$\delta \int_{\tau_0}^{\tau_1} \sum_{i=1}^{n+1} P_i Q_i' \, d\tau = 0 \qquad (6\text{-}294)$$

where

$$Q_{n+1} = T, \qquad P_{n+1} = -K \qquad (6\text{-}295)$$

and $K(Q, P, T)$ is the new Hamiltonian function.

A comparison of Eqs. (6-289) and (6-293) indicates that the differential form associated with the canonical transformation is

$$\sum_{i=1}^{n} p_i \, dq_i - H \, dt - \sum_{i=1}^{n} P_i \, dQ_i + K \, dT = d\phi \qquad (6\text{-}296)$$

where $\phi(q, Q, t, T)$ is the generating function for the transformation. Note that terms in $d\tau$ are missing from the differential form, and therefore ϕ is not a function of τ. Furthermore, this indicates that the Hamiltonian functions describing the system in extended phase space are both *identically zero*.† At first glance, this might seem to imply (because of the canonical equations) that the canonical variables are all constant, but this will be shown not to be the case.

Suppose we write Eq. (6-296) in the condensed variational form

$$\sum_{i=1}^{n+1} p_i \, \delta q_i - \sum_{i=1}^{n+1} P_i \, \delta Q_i = \delta \phi \qquad (6\text{-}297)$$

which is similar to Eq. (6-22). Assuming a nonzero determinant

$$\left| \frac{\partial^2 \phi}{\partial q_i \, \partial Q_j} \right| \neq 0 \qquad (i, j, = 1, 2, \ldots, n+1) \qquad (6\text{-}298)$$

we are assured of the independence of the q's and Q's. Then we expand $\delta\phi$ and equate the coefficients on either side of Eq. (6-297), obtaining

$$p_i = \frac{\partial \phi}{\partial q_i} \qquad (i = 1, 2, \ldots, n+1) \qquad (6\text{-}299)$$

$$P_i = -\frac{\partial \phi}{\partial Q_i} \qquad (i = 1, 2, \ldots, n+1) \qquad (6\text{-}300)$$

These equations can be solved for the functions $Q(q, p, t)$, $P(q, p, t)$, $T(q, p, t)$ which specify the transformation.

Let us return now to a consideration of the canonical equations of motion. Although the Hamiltonian function in extended phase space is identically zero, we must account for the constraint expressed in Eq. (6-292). This is accomplished by using the Lagrange multiplier method, resulting in an augmented Hamiltonian

$$\mathcal{H} = \lambda[p_t + H(q, p, t)] \qquad (6\text{-}301)$$

†Note that these functions are different from $H(q, p, t)$ and $K(Q, P, T)$.

where p_t is identical with p_{n+1}. The corresponding canonical equations are

$$q'_i = \frac{\partial \mathfrak{K}}{\partial p_i} = \lambda \frac{\partial H}{\partial p_i} \qquad (i = 1, 2, \ldots, n) \tag{6-302}$$

$$t' = \frac{\partial \mathfrak{K}}{\partial p_t} = \lambda \tag{6-303}$$

$$p'_i = -\frac{\partial \mathfrak{K}}{\partial q_i} = -\lambda \frac{\partial H}{\partial q_i} \qquad (i = 1, 2, \ldots, n) \tag{6-304}$$

$$p'_t = -\frac{\partial \mathfrak{K}}{\partial t} = -\lambda \frac{\partial H}{\partial t} \tag{6-305}$$

Using Eq. (6-303), we see that

$$\frac{1}{\lambda} q'_i = \dot{q}_i \tag{6-306}$$

and therefore Eq. (6-302) becomes

$$\dot{q}_i = \frac{\partial H}{\partial p_i} \qquad (i = 1, 2, \ldots, n) \tag{6-307}$$

Similarly, from Eq. (6-304) we have

$$\dot{p}_i = -\frac{\partial H}{\partial q_i} \qquad (i = 1, 2, \ldots, n) \tag{6-308}$$

Thus we obtain the usual canonical equations for the system. In addition, we note that $p_t = -H$, and therefore Eq. (6-305) reduces to the familiar equation

$$\dot{H} = \frac{\partial H}{\partial t} \tag{6-309}$$

which is identical with Eq. (4-116).

A similar set of canonical equations can be obtained from the new augmented Hamiltonian $\mathfrak{K}$, where we note that

$$\mathcal{K} = \mathfrak{K} \tag{6-310}$$

This equality occurs because the generating function ϕ is not an explicit function of τ. Of course, the new Hamiltonian is expressed in terms of the new variables.

Now let us summarize what has been accomplished. We have considered time as an additional dimension in configuration space and increased the number of dimensions of the phase space to $(2n + 2)$. Since the Hamiltonian functions and the generating function do not contain τ explicitly, the formal analysis resembles that for a conservative scleronomic system; that is, the augmented Hamiltonian functions remain constant in value and, in fact, are equal to zero.

Example 6-9. Consider the canonical transformation generated by

$$\phi = qQ + tT \tag{6-311}$$

Find the transformation equations, the new augmented Hamiltonian, and the canonical equations for the case in which

$$H = \frac{p^2}{2m} + \frac{1}{2}k(t - q)^2 \qquad (6\text{-}312)$$

First let us obtain the transformation equations. From Eqs. (6-299) and (6-300) we have

$$p = \frac{\partial \phi}{\partial q} = Q, \qquad p_t = \frac{\partial \phi}{\partial t} = T \qquad (6\text{-}313)$$

and

$$P = -\frac{\partial \phi}{\partial Q} = -q, \qquad P_T = -\frac{\partial \phi}{\partial T} = -t \qquad (6\text{-}314)$$

The canonical equations are obtained by using Eqs. (6-302) and (6-304) which yield

$$q' = \frac{\lambda p}{m}, \qquad p' = \lambda k(t - q) \qquad (6\text{-}315)$$

But the multiplier λ is equal to $dt/d\tau$, so we can write

$$\dot{q} = \frac{p}{m}, \qquad \dot{p} = k(t - q) \qquad (6\text{-}316)$$

which are the usual equations of motion in terms of the original coordinates.

With the aid of Eqs. (6-301) and (6-310), we see that the new augmented Hamiltonian is

$$\mathcal{K} = \lambda(p_t + H)$$
$$= \lambda\left[T + \frac{Q^2}{2m} + \frac{1}{2}k(P - P_T)^2 \right] \qquad (6\text{-}317)$$

The canonical equations in terms of the new variables are

$$Q' = \frac{\partial \mathcal{K}}{\partial P} = \lambda k(P - P_T) \qquad (6\text{-}318)$$

$$T' = \frac{\partial \mathcal{K}}{\partial P_T} = -\lambda k(P - P_T) \qquad (6\text{-}319)$$

$$P' = -\frac{\partial \mathcal{K}}{\partial Q} = -\frac{\lambda Q}{m} \qquad (6\text{-}320)$$

$$P_T' = -\frac{\partial \mathcal{K}}{\partial T} = -\lambda \qquad (6\text{-}321)$$

Equations (6-318) and (6-320) correspond to the canonical equations of Eq. (6-316). Equation (6-321) is equivalent to $\lambda = dt/d\tau$, and Eq. (6-319) is essentially the statement that $\dot{H} = \partial H/\partial t$.

A check of the transformation equations (6-313) and (6-314) using the Poisson bracket criteria confirms that the transformation is canonical.

6-5. MATRIX FORMULATIONS

Hamilton's Equations. We have been concerned with holonomic systems for which the standard form of the canonical equations applies, namely,

$$\dot{q}_i = \frac{\partial H}{\partial p_i}, \qquad \dot{p}_i = -\frac{\partial H}{\partial q_i} \qquad (i = 1, 2, \ldots, n) \tag{6-322}$$

where the Hamiltonian function has the form $H(q, p, t)$. Here we have $2n$ first-order ordinary differential equations. It is convenient to consider a vector $\mathbf{x}$ in phase space and use the notation

$$\{x_j\} \equiv \{q_1, q_2, \ldots, q_n, p_1, p_2, \ldots, p_n\} \tag{6-323}$$

where $j = 1, 2, \ldots, 2n$. Then the canonical equations can be written in the matrix form

$$\dot{\mathbf{x}} = \mathbf{Z}\mathbf{H}_x \tag{6-324}$$

or, in detail,

$$
\begin{Bmatrix} \dot{q}_1 \\ \cdot \\ \cdot \\ \cdot \\ \dot{q}_n \\ \dot{p}_1 \\ \cdot \\ \cdot \\ \cdot \\ \dot{p}_n \end{Bmatrix} =
\left[\begin{array}{c|c} \mathbf{0} & \mathbf{1} \\ \hline -\mathbf{1} & \mathbf{0} \end{array} \right]
\begin{Bmatrix} \dfrac{\partial H}{\partial q_1} \\ \cdot \\ \cdot \\ \cdot \\ \cdot \\ \cdot \\ \dfrac{\partial H}{\partial p_n} \end{Bmatrix} \tag{6-325}
$$

where the skew-symmetric matrix $\mathbf{Z}$ is partitioned into four $n \times n$ submatrices. The symbols $\mathbf{0}$ and $\mathbf{1}$ represent null and unit matrices, respectively.

Note that

$$\mathbf{Z}^T = \mathbf{Z}^{-1} = -\mathbf{Z} \tag{6-326}$$

Also, we see that $\mathbf{H}_x$ can be considered as the gradient of the Hamiltonian function in a phase space of $2n$ dimensions, since its components consist of the first partial derivatives of H with respect to the q's and p's.

Symplectic Matrices. Let us consider next the matrix formulation of the necessary and sufficient conditions for a canonical transformation. Suppose we are given transformation equations of the form

$$X_i = X_i(x, t) \qquad (i = 1, 2, \ldots, 2n) \tag{6-327}$$

where

$$\{X_i\} \equiv \{Q_1, Q_2, \ldots, Q_n, P_1, P_2, \ldots, P_n\} \tag{6-328}$$

The Jacobian matrix for this transformation is the $2n \times 2n$ matrix having a typical element

$$M_{ij} = \frac{\partial X_i}{\partial x_j} \qquad (6\text{-}329)$$

More specifically, we can write

$$\mathbf{M} = \left[\begin{array}{c:c} \mathbf{A} & \mathbf{B} \\ \hdashline \mathbf{C} & \mathbf{D} \end{array} \right] \qquad (6\text{-}330)$$

where $\mathbf{A}$, $\mathbf{B}$, $\mathbf{C}$, and $\mathbf{D}$ are $n \times n$ submatrices whose typical elements are

$$A_{ij} = \frac{\partial Q_i}{\partial q_j}, \qquad B_{ij} = \frac{\partial Q_i}{\partial p_j}, \qquad C_{ij} = \frac{\partial P_i}{\partial q_j}, \qquad D_{ij} = \frac{\partial P_i}{\partial p_j} \qquad (6\text{-}331)$$

The Poisson bracket formulation of the necessary and sufficient conditions for a canonical transformation were given in Eq. (6-275). We can write them in the form

$$(Q_j, Q_k) = 0, \qquad (P_j, P_k) = 0, \qquad (Q_j, P_k) = -(P_k, Q_j) = \mu \delta_{jk} \qquad (6\text{-}332)$$

where μ is a nonzero constant which, we recall, is equal to $+1$ in the usual, more restrictive definition of a canonical transformation. Using the $\mathbf{A}$, $\mathbf{B}$, $\mathbf{C}$, $\mathbf{D}$ matrices, we can express these conditions as follows:

$$\mathbf{AB}^T - \mathbf{BA}^T = \mathbf{0}, \qquad \mathbf{CD}^T - \mathbf{DC}^T = \mathbf{0} \qquad (6\text{-}333)$$
$$\mathbf{AD}^T - \mathbf{BC}^T = \mu\mathbf{1}, \qquad \mathbf{CB}^T - \mathbf{DA}^T = -\mu\mathbf{1}$$

The first two equations imply that $\mathbf{AB}^T$ and $\mathbf{CD}^T$ are symmetric matrices, while the last two equations are essentially identical statements of a single condition on the matrices. All four equations are combined in the single expression

$$\mathbf{MZM}^T = \mu\mathbf{Z} \qquad (6\text{-}334)$$

where μ is a nonzero scalar constant. If this equation is premultiplied by $\mathbf{Z}^{-1}\mathbf{M}^{-1}$ and postmultiplied by $\mathbf{ZM}$, we obtain the alternate form

$$\mathbf{M}^T\mathbf{ZM} = \mu\mathbf{Z} \qquad (6\text{-}335)$$

Here we have used Eq. (6-326) and the fact that

$$\mathbf{ZZ} = -\mathbf{1} \qquad (6\text{-}336)$$

A $2n \times 2n$ matrix $\mathbf{M}$ which satisfies Eq. (6-335), or Eq. (6-334), is known as a *symplectic matrix*. Hence we can state that the necessary and sufficient condition for a canonical transformation is that its Jacobian matrix be symplectic.

Now suppose we consider the inverse transformation to Eq. (6-327), namely,

$$x_i = x_i(X, t) \qquad (i = 1, 2, \ldots, 2n) \qquad (6\text{-}337)$$

Let its Jacobian matrix be **m**, where

$$m_{ij} = \frac{\partial x_i}{\partial X_j} \tag{6-338}$$

The matrices **M** and **m** are inverses of each other, that is,

$$\mathbf{M} = \mathbf{m}^{-1} \tag{6-339}$$

If the original transformation is canonical, we know that the inverse transformation must also be canonical. Hence if **M** is symplectic, then **m** is also symplectic. To show this, we can premultiply each side of Eq. (6-335) by $\mathbf{m}^T$ and postmultiply each side by **m**. Then, using Eq. (6-339), we obtain

$$\mathbf{m}^T \mathbf{Z} \mathbf{m} = \frac{1}{\mu} \mathbf{Z} \tag{6-340}$$

where $1/\mu$ is a nonzero scalar constant. An alternate form is

$$\mathbf{m} \mathbf{Z} \mathbf{m}^T = \frac{1}{\mu} \mathbf{Z} \tag{6-341}$$

Eqs. (6-334), (6-335), (6-340), and (6-341) constitute four equivalent statements of the necessary and sufficient conditions for a canonical transformation. Any one implies the other three.

Example 6-10. Suppose we are given the transformation

$$Q = aq, \qquad P = bp \tag{6-342}$$

Let us employ the matrix method to find any conditions on the constants a and b which are required for a canonical transformation.

In this case the Jacobian matrix is

$$\mathbf{M} = \begin{bmatrix} a & 0 \\ 0 & b \end{bmatrix} \tag{6-343}$$

Upon substituting into Eq. (6-335), we obtain

$$\mathbf{M}^T \mathbf{Z} \mathbf{M} = \begin{bmatrix} 0 & ab \\ -ab & 0 \end{bmatrix} = \mu \mathbf{Z} \tag{6-344}$$

where $\mu = ab$. Thus the given transformation is canonical for all nonzero values of ab, but is canonical in the usual, more restricted, sense only for $ab = 1$.

6-6. FURTHER TOPICS

Infinitesimal Canonical Transformations. Suppose we consider a canonical transformation in which the new q's and p's are only slightly different from the old values. Let us choose a generating function $F_2(q, P, t)$ which differs by an infinitesimal amount from the generating function for an identity transformation. Referring to Eq. (6-94), for example, we can take

$$F_2(q, P, t) = \sum_{i=1}^{n} q_i P_i + \epsilon G(q, P, t) \qquad (6\text{-}345)$$

where ϵ is a small parameter and $G(q, P, t)$ is assumed to be twice differentiable. Then, with the aid of Eqs. (6-47) and (6-48), we obtain

$$p_i = \frac{\partial F_2}{\partial q_i} = P_i + \epsilon \frac{\partial G}{\partial q_i}$$

$$Q_i = \frac{\partial F_2}{\partial P_i} = q_i + \epsilon \frac{\partial G}{\partial P_i}$$

or

$$Q_i - q_i = \epsilon \frac{\partial G}{\partial P_i}$$

$$P_i - p_i = -\epsilon \frac{\partial G}{\partial q_i} \qquad (6\text{-}346)$$

These equations represent an *infinitesimal canonical transformation*. Notice that the small changes in the coordinates and momenta are of first order in ϵ. So if we neglect terms of order ϵ^2 or higher, we can replace each P_i by p_i in G. The resulting function $G(q, p, t)$ can be considered as the *generator* of the infinitesimal canonical transformation.

Now suppose we denote $Q_i - q_i$ by δq_i and $P_i - p_i$ by δp_i. Then, from Eq. (6-346), we see that the infinitesimal changes in the q's and p's generated by $G(q, p, t)$ are given by

$$\delta q_i = \epsilon \frac{\partial G}{\partial p_i}$$

$$\delta p_i = -\epsilon \frac{\partial G}{\partial q_i} \qquad (6\text{-}347)$$

These changes may or may not be considered to take place with the passage of time. For the particular case where we choose the Hamiltonian $H(q, p, t)$ as the generator of an infinitesimal canonical transformation, we can assume that the transformation takes place with a time increment dt. Then, letting $\epsilon = dt$, Eq. (6-347) becomes

$$\dot{q}_i = \frac{\partial H}{\partial p_i}$$

$$\dot{p}_i = -\frac{\partial H}{\partial q_i} \qquad (6\text{-}348)$$

which we recognize as Hamilton's canonical equations. Hence we conclude that the Hamiltonian function can be considered to generate an infinitesimal canonical transformation at each instant of time. This sequence of transformations results in the evolution of the dynamical system with time. Since a sequence of canonical transformations is equivalent to a single canonical transformation, we see that any two points on a given trajectory in phase space are connected by a canonical transformation.

Next let us consider the variation in the value of a function $f(q, p, t)$ due to changes in the q's and p's resulting from an infinitesimal canonical transformation. Assuming a fixed time, we have

$$\delta f = \sum_{i=1}^{n} \left(\frac{\partial f}{\partial q_i} \delta q_i + \frac{\partial f}{\partial p_i} \delta p_i \right)$$

which, with the aid of Eq. (6-347), becomes

$$\delta f = \epsilon \sum_{i=1}^{n} \left(\frac{\partial f}{\partial q_i} \frac{\partial G}{\partial p_i} - \frac{\partial f}{\partial p_i} \frac{\partial G}{\partial q_i} \right) = \epsilon(f, G) \tag{6-349}$$

More generally, if we allow the time to vary and let $\epsilon = dt$, we obtain

$$df = \left[(f, G) + \frac{\partial f}{\partial t} \right] dt \tag{6-350}$$

For the case where the generating function G is the Hamiltonian, we see that

$$\dot{f} = (f, H) + \frac{\partial f}{\partial t} \tag{6-351}$$

in agreement with Eq. (6-235).

In general, an infinitesimal canonical transformation changes the Hamiltonian function in both form and value. From Eq. (6-49) we have

$$K(Q, P, t) = H(q, p, t) + \epsilon \frac{\partial G}{\partial t} \tag{6-352}$$

But, for a given time t, we see from Eq. (6-349) that

$$H(Q, P, t) = H(q, p, t) + \delta H$$
$$= H(q, p, t) + \epsilon(H, G) \tag{6-353}$$

where $H(Q, P, t)$ is the original Hamiltonian function with Q_i, P_i substituted for q_i, p_i. Hence

$$K(Q, P, t) = H(Q, P, t) + \epsilon \left[(G, H) + \frac{\partial G}{\partial t} \right] \tag{6-354}$$

In obtaining this result, note that the Poisson bracket $(G, H) = -(H, G)$.

Now suppose that the generating function is a constant of the motion. Then, in accordance with Eq. (6-351),

$$(G, H) + \frac{\partial G}{\partial t} = 0 \tag{6-355}$$

and we see that the new Hamiltonian is merely the old function with the new Q's and P's substituted for the old.

As an illustration of this point, suppose we describe a system of N particles in terms of a Cartesian coordinate system and take the total angular momentum about the x axis as the generator of an infinitesimal canonical transformation. Hence we can write

$$G = \sum_{i=1}^{N} m_i(y_i \dot{z}_i - z_i \dot{y}_i)$$

$$= \sum_{i=1}^{N} (y_i p_{zi} - z_i p_{yi}) \qquad (6\text{-}356)$$

and, using Eq. (6-347), we find that

$$\delta x_i = 0, \qquad \delta y_i = -\epsilon z_i, \qquad \delta z_i = \epsilon y_i \qquad (6\text{-}357)$$

and

$$\delta p_{xi} = 0, \qquad \delta p_{yi} = -\epsilon p_{zi}, \qquad \delta p_{zi} = \epsilon p_{yi} \qquad (6\text{-}358)$$

This transformation corresponds to an infinitesimal rotation $d\theta = \epsilon$ of the system of particles about the x axis. A similar rotation applies to the momentum components. More generally, the use of a momentum p_i as a generating function will result in an infinitesimal change δq_i in the conjugate coordinate. This applies, as in the given illustration, even if p_i is expressed in terms of other variables.

In order to show the effect of the transformation on the Hamiltonian, let us assume a specific case in which each particle is attracted toward the origin by a force proportional to its distance. Then the generating function is a constant of the motion and, furthermore, we have a natural system with the Hamiltonian function equal to the constant total energy.

$$H = \frac{1}{2} \sum_{i=1}^{N} \left(\frac{p_{xi}^2}{m_i} + \frac{p_{yi}^2}{m_i} + \frac{p_{zi}^2}{m_i} \right) + \frac{1}{2} \sum_{i=1}^{N} k_i(x_i^2 + y_i^2 + z_i^2) \qquad (6\text{-}359)$$

Next use the transformation equations

$$x_i = X_i, \qquad y_i = Y_i + \epsilon Z_i, \qquad z_i = Z_i - \epsilon Y_i \qquad (6\text{-}360)$$

$$p_{xi} = P_{xi}, \qquad p_{yi} = P_{yi} + \epsilon P_{zi}, \qquad p_{zi} = P_{zi} - \epsilon P_{yi} \qquad (6\text{-}361)$$

and, applying Eq. (6-352) and neglecting terms of order ϵ^2, we obtain

$$K = \frac{1}{2} \sum_{i=1}^{N} \left(\frac{P_{xi}^2}{m_i} + \frac{P_{yi}^2}{m_i} + \frac{P_{zi}^2}{m_i} \right) + \frac{1}{2} \sum_{i=1}^{N} (X_i^2 + Y_i^2 + Z_i^2) \qquad (6\text{-}362)$$

We conclude from this example that the infinitesimal rotation generated by the angular momentum function leaves the Hamiltonian unchanged in both form and value, indicating a symmetry in this function. This can be seen by noting that the sums of the squares of the components of momentum or position are invariant under a rotation about the origin. Conversely, the Hamiltonian function generates a motion of the system in time which leaves the angular momentum invariant.

Liouville's Theorem. Once again let us consider a holonomic system which is described by the standard form of Hamilton's equations. With the passage of time the system traces out a trajectory in phase space. Any two points on a given trajectory are connected by a canonical transformation, and, in fact, any two adjacent points are connected by an infinitesimal canonical transformation. These transformations must approach the identity transformation as the time interval approaches zero, and therefore we consider only the more restricted case of canonical transformations in which $\mu = 1$.

Liouville's theorem was discussed in Sec. 4-4, and we found that the Jacobian determinant is equal to unity for a canonical transformation relating an initial point on a trajectory in phase space to any later point on the trajectory. Using the terminology of this chapter, we see that

$$|\mathbf{M}| = \frac{\partial(Q_1, \ldots, P_n)}{\partial(q_1, \ldots, p_n)} = 1 \qquad (6\text{-}363)$$

where M_{ij} is given by Eq. (6-329).

Another viewpoint of Liouville's theorem, which is particularly applicable to statistical mechanics, is to consider the motion of a group or *ensemble* of identical systems whose phase points lie within a volume element of phase space at some initial time. As these points proceed along their individual trajectories, the corresponding volume element may change its shape, but, in accordance with Liouville's theorem, its volume does not change. In other words, *the phase fluid is incompressible.* If we define a density $\rho(q, p, t)$ for these phase points, Liouville's theorem then implies that ρ is a constant of the motion; that is, using the Poisson bracket notation, we have

$$\dot{\rho} = (\rho, H) + \frac{\partial \rho}{\partial t} = 0 \qquad (6\text{-}364)$$

Liouville's theorem can also be stated in terms of integral invariants, and this is the next topic for discussion.

Integral Invariants. Consider a system of differential equations of the form

$$\dot{x}_i = X_i(x_1, x_2, \ldots, x_m, t) \qquad (i = 1, 2, \ldots, m) \qquad (6\text{-}365)$$

We can consider the x's to be the components of an m-dimensional vector $\mathbf{x}$. A solution point of this system traces a curve in the $(m + 1)$-space $(\mathbf{x}, t)$. We shall be interested, however, not so much in a single solution curve as in the totality of these curves. Furthermore, we shall assume that they fill the space in such a way that only one curve passes through any given point.

Let us assume the existence of a vector field $\mathbf{F}(\mathbf{x}, t)$ in this space and consider the line integral

$$I = \int_\gamma \mathbf{F} \cdot d\mathbf{x} = \int_\gamma \sum_{i=1}^m F_i \, dx_i \qquad (6\text{-}366)$$

where the curve γ consists of solution points at the same instant t. Since each point of γ is also a point on a solution curve C, we see that as the solution proceeds these points move along their respective curves, causing γ to change shape (Fig. 6-3). It may turn out that the value of the integral I does

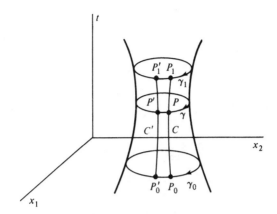

Figure 6-3. The curve of integration γ at various times.

not change with time; if so, it is known as an *integral invariant*. Since γ represents a one-dimensional domain, the integral invariant is of the *first order*. If the condition holds for all curves, open or closed, the integral is called an *absolute* integral invariant. If, on the other hand, the integral is constant only for closed curves, it is a *relative* integral invariant.

Now let us consider the conditions under which this invariance occurs. Suppose we introduce a parameter u which varies from 0 to 1 as the integration proceeds along γ. For a fixed value of u, the corresponding point traces a single solution curve C with increasing time. Therefore those solution curves of Eq. (6-365) which intersect γ can be written as functions of u and t, that is,

$$x_i = x_i(u, t) \qquad (i = 1, 2, \ldots, m) \qquad (6\text{-}367)$$

where these functions are twice differentiable. Hence we obtain

$$\frac{\partial x_i}{\partial t} = X_i(\mathbf{x}, t) \qquad (i = 1, 2, \ldots, m) \qquad (6\text{-}368)$$

With this notation, we can write the integral of Eq. (6-366) in the alternate forms

$$I = \int_\gamma \sum_{i=1}^{m} F_i \, dx_i = \int_0^1 \sum_{i=1}^{m} F_i \frac{\partial x_i}{\partial u} du \qquad (6\text{-}369)$$

The invariance of I requires that dI/dt be zero as we follow the set of solution points comprising γ. It is convenient to differentiate the second form since it has fixed limits. First we note that u remains constant for this differentiation within the integral, and we obtain

$$\frac{dF_i}{dt} = \sum_{k=1}^{m} \frac{\partial F_i}{\partial x_k} \frac{\partial x_k}{\partial t} + \frac{\partial F_i}{\partial t}$$

$$= \sum_{k=1}^{m} \frac{\partial F_i}{\partial x_k} X_k + \frac{\partial F_i}{\partial t} \qquad (6\text{-}370)$$

Also,

$$\frac{\partial}{\partial t}\left(\frac{\partial x_i}{\partial u}\right) = \frac{\partial}{\partial u}\left(\frac{\partial x_i}{\partial t}\right) = \sum_{k=1}^{m} \frac{\partial X_i}{\partial x_k} \frac{\partial x_k}{\partial u} \tag{6-371}$$

Then, upon interchanging the indices in the term $F_i(\partial X_i/\partial x_k)(\partial x_k/\partial u)$, we obtain

$$\frac{dI}{dt} = \int_0^1 \sum_{i=1}^{m} \left[\sum_{k=1}^{m} \left(\frac{\partial F_i}{\partial x_k} X_k + F_k \frac{\partial X_k}{\partial x_i} \right) + \frac{\partial F_i}{\partial t} \right] \frac{\partial x_i}{\partial u} \, du$$

$$= \int_\gamma \sum_{i=1}^{m} \left[\sum_{k=1}^{m} \left(\frac{\partial F_i}{\partial x_k} X_k + F_k \frac{\partial X_k}{\partial x_i} \right) + \frac{\partial F_i}{\partial t} \right] dx_i \tag{6-372}$$

In order for this integral to be zero for an arbitrary γ, that is, for independent dx's, each coefficient must be zero. Hence

$$\sum_{k=1}^{m} \left(\frac{\partial F_i}{\partial x_k} X_k + F_k \frac{\partial X_k}{\partial x_i} \right) + \frac{\partial F_i}{\partial t} = 0 \qquad (i = 1, 2, \ldots, m) \tag{6-373}$$

is the necessary condition for an *absolute* integral invariant. It is also sufficient.

It is interesting to note at this point that there is an absolute integral invariant corresponding to each integral of the set of differential equations (6-365). For example, suppose

$$\phi(x_1, x_2, \ldots, x_m, t) = \alpha \tag{6-374}$$

is such an integral. Let us choose a vector field **F** such that

$$F_i = \frac{\partial \phi}{\partial x_i} \qquad (i = 1, 2, \ldots, m) \tag{6-375}$$

Then we can apply Eq. (6-373) to obtain

$$\sum_{k=1}^{m} \left(\frac{\partial^2 \phi}{\partial x_k \, \partial x_i} X_k + \frac{\partial \phi}{\partial x_k} \frac{\partial X_k}{\partial x_i} \right) + \frac{\partial^2 \phi}{\partial t \, \partial x_i} = \frac{\partial}{\partial x_i}\left(\frac{d\phi}{dt}\right) = 0 \tag{6-376}$$

since $d\phi/dt$ is identically zero. Therefore we see that

$$I = \int_\gamma \sum_{i=1}^{m} \frac{\partial \phi}{\partial x_i} \, dx_i \tag{6-377}$$

is an absolute integral invariant for this system.

Now let us return to a further consideration of Eq. (6-372) which can be written in the form

$$\frac{dI}{dt} = \int_\gamma \sum_{i=1}^{m} \left\{ \sum_{k=1}^{m} \left[\frac{\partial}{\partial x_i} (F_k X_k) + \left(\frac{\partial F_i}{\partial x_k} - \frac{\partial F_k}{\partial x_i} \right) X_k \right] + \frac{\partial F_i}{\partial t} \right\} dx_i \tag{6-378}$$

But we recall that time is held constant for this integration, so we can write

$$\sum_{i=1}^{m} \frac{\partial}{\partial x_i} (F_k X_k) \, dx_i = d(F_k X_k) \tag{6-379}$$

Then we obtain another form for Eq. (6-372), namely,

$$\frac{dI}{dt} = \sum_{k=1}^{m} \left[F_k X_k \right]_{u=0}^{1} + \int_{\gamma} \sum_{i=1}^{m} \left[\sum_{k=1}^{m} \left(\frac{\partial F_i}{\partial x_k} - \frac{\partial F_k}{\partial x_i} \right) X_k + \frac{\partial F_i}{\partial t} \right] dx_i \qquad (6\text{-}380)$$

The first term on the right vanishes for all closed curves because the values of $\mathbf{x}$ corresponding to $u = 0$ and $u = 1$ are identical, and $F_k X_k$ is a function of $\mathbf{x}$ only at any given t. We conclude, then, that the line integral I of Eq. (6-366) is a *relative* integral invariant if the integrand of Eq. (6-380) is a perfect *space differential*, since, once again, we consider time as a constant parameter during the integration.

So far we have been concerned with integral invariants of the first order. One can show, however, that each relative integral invariant of the first order is equivalent to an absolute integral invariant of the second order. This follows from an application of Stokes' theorem which states that the line integral of Eq. (6-366) taken around a closed curve is equivalent to the integral

$$I = \int \int \sum_{i=1}^{m} \sum_{j=1}^{m} \left(\frac{\partial F_i}{\partial x_j} - \frac{\partial F_j}{\partial x_i} \right) dx_i \, dx_j \qquad (6\text{-}381)$$

taken over a surface bounded by the curve γ. We might, for example, follow the motion of any surface which is bounded by γ at the initial instant, with care being taken to properly correlate the positive surface direction and the direction of integration around γ. This results in an *absolute* integral invariant of second order because the surface is *not closed*.

A similar argument can be used to show that any relative integral invariant of order r is equivalent to an absolute integral invariant of order $(r + 1)$. We infer that any integral invariant of the maximum order m must be absolute.

Now let us consider integral invariants of order m. In this case we have the multiple integral

$$I = \int \cdots \int M(\mathbf{x}, t) \, dx_1 \, dx_2 \cdots dx_m \qquad (6\text{-}382)$$

and follow a set of system points within a certain initial region as they move in accordance with Eq. (6-365). We seek the conditions on M which must apply in order for I to be an integral invariant.

Suppose we consider the m integrals of the motion

$$\phi_i(x_1, x_2, \ldots, x_m, t) = \alpha_i \qquad (i = 1, 2, \ldots, m) \qquad (6\text{-}383)$$

We can regard these equations as a transformation from the x's to α's, where the α's are constant along any solution curve. Then we can write the integral in terms of the Jacobian of the transformation as follows:

$$I = \int \cdots \int M \frac{\partial(x_1, \ldots, x_m)}{\partial(\alpha_1, \ldots, \alpha_m)} \, d\alpha_1 \, d\alpha_2 \cdots d\alpha_m \qquad (6\text{-}384)$$

The limits on the α's are not functions of time, and the region of integration is arbitrary, so the required condition is that

$$\frac{d}{dt}\left[M \frac{\partial(x_1, \ldots, x_m)}{\partial(\alpha_1, \ldots, \alpha_m)} \right] = 0 \tag{6-385}$$

The derivative with respect to time of a typical element of the Jacobian is

$$\frac{d}{dt}\left(\frac{\partial x_i}{\partial \alpha_j} \right) = \frac{\partial \dot{x}_i}{\partial \alpha_j} = \sum_{k=1}^{m} \frac{\partial X_i}{\partial x_k} \frac{\partial x_k}{\partial \alpha_j} \tag{6-386}$$

and the derivative of the determinant is found as the sum of m determinants, with the elements of successive rows being differentiated, one row to each determinant. We find, then, that the time derivative of the Jacobian is

$$\frac{d}{dt}\left[\frac{\partial(x_1, x_2, \ldots, x_m)}{\partial(\alpha_1, \alpha_2, \ldots, \alpha_m)} \right] = \left(\frac{\partial X_1}{\partial x_1} + \frac{\partial X_2}{\partial x_2} + \cdots + \frac{\partial X_m}{\partial x_m} \right) \frac{\partial(x_1, x_2, \ldots, x_m)}{\partial(\alpha_1, \alpha_2, \ldots, \alpha_m)} \tag{6-387}$$

The necessary condition on M, obtained from Eqs. (6-385) and (6-387), can therefore be written as

$$\frac{dM}{dt} + M \sum_{i=1}^{m} \frac{\partial X_i}{\partial x_i} = 0 \tag{6-388}$$

or

$$\sum_{i=1}^{m} \frac{\partial}{\partial x_i} (MX_i) + \frac{\partial M}{\partial t} = 0 \tag{6-389}$$

Thus, we see that the functions $M(x_1, x_2, \ldots, x_m, t)$, which are known as Jacobi multipliers, must satisfy a linear partial differential equation. The corresponding integral I, given by Eq. (6-382), is an *absolute* integral invariant.

Now let us obtain some integral invariants which are associated with Hamiltonian systems. First, recall from Eq. (5-15) that

$$\sum_{i=1}^{n} p_{i1} \, dq_{i1} - \sum_{i=1}^{n} p_{i0} \, dq_{i0} - H_1 \, dt_1 + H_0 \, dt_0 = dS(q_1, q_0, t_1, t_0) \tag{6-390}$$

Next, suppose we consider a closed curve γ_0 in an extended configuration space of $(n + 1)$ dimensions. It consists of a set of solution points in this space at time t_0. We assume that one solution curve C passes through each point of γ_0, but γ_0 is nowhere tangent to any solution curve. As in Fig. 6-3, the curve γ indicates the locations at time t of the original set of solution points. Hence, at some final time t_1, a typical point P_1 on γ_1 corresponds to a point P_0 on γ_0.

Now let us examine the line integral

$$I = \oint_\gamma \sum_{i=1}^{n} p_i \, dq_i \tag{6-391}$$

where we consider that $p_i = p_i(q, t)$ since the initial conditions are given along γ_0, and therefore the solution curves are specified. We wish to compare the values of this integral taken over the closed curves γ_0 and γ_1. As the integration proceeds along γ_0 we assume that another integration is taking place along the corresponding points of γ_1. Since each of these integrations takes place at a fixed time, we can set dt_0 and dt_1 equal to zero in Eq. (6-390). Then, noting that the end-points of the integrations occur at the same values of the variables $(\mathbf{q}_0, \mathbf{q}_1, t_0, t_1)$, we see that

$$\int dS = 0 \tag{6-392}$$

and we obtain

$$\oint_{\gamma_0} \sum_{i=1}^{n} p_i \, dq_i = \oint_{\gamma_1} \sum_{i=1}^{n} p_i \, dq_i \tag{6-393}$$

In other words, the integral I of Eq. (6-391) is a *relative* integral invariant whose constant value depends upon the family of solution curves intersected by γ.

This result can be generalized easily to include a continuous closed curve of integration γ in extended configuration space, without the restriction that time is constant along the curve. Since the initial and final points of the integration are identical, Eq. (6-392) still applies, and we obtain from Eq. (6-390) that

$$\oint_{\gamma_0} \left(\sum_{i=1}^{n} p_i \, dq_i - H \, dt \right) = \oint_{\gamma_1} \left(\sum_{i=1}^{n} p_i \, dq_i - H \, dt \right) \tag{6-394}$$

Thus we obtain the result that

$$I = \oint_{\gamma} \left(\sum_{i=1}^{n} p_i \, dq_i - H \, dt \right) \tag{6-395}$$

is a relative integral invariant of the first order. As before, each curve of integration γ is characterized by the fact that it is intersected by the same family of solution curves, and this family determines the value of I. But a single value of time is no longer associated with each γ.

As a final example of an integral invariant which is associated with a Hamiltonian system, consider an integral in *phase space* of the form

$$I = \int \cdots \int dq_1 \cdots dq_n \, dp_1 \cdots dp_n \tag{6-396}$$

This is an example of an integral of maximum order $2n$. A comparison with Eq. (6-382) shows that the vector $\mathbf{x}$ has the components $(q_1, \ldots, q_n, p_1, \ldots, p_n)$ and the Jacobi multiplier M is in this case equal to unity. Furthermore, from a comparison of the canonical equations and Eq. (6-365), we see that

$$X_i = \frac{\partial H}{\partial p_i} \qquad (i = 1, 2, \ldots, n)$$

$$X_i = -\frac{\partial H}{\partial q_{i-n}} \qquad (i = n+1, \ldots, 2n) \tag{6-397}$$

A substitution of these expressions for X_i into Eq. (6-389) leads to

$$\sum_{i=1}^{n} \left(\frac{\partial^2 H}{\partial q_i \, \partial p_i} - \frac{\partial^2 H}{\partial p_i \, \partial q_i} \right) = 0 \qquad (6\text{-}398)$$

Hence Eq. (6-389) is satisfied identically, and we conclude that I of Eq. (6-396) is an *absolute* integral invariant of order $2n$. Here the integration is over a finite volume enclosing a certain set of phase points whose motion is followed as time proceeds The invariance of the integral indicates that the volume of an arbitrary portion of the so-called phase fluid is constant as each point traces out its trajectory in accordance with the canonical equations. This is another statement of *Liouville's theorem*.

REFERENCES

1. CARATHÉODORY, C., *Calculus of Variations and Partial Differential Equations of the First Order, Part I.* (English translation by R. B. Dean and J. J. Brandstatter). San Francisco: Holden-Day, Inc., 1965. An excellent reference on canonical transformation theory. Careful reading is necessary with respect to the author's sign conventions.

2. GOLDSTEIN, H., *Classical Mechanics.* Reading, Mass.: Addison-Wesley Press, Inc., 1950. The principal topics of the theory are clearly explained.

3. RUND, H., *The Hamilton-Jacobi Theory in the Calculus of Variations.* London: D. Van Nostrand Company, Ltd., 1966. A particularly good reference on the theory of integral invariants.

4. LANCZOS, C., *The Variational Principles of Mechanics.* Toronto: University of Toronto Press, 1949. A good general treatment of the subject. Time transformations are well done.

5. PARS, L. A., *A Treatise on Analytical Dynamics.* London: Heinemann, 1965. This reference presents a broad and thorough treatment of the material of this chapter.

PROBLEMS

6-1. For a certain canonical transformation it is known that

$$Q = \sqrt{q^2 + p^2}$$

$$\psi = \frac{1}{2}(q^2 + p^2) \tan^{-1} \frac{q}{p} + \frac{1}{2}qp$$

Find $P(q, p)$ and $\phi(q, Q)$.

6-2. Given the canonical transformation equations

$$Q = \sin q, \qquad P = \frac{p - mv_0}{\cos q}$$

Find $\psi, F_1, F_2, F_3,$ and F_4.

6-3. A particle of mass m moves in the xy plane under the action of a potential function $V = ky$. For a homogeneous point transformation

$$Q_1 = xy, \qquad Q_2 = \frac{1}{2}(x^2 - y^2)$$

find the expressions for P_1 and P_2 and the generating function $F_2(q, P)$. What is the new Hamiltonian function $K(Q, P)$?

6-4. Given the canonical transformation

$$Q_1 = q_1 - v_0 t \qquad\qquad P_1 = p_1 - v_0$$
$$Q_2 = \sqrt{2p_2}\, e^{-t} \sin q_2 \qquad P_2 = \sqrt{2p_2}\, e^t \cos q_2$$

and the Hamiltonian function $H = \frac{1}{2}(p_1^2 + p_2^2 + q_1^2)$. Find the generating function $F_2(q, P, t)$, the new Hamiltonian $K(Q, P, t)$, and the canonical equations in terms of the new variables.

6-5. Given a standard holonomic system with $H = \frac{1}{2}(q^2 + p^2)$. Assuming the canonical transformation of Example 6-2, find $K(Q, P, t)$ and the corresponding canonical equations.

6-6. Consider a canonical transformation for which $Q = q \cos^2 2p$ and $\psi = qp - \frac{1}{4}q \sin 4p$. Find expressions for $P(q, p)$, $F_1(q, Q)$, and $F_4(p, P)$.

6-7. For the case in which the generating function $F_4(p, P, t)$ is identically zero, show that

$$\sum_{i=1}^{n} q_i \frac{\partial P_j}{\partial q_i} = 0 \qquad (j = 1, 2, \ldots, n)$$

6-8. Given a system with $H = \frac{1}{2}(q^2 + p^2)$. Employ the Hamilton-Jacobi method to solve for the motion, using first $F_2(q, P, t)$, and then $F_4(p, P, t)$. Note that P is identical with α and Q with β in each case.

6-9. Consider a homogeneous canonical transformation having $n = 2$ and an auxiliary condition

$$\Omega(q, Q) = \sin(q_1 + Q_1) - \sqrt{q_2^2 + Q_2^2} = 0$$

where $0 \leq q_1 + Q_1 \leq \pi/2$ and $0 \leq q_2^2 + Q_2^2 \leq 1$. Obtain the transformation equations and use the differential form to confirm that they are canonical. Show that the matrix $(\partial Q_i / \partial p_j)$ has rank one.

6-10. In the usual case we require that the determinant $|\partial^2 \phi / \partial q_i \, \partial Q_j|$ be nonzero. As a counterexample, suppose we are given the generating function

$$F_1(q, Q, t) = v_0 q_1 + q_2 \cos^{-1} \frac{Q_2 e^{-t}}{\sqrt{2q_2}} - \frac{1}{2} Q_2 e^{-t} \sqrt{2q_2 - Q_2^2 e^{-2t}}$$

and the auxiliary condition $\Omega = Q_1 - q_1 + v_0 t = 0$. Show that an augmented generating function can be used to obtain the complete canonical transformation equations. Solve for $(K - H)$ in terms of the old variables.

6-11. Given the canonical transformation equations

$$Q_i = \sum_{j=1}^{n} a_{ij} q_j, \qquad P_i = \sum_{j=1}^{n} a_{ij} p_j$$

where $\sum_{i=1}^{n} a_{ij}a_{ik} = \delta_{jk}$ and $a_{ij} = a_{ij}(t)$. Assuming that $H = \frac{1}{2} \sum_{k=1}^{n} p_k^2$ find $K(Q, P, t)$ and the canonical equations in terms of the new variables.

6-12. Using the notation of Example 6-4, obtain a generating function $F_2^*(q, P^*)$ for a rotation followed by a translation.

6-13. A system with one degree of freedom is described by $T = \frac{1}{2}m(\dot{q} + t)^2$ and $V = \frac{1}{2}kq^2$. Now consider a canonical transformation in extended phase space generated by $F_2 = qP_T + tP$. Write the transformation equations and use the Poisson bracket expressions to check that they are canonical. Find the new augmented Hamiltonian and obtain the equations of motion in the new variables. Show that these equations are equivalent to the equations of motion in terms of the original variables.

6-14. Given the transformation equations

$$Q_1 = q_1 q_2, \qquad\qquad Q_2 = \frac{1}{2}(q_1^2 - q_2^2)$$

$$P_1 = \frac{q_2 p_1 + q_1 p_2}{q_1^2 + q_2^2}, \qquad P_2 = \frac{q_1 p_1 - q_2 p_2}{q_1^2 + q_2^2}$$

Use the symplectic matrix method to show that these equations represent a canonical transformation.

6-15. Consider a system described by the differential equations

$$\dot{x}_i = X_i(x_1, x_2, x_3, t) \qquad (i = 1, 2, 3)$$

with independent integrals of the motion of the form

$$\phi_i(x_1, x_2, x_3, t) = \alpha_i \qquad (i = 1, 2, 3)$$

where the α's are constants. Give a detailed derivation of the time derivative of the Jacobian $\partial(x_1, x_2, x_3)/\partial(\alpha_1, \alpha_2, \alpha_3)$ and compare with the result given in Eq. (6-387).

6-16. Given a canonical transformation relating (q_i, p_i) and (Q_i, P_i) for $i = 1, 2, \ldots, n$. Use the bilinear covariant criterion of Eq. (6-263) to show that

$$\frac{\partial Q_i}{\partial q_k} = \frac{\partial p_k}{\partial P_i}, \qquad \frac{\partial P_i}{\partial p_k} = \frac{\partial q_k}{\partial Q_i}$$

and

$$\frac{\partial Q_i}{\partial p_k} = -\frac{\partial q_k}{\partial P_i}, \qquad \frac{\partial P_i}{\partial q_k} = -\frac{\partial p_k}{\partial Q_i}$$

for all (i, k). This can be accomplished by expanding in terms of mixed (old and new) variables, such as $\delta q_k \, dP_i$, and considering the case where the δ-variation involves only q_k from the old variables, and the d-variation involves only P_i from the new variables.

6-17. Consider a standard Hamiltonian system. Use Eq. (6-380) to show that the line integral

$$I = \int_\gamma \sum_{i=1}^{n} q_i \, dp_i$$

is a relative integral invariant.

6-18. Given a system with $T = \frac{1}{2}m(\dot{q}_1^2 + \dot{q}_2^2)$ and $V = \frac{1}{2}kq_2^2$. Use Eq. (6-380) to show that the line integral

$$I = \int_\gamma (p_1 \, dq_1 + p_2 \, dq_2)$$

is a relative integral invariant and the line integral

$$I = \int_\gamma \left[kq_2 \, dq_2 + \left(1 + \frac{p_1}{m}\right) dp_1 + \frac{p_2}{m} \, dp_2 \right]$$

is an absolute integral invariant.

6-19. For the system of the previous problem, show that

$$I = \int \int \int \int (mkq_2^2 + p_2^2) dq_1 \, dq_2 \, dp_1 \, dp_2$$

is an absolute integral invariant.

6-20. The polar coordinates (r, θ) specify the location of a particle of mass m as it moves in the xy plane. Consider an infinitesimal canonical transformation corresponding to a small translation ϵ parallel to the x axis. Find the generating function $G(q, p)$ of this transformation and the expressions for δr, $\delta\theta$, δp_r, and δp_θ.

7

INTRODUCTION TO RELATIVITY

The basic idea of a principle of relativity in mechanics did not originate with Einstein, but rather it dates back to the times of Galileo and Newton. Recall that Newton's first law of motion states that a body with no forces acting on it remains in its state of rest or of uniform motion in a straight line. Now this law presupposes a frame of reference from which the path of the body is measured, and only for certain choices does Newton's law apply. Newton tended to think in terms of an absolute or primary inertial frame such as a frame fixed with respect to the distant stars, but he was aware that other inertial frames exist in which an unforced body moves uniformly in a straight line. Since each of these frames is equally valid as a reference for writing the equations of motion in accordance with Newton's laws, we have a principle of relativity even in Newtonian mechanics.

In his special theory of relativity, Einstein extended the principle of relativity from purely mechanical systems to electromagnetic phenomena and, by hypothesis, to all of physics. This postulate had surprising consequences with respect to the prevailing common-sense notions concerning distance and time, and had a profound effect upon the further development of physical theory. In this chapter we shall present the principal ideas of Einstein's special theory of relativity.

7-1. INTRODUCTION

Galilean Transformations. First let us define an *inertial frame* as any frame of reference in which a free particle remains at rest or moves uniformly in a straight line. If we are given an inertial frame, then a second frame which translates with a constant velocity relative to the given frame is itself inertial, since the same free particle moves uniformly in a straight line relative to this frame as well. By similar reasoning, we can show that an infinite number of inertial frames exist, each translating uniformly with respect to the others.

Now consider two inertial frames I and I'. For convenience we can choose their orientations such that the direction of the relative velocity lies along the common x axes. Suppose, then, that frame I' translates with a constant velocity V relative to I, as shown in Fig. 7-1. Also, let us assume that the origins O and O' coincide at time $t = 0$.

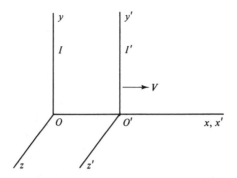

Figure 7-1. Two inertial reference frames in relative motion.

If we adopt the viewpoint of Newtonian mechanics, the transformation equations relating position and time in the two inertial frames are

$$x' = x - Vt$$
$$y' = y$$
$$z' = z \qquad\qquad (7\text{-}1)$$
$$t' = t$$

These equations describe a *Galilean transformation*. Note that time is considered to be an absolute quantity which is the same in both frames and, furthermore, the concepts of space and time are completely separable. The corresponding Cartesian components of velocity are related by

$$v_x' = v_x - V$$
$$v_y' = v_y \qquad\qquad (7\text{-}2)$$
$$v_z' = v_z$$

and the acceleration components in the two inertial frames are identical.

$$a_x' = a_x$$
$$a_y' = a_y \qquad\qquad (7\text{-}3)$$
$$a_z' = a_z$$

The equations which we have presented so far are concerned with the kinematics of a point; that is, they state the relationships of the position, velocity, and acceleration of a point relative to the two frames I and I' at a given time t.

If we next consider the motion of a particle of mass m which is acted upon by a force $\mathbf{F}$, we can apply Newton's second law of motion and obtain

$$\mathbf{F} = m\mathbf{a} \qquad\qquad (7\text{-}4)$$

where **a** is the instantaneous acceleration of the particle relative to the inertial frame I. Similarly, the equation of motion relative to I' is

$$\mathbf{F}' = m'\mathbf{a}' \tag{7-5}$$

Now, in accordance with the viewpoint of Newtonian mechanics, force and mass are absolute quantities, so $\mathbf{F} = \mathbf{F}'$ and $m = m'$. From Eq. (7-3) we see that $\mathbf{a} = \mathbf{a}'$. Hence we conclude that a Galilean transformation leaves the basic equation of motion unchanged, not only in its mathematical form, but also in the numerical values of the force, mass, and acceleration. Since all of Newton's laws of motion retain the same form, that is, are *covariant*, under a Galilean transformation, we can state the *Newtonian principle of relativity: All the laws of mechanics have the same form in every inertial frame of reference.* There are no preferred inertial frames.

Maxwell's Equations. Although we have seen that the laws of Newtonian mechanics apply equally in every inertial frame, one might ask whether the laws governing other areas of physics have the same form in these frames. As an example, let us consider Maxwell's equations which apply to electromagnetic phenomena. They are

$$\nabla \times \mathbf{H} = \mathbf{i} + \frac{\partial \mathbf{D}}{\partial t} \tag{7-6}$$

$$\nabla \times \mathbf{E} = -\frac{\partial \mathbf{B}}{\partial t} \tag{7-7}$$

$$\nabla \cdot \mathbf{D} = \rho \tag{7-8}$$

$$\nabla \cdot \mathbf{B} = 0 \tag{7-9}$$

where

$$\mathbf{D} = \epsilon \mathbf{E} \tag{7-10}$$

$$\mathbf{B} = \mu \mathbf{H} \tag{7-11}$$

We are particularly interested in the case of the propagation of electromagnetic waves in *free space* where the current density **i** and the charge density ρ are both zero. Hence we can combine these equations to obtain

$$\nabla^2 \mathbf{B} = \frac{1}{c^2} \frac{\partial^2 \mathbf{B}}{\partial t^2} \tag{7-12}$$

$$\nabla^2 \mathbf{E} = \frac{1}{c^2} \frac{\partial^2 \mathbf{E}}{\partial t^2} \tag{7-13}$$

where **B** is the magnetic flux density and **E** is the electric field strength. Equations (7-12) and (7-13) have the form of *wave equations* with a propagation velocity

$$c = \frac{1}{\sqrt{\mu_0 \epsilon_0}} \tag{7-14}$$

where μ_0 is the magnetic permeability of a vacuum and ϵ_0 is the corresponding dielectric constant. Measurements of these constants result in $\mu_0 = 1.257 \times 10^{-6}$ henries per meter and $\epsilon_0 = 8.86 \times 10^{-12}$ farads per meter, which yield a velocity $c = 3.00 \times 10^8$ meters per second. This propagation velocity applies to all electromagnetic waves, including ordinary visible light, and from the form of the equations is independent of the frequency and motion of the source.

Now if a relativity principle applies to electromagnetic phenomena, and therefore Maxwell's equations are valid in every inertial frame, then the Galilean transformation equations cannot be valid, for they would predict that different inertial observers would measure different velocities of light. For example, if a pulse of light is sent with velocity c along the x axis of inertial frame I in Fig. 7-1, the velocity of this light pulse as viewed by an observer translating with the inertial frame I' would be $(c - V)$ in accordance with Eq. (7-2). Furthermore, if we assume that light travels with the same speed in all directions relative to frame I, it would have different speeds in different directions relative to I'. So, if Maxwell's equations are valid at all, it is apparent that a choice must be made between the existence of a single preferred inertial frame and therefore no relativity principle, or else the Galilean transformation equations must be discarded.

The Ether Theory. Before abandoning the Galilean transformation which had so much intuitive appeal, the scientists of the late nineteenth century sought by experimental means to detect an "ether" through which light propagated in accordance with Maxwell's equations. This ether was considered to be a medium for the transmission of electromagnetic vibrations in a manner analogous to the transmission of sound waves through an elastic medium. The ether would then represent a preferred inertial frame, and light would be propagated in all directions with a constant speed c relative to this frame. Relative to the earth, however, the velocity of light would be different in different directions because of the motion of the earth in its orbit around the sun. This orbital velocity is approximately 3×10^4 meters per second, or 10^{-4} times the velocity of light. It is of sufficient magnitude that the presence of an ether should be detectable by measurements of the velocity of light.

The most important of the experiments which ultimately served to deny the existence of an ether was reported by Michelson and Morley in 1887. An interferometer was set up, as shown in Fig. 7-2, with the path lengths of the orthogonal beams 1 and 2 approximately equal. If we suppose that beam 1 is oriented in the direction of the earth's orbital velocity around the sun, and the sun is fixed relative to the ether, then in accordance with the ether theory the round trip time for path 2 is slightly less than that for path 1. On the other hand, the situation is reversed if path 2 lies along the direction

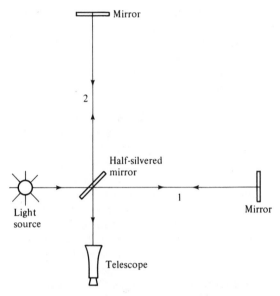

Figure 7-2. The Michelson-Morley interferometer.

of the earth's motion. Michelson and Morley first observed the fringe locations for one orientation, and then looked for any shifts of the fringes as the whole apparatus was rotated 90 degrees in its own plane. But despite measurements conducted at various times during the year, involving different directions of the earth's orbital velocity, no shifts were noted that were not within the expected limits of experimental error. In fact, the speed of light in the two perpendicular directions was shown to be the same to within approximately 17 parts per million, a precision considerably better than that required to detect the expected velocity difference of 100 parts per million. In other words, no velocity of the earth relative to a hypothetical ether could be found.

The Principle of Relativity. With the null result of the Michelson-Morley experiment, the retention of an absolute ether and the Galilean transformation equations seemed to require an additional hypothesis which was provided separately by Lorentz and FitzGerald. It was the contraction hypothesis whereby a rigid body moving with a velocity v relative to the ether contracts in the direction of its motion according to the equation

$$l = \sqrt{1 - \frac{v^2}{c^2}} l_0 \qquad (7\text{-}15)$$

where l is its length in the direction of its motion and l_0 is the corresponding length of the same body when it is at rest. This contraction factor was of such a magnitude that the effects of a fixed ether and the shortening of the apparatus in the direction of motion exactly offset each other, with the result that no fringe shifts would occur.

To Einstein, however, it seemed preferable to dispense with the ether theory once and for all, and instead to extend the Newtonian principle of relativity from mechanics, first to electromagnetic phenomena, and then to *all* of physics. Thus, in 1905 he proposed that the laws of physics are the same in every inertial frame; that is, there are no preferred inertial frames. This is known as *Einstein's principle of relativity*.

With the acceptance of the principle of relativity, and therefore with the conclusion that Maxwell's equations are valid in any inertial frame, we can now state the two basic postulates on which the special theory of relativity is based, namely, (1) the laws of physics are the same in all inertial frames and (2) the speed of light in free space, measured relative to any inertial frame, is a universal constant which is independent of the motion of the source.

7-2. RELATIVISTIC KINEMATICS

The Lorentz Transformation Equations. With the acceptance of the second basic postulate of relativity, it is apparent that the Galilean transformation equations must be replaced, for they indicate that a given light wavefront propagates with unequal speeds relative to different inertial frames. What is desired is a set of equations giving (x', y', z', t') in terms of (x, y, z, t), with the properties that the basic laws of electromagnetism and mechanics retain the same *form* in each inertial frame.

Rather than considering the most general case of two Cartesian coordinate systems in uniform relative motion, let us return to the inertial frames I and I' of Fig. 7-1. Let I' translate with a uniform velocity V relative to I in the direction of the positive x axes. Assume that the origins O and O' coincide at the instant $t = t' = 0$. As a first condition on the transformation equations, let us require that the speed of light have the same value c in all directions with respect to each inertial frame. Therefore if a flash of light is emitted at the common origin at $t = t' = 0$, an expanding spherical wavefront must propagate radially outward with a speed c relative to both I and I'. Relative to I the wavefront location is therefore

$$x^2 + y^2 + z^2 = c^2 t^2 \tag{7-16}$$

and, relative to I',

$$x'^2 + y'^2 + z'^2 = c^2 t'^2 \tag{7-17}$$

Thus the transformation equations must be consistent with Eqs. (7-16) and (7-17).

A second condition on the transformation equations is obtained from the definition of an inertial frame. Recall that a free particle moves with a constant velocity in a straight line relative to any inertial frame. This implies that the four-dimensional trajectories in (x, y, z, t) and also in (x', y', z', t') are straight lines. Hence the transformation equations must be *linear*.

Let us assume that distances are measured in frames I and I' by means of measuring sticks having the same length when at rest in a given inertial frame. Also, clocks of identical construction are used in measuring the times in I and I'.

We have chosen a set of mutually orthogonal axes in each frame, and when $t = 0$ in I or when $t' = 0$ in I' the corresponding axes coincide. We note further that the relative translational velocity is such that the x axes of the two systems always coincide. Therefore it follows that all points in a plane having a certain constant value of y must also have a constant value of y'. Furthermore, a symmetry argument can be used to show that these two constant values must be equal. For example, let us suppose that y' is smaller in magnitude than y, presumably because of the motion of I' relative to I. Then, by reversing the positive directions of the x and z axes in each system, and observing the motion of I relative to I', we have the same situation as previously except that the primed and unprimed systems are interchanged. Hence the same argument could be used to show that y is smaller than y', creating an inconsistency. This inconsistency is removed only if

$$y' = y \qquad (7\text{-}18)$$

A similar symmetry argument shows that

$$z' = z \qquad (7\text{-}19)$$

There remains the problem of relating the x and t variables in the two systems. Let us assume transformation equations of the form

$$\begin{aligned} x' &= ax + bt \\ t' &= ex + ft \end{aligned} \qquad (7\text{-}20)$$

where a, b, e, and f are constants to be determined.†

Using the condition that the transformation equations are consistent with Eqs. (7-16) and (7-17), and with the aid of Eqs. (7-18) and (7-19), we obtain

$$\begin{aligned} x^2 - c^2 t^2 &= x'^2 - c^2 t'^2 \\ &= a^2 x^2 + 2abxt + b^2 t^2 - c^2(e^2 x^2 + 2efxt + f^2 t^2) \end{aligned} \qquad (7\text{-}21)$$

This equation is valid for independent values of x and t, so let us equate the coefficients of the x^2 terms and obtain

$$a^2 - c^2 e^2 = 1 \qquad (7\text{-}22)$$

---

†These parameters actually will be functions of V, the translational velocity of I' relative to I. But for any two given inertial frames having a certain relative velocity, they are constants since we know that the transformation equations must be linear in x and t.

Similarly, by equating the coefficients of the t^2 terms, we find that

$$b^2 - c^2 f^2 = -c^2 \tag{7-23}$$

and, for the xt terms, we have

$$ab - c^2 ef = 0 \tag{7-24}$$

Thus far we have three equations from which to solve for the four coefficients. A fourth equation which serves to introduce the relative velocity V is obtained by noting that the position of the origin O' is given by $x = Vt$ or $x' = 0$. Hence, from the first equation of (7-20), we have

$$aVt + bt = 0 \tag{7-25}$$

or

$$b = -Va \tag{7-26}$$

Similarly, the position of O is at $x = 0$, or $x' = -Vt'$, and Eq. (7-20) yields

$$b = -Vf \tag{7-27}$$

Comparing Eqs. (7-26) and (7-27), we see that

$$a = f \tag{7-28}$$

Then, using Eqs. (7-23) and (7-27), we obtain

$$f^2 = \frac{c^2}{c^2 - V^2} \tag{7-29}$$

or

$$a = f = \frac{1}{\sqrt{1 - \left(\dfrac{V}{c}\right)^2}} \tag{7-30}$$

Here we choose the positive root because we know that the transformation equations must approach the Galilean transformation for small V; that is, a and f must approach one. Next, from Eqs. (7-26) and (7-30), we have

$$b = \frac{-V}{\sqrt{1 - \left(\dfrac{V}{c}\right)^2}} \tag{7-31}$$

Finally, from Eqs. (7-24) and (7-30), we obtain

$$e = \frac{b}{c^2} = \frac{-V/c^2}{\sqrt{1 - \left(\dfrac{V}{c}\right)^2}} \tag{7-32}$$

Thus the four coefficients are found to be functions of V, as expected, and, furthermore, this transformation approaches the Galilean transformation for small V. It is convenient to use the notation

$$\beta \equiv \frac{V}{c} \tag{7-33}$$

Then the transformation equations are

$$x' = \frac{x - Vt}{\sqrt{1 - \beta^2}}$$

$$y' = y$$

$$z' = z \tag{7-34}$$

$$t' = \frac{t - \dfrac{Vx}{c^2}}{\sqrt{1 - \beta^2}}$$

These are the *Lorentz transformation equations*. They form the basis of the kinematics of special relativity. Note, in particular, that the time t' is a function of the spatial coordinate x, as well as the time t. Thus, the time can no longer be considered as an absolute quantity which has the same value for all observers. Rather, it is dependent upon the position and velocity of the observer. This mingling of the coordinates of space and time in the Lorentz transformation equations gives rise to the idea of a four-dimensional *space-time* in which x, y, z, and t have comparable but not exactly similar roles.

The Lorentz transformation equations can be expressed in an alternate form by interchanging the primed and unprimed quantities and changing the sign of V. This is equivalent to solving Eq. (7-34) for the unprimed variables and leads to

$$x = \frac{x' + Vt'}{\sqrt{1 - \beta^2}}$$

$$y = y'$$

$$z = z' \tag{7-35}$$

$$t = \frac{t' + \dfrac{Vx'}{c^2}}{\sqrt{1 - \beta^2}}$$

Perhaps the most symmetrical form of the Lorentz transformation equations arises by multiplying the time equation by c, with the result that we can consider the time variable to be ct or ct'. Thus, we measure time in units of *length*, just as for x, y, or z. In other words, a certain interval of time can be expressed in terms of the *distance* that light would travel during the given time.

Now let us define the scalar parameter γ by

$$\gamma \equiv \frac{1}{\sqrt{1 - \beta^2}} \tag{7-36}$$

Then we can write the matrix equation

$$\begin{Bmatrix} x' \\ ct' \end{Bmatrix} = \gamma \begin{bmatrix} 1 & -\beta \\ -\beta & 1 \end{bmatrix} \begin{Bmatrix} x \\ ct \end{Bmatrix} \tag{7-37}$$

or, conversely,

$$\begin{Bmatrix} x \\ ct \end{Bmatrix} = \gamma \begin{bmatrix} 1 & \beta \\ \beta & 1 \end{bmatrix} \begin{Bmatrix} x' \\ ct' \end{Bmatrix} \tag{7-38}$$

A more general form of the Lorentz transformation equations occurs when I' translates relative to I with a constant velocity $\mathbf{V}$ which may be in any direction. In order to simplify the notation, suppose we designate the position vectors relative to I and I' by $\mathbf{x}$ and $\mathbf{x}'$, respectively, where

$$\mathbf{x} = x\mathbf{i} + y\mathbf{j} + z\mathbf{k} \tag{7-39}$$

$$\mathbf{x}' = x'\mathbf{i}' + y'\mathbf{j}' + z'\mathbf{k}' \tag{7-40}$$

Once again we assume that the origins O and O' coincide at $t = t' = 0$, and corresponding axes are parallel in the limiting case in which $\mathbf{V}$ approaches zero. For these assumptions, one can show that†

$$\mathbf{x}' = \mathbf{x} + \left[\frac{\mathbf{x} \cdot \mathbf{V}}{V^2}(\gamma - 1) - \gamma t \right] \mathbf{V} \tag{7-41}$$

$$t' = \gamma \left(t - \frac{\mathbf{x} \cdot \mathbf{V}}{c^2} \right) \tag{7-42}$$

Even more general Lorentz transformation equations can be obtained if, for example, the primed system undergoes a certain fixed rotation such that corresponding axes are not parallel, or perhaps O and O' do not coincide at time zero. We shall not concern ourselves further with these possibilities, however.

Events and Simultaneity. The kinematics of relativity is concerned with the motion of a point in space and time. Thus, a moving particle traces out a trajectory in the four-dimensional space (x, y, z, t). Now if an event such as a flash of light occurs at a certain time and place, it can be specified by giving the values of x, y, z, and t relative to some inertial frame. In other words, we can say that *an event is a point in spacetime.* Furthermore, we see that the Lorentz transformation equations deal with *events*; that is, they relate the values of x, y, z, and t relative to two different inertial frames.

Now if we suppose that a certain event such as a light flash occurs, how are the corresponding values (x, y, z, t) obtained? First we can choose an inertial frame I as a reference. Now suppose we imagine that an array of signs is constructed, and these signs are fixed at regular points throughout the reference frame. The *calibration* giving the spatial coordinates (x, y, z) of each sign is a straightforward procedure because these signs are not moving, and any measuring sticks would also be fixed during this calibration process. The location of a given event is found by interpolation from the signs closest to the event.

†C. Møller, *The Theory of Relativity* (New York: Oxford University Press, 1952), p. 41.

Next, let us suppose that a clock of standard construction is associated with each sign. The question immediately arises concerning how this array of clocks is to be *synchronized*. One possibility is to assume that the clock located at the origin O is designated as the master clock. In order to synchronize all the other identical clocks with the master clock, we assume that a flash of light is emitted at the origin at time t_0. This flash propagates as a spherical wavefront which expands at a constant speed $c = (\mu_0 \epsilon_0)^{-1/2}$, where μ_0 and ϵ_0 are constants representing properties of a vacuum, as we have seen. Hence if a certain clock is located at a distance d from the origin, it should be set to the time $t_0 + d/c$ at the instant when the synchronizing light flash reaches it. By using this procedure, allowance is made for the finite propagation speed of light in the synchronizing of this clock.

A group of standard clocks, fixed in I and synchronized using a flash from the origin O, can be shown to be synchronized with respect to any other clock of the group which might be considered as a master. In other words, the synchronization obtained by this process is independent of which clock is chosen as the master clock.

Assuming, then, that each inertial frame has an array of signs and synchronized clocks which are fixed in the given frame and move with it, the spacetime coordinates of an event can be clearly defined. Two events are defined to be *simultaneous* with respect to a given inertial frame if they have the same value of time in that frame. But two events which are found to be simultaneous by an observer in one inertial frame are not necessarily simultaneous when viewed by an observer in some other inertial frame. For example, a set of clocks which are synchronized in frame I are not synchronized relative to I'. In general, the simultaneity of two spatially separated events is not an absolute relationship, but it depends upon the frame of reference.

At this point, a little more explanation is in order concerning the notion of an observer. An *observer* is always associated with a *definite inertial frame* and acts as a *recorder of events* using the spacetime coordinates of that frame. Hence one does not think of an observer as being located at a certain point and looking at a sequence of physical events from the perspective of that position. Rather, an observer in our definition is more like a distributed set of devices which record the values (x, y, z, t) for all events of interest by using the readings on the local signs and clocks. Each inertial frame can be considered to have an observer associated with it, but it is superfluous to consider that any given inertial frame has more than one observer.

Example 7-1. Let us consider an example, known as *Einstein's train*, which illustrates the relativity of simultaneity. Suppose a train (Fig. 7-3) moves with a constant relativistic velocity V along a straight track fixed in the inertial frame I. A second inertial frame I' is attached to the train and translates with it. Now suppose that lightning strikes each end of the train, making

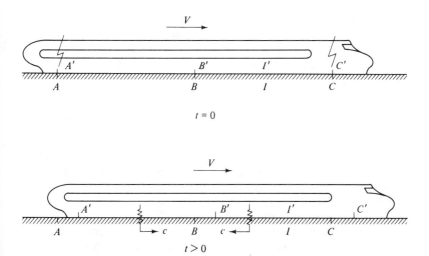

Figure 7-3. Einstein's train, showing the lightning striking at time zero, and also the wavefronts approaching each other.

marks on both the train and the ground. A man happens to be located at B, the point in frame I which is midway between the marks at A and C. He notes that the flashes from A and C arrive at B at the same instant. Since the two flashes travel with the same speed c relative to I, and he can confirm that the distances AB and CB are equal, he concludes that the flashes at the points A and C are *simultaneous*.

On the other hand, a man riding on the train at the midpoint B' finds that the lightning flash from the front end C' of the train arrives at B' before the flash from the rear end A'. Therefore, since the speed of light has the same value c relative to I' for both flashes, he concludes that the flash at C' occurs *before* that at A'. Furthermore, he notes that the two wavefronts meet at a point between B' and the rear of the train, again indicating that the flash at the front end occurs first.

Although there is an apparent disagreement between the men at B and B' concerning whether or not the lightning flashes are simultaneous, the truth of the matter is that both are correct. The two flashes occur at the same time t in frame I, but not at the same time t' in frame I'. Furthermore, if we should consider a second train passing the first and moving in the opposite direction, a passenger on this second train would find that the flash at A' occurs before that at C'.

We can conclude that there is no absolute simultaneity for two spatially separated events. If the events are simultaneous in a certain inertial frame, one can always find another inertial frame in which one of the events precedes the other.

Time Dilation. Now let us turn directly to the Lorentz transformation equations and consider the time interval between two events that occur at the same spatial location in I', namely, at the location of a certain clock. This time interval is registered by a clock fixed in I' and translating with it. Hence, for the events 1 and 2, we take

$$x_1' = x_2' \tag{7-43}$$

Then, using the Lorentz transformation for the time given by Eq. (7-35), we obtain

$$t_2' - t_1' = \sqrt{1 - \beta^2}\,(t_2 - t_1) \tag{7-44}$$

Thus we find that the time interval $(t_2' - t_1')$ registered by a single moving clock is less than the interval $(t_2 - t_1)$ between the same two events registered on separate but synchronized clocks in I. Hence an observer in I would say that the rate of the moving I' clock is slow. Conversely, if an observer in I' compares the reading of his single clock against the sequence of clocks in I that he passes, he finds that these clocks run faster than his own. This is the *time dilation* effect.

Using a similar line of reasoning, an observer fixed in I', and watching a certain clock of the I system go by with a relative speed V, will find that this clock runs more slowly than the synchronized clocks of his system. We can summarize, then, by noting that every clock appears to run at its fastest rate when it is at rest relative to the observer. If it moves with a relative speed V, its rate measured in the observer's frame is slowed by a factor $(1 - \beta^2)^{1/2}$, where $\beta = V/c$.

Longitudinal Contraction. In the theory of special relativity, the distance between two points in a rigid body is found by measuring the spatial separation of the two points *at the same instant* in the given inertial frame. We have seen that the simultaneity implied by the phrase "at the same instant" is not absolute, so one might expect different inertial observers to obtain different dimensions for the same rigid body. This is actually the case.

As an example, suppose we consider a rigid rod which has an orientation parallel to the x axis of the inertial frame I (Fig. 7-4) and translates with a velocity V in the x direction relative to this frame. We assume that it remains

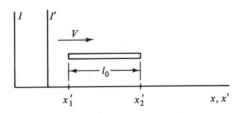

Figure 7-4. A rod translating with frame I'.

at rest relative to I' and has a *rest length* l_0, where

$$x_2' - x_1' = l_0 \qquad (7\text{-}45)$$

In order to determine its length with respect to the frame I, we use the Lorentz equations of (7-34) and assume that the locations of the ends of the rod are found to be at x_1 and x_2 at the same time t. We obtain

$$x_1' = \frac{x_1 - Vt}{\sqrt{1 - \beta^2}}$$

$$x_2' = \frac{x_2 - Vt}{\sqrt{1 - \beta^2}}$$

or

$$x_2' - x_1' = \frac{x_2 - x_1}{\sqrt{1 - \beta^2}} \qquad (7\text{-}46)$$

where $\beta = V/c$. If we let l be the length of the rod in frame I, we see from Eqs. (7-45) and (7-46) that

$$l = x_2 - x_1 = \sqrt{1 - \beta^2}\, l_0 \qquad (7\text{-}47)$$

This result indicates that the length of the moving rod as observed from I is *shorter* than its rest length in I'.

Conversely, if we assume that a rod of length l_0 is fixed in frame I and is observed from I', then a similar analysis will show that the length l' measured in I' is shorter than l_0 by the same factor $(1 - \beta^2)^{1/2}$. In this case the locations of the end-points of the rod are determined at the same time t' in the observer's frame. The seeming paradox that the rod fixed in the other frame seems shorter to each observer is explained by noting that each observer measures the distance between the *simultaneous* locations of the ends of the rod, as determined by the clocks of his own frame.

The longitudinal contraction effect can be summarized by noting that every rigid body appears to be longest when it is at rest relative to the observer. When it is not at rest it appears to be contracted in the direction of its motion by the factor $(1 - \beta^2)^{1/2}$. The dimensions perpendicular to the direction of motion are unchanged since $y = y'$ and $z = z'$ in accordance with the Lorentz transformation equations.

The Invariant Interval. In the Galilean transformations which are associated with nonrelativistic kinematics, time is considered to be an *absolute* quantity. In other words, the time interval between a given pair of events is independent of the frame of reference. Also, according to nonrelativistic theory, the spatial interval, that is, the *distance* separating any two *simultaneous* events, does not depend upon the observer's reference frame. So we can say that these intervals are *invariant* with respect to a Galilean transformation.

When we turn to relativistic kinematics, however, we find that the Lorentz transformation equations are not consistent with these invariant intervals in time and space. Instead, there is a single invariant interval between the arbitrary events 1 and 2 which combines the measurements of space and time. This invariant interval s is given by the equation

$$s^2 = c^2(t_2 - t_1)^2 - (x_2 - x_1)^2 - (y_2 - y_1)^2 - (z_2 - z_1)^2 \quad (7\text{-}48)$$

for the case where the two events are recorded by an observer in the inertial frame I. If the spacetime coordinates of the same two events are given relative to the inertial frame I', then the corresponding squared interval is

$$s'^2 = c^2(t_2' - t_1')^2 - (x_2' - x_1')^2 - (y_2' - y_1')^2 - (z_2' - z_1')^2 \quad (7\text{-}49)$$

A direct substitution from Eq. (7-34) for the primed quantities shows that this interval is indeed invariant under a Lorentz transformation, that is,

$$s^2 = s'^2 \quad (7\text{-}50)$$

for any two events.

The interval s can be visualized for the case of the spacetime coordinates (x, y, t) by using a set of three mutually orthogonal axes, as shown in Fig. 7-5. Let us suppose that a flash of light occurs at the origin at time $t = 0$. In ordinary two-dimensional space a circular wavefront is formed which increases in radius at a uniform rate c. Therefore, in the three-dimensional spacetime diagram of Fig. 7-5, the wavefront generates a conical surface whose vertex is at the origin. This is the *light cone* for positive t. This cone can also be extended backward in time to include those events whose light flashes arrive at the origin at $t = 0$. These two cones are easily generalized for the case of four-dimensional spacetime and are described by the equation

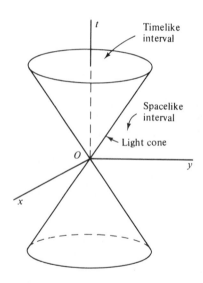

Figure 7-5. Light cones for an event at the origin.

$$x^2 + y^2 + z^2 = c^2 t^2 \quad (7\text{-}51)$$

Comparing this equation with Eq. (7-48), we see that it represents the surface of all events whose interval s relative to the origin is zero.

So far we have considered light cones relative to the origin. But additional light cones can be imagined to originate at other points in spacetime. For example, the light cones relative to the reference point (x_1, y_1, z_1, t_1)

are given by

$$(x - x_1)^2 + (y - y_1)^2 + (z - z_1)^2 = c^2(t - t_1)^2 \tag{7-52}$$

In general, the interval between any two events can be classified into one of three types:

1. *Timelike interval, $s^2 > 0$*. One event lies inside the light cone of the other. The two events can be causally linked, but are not necessarily so related. In other words, it is possible for a signal travelling with a speed less than c to connect the events, and therefore the earlier event could have caused the later event.

2. *Spacelike interval, $s^2 < 0$*. One event lies outside the light cone of the other. The two events cannot be connected by a light signal; hence they are not causally related. Also, for events separated by a spacelike interval, it is always possible to find two inertial frames in which the two events occur in the opposite time sequence.

3. *Lightlike interval, $s^2 = 0$*. One event lies on the light cone of the other; that is, it is possible for a light signal to connect the two events.

Proper Time and Proper Distance. Let us consider two events which are separated by a *timelike interval* which we shall designate by $\Delta\tau$. From Eq. (7-48) we can write

$$\Delta\tau = \frac{1}{c}[c^2\,\Delta t^2 - \Delta x^2 - \Delta y^2 - \Delta z^2]^{1/2} \tag{7-53}$$

where Δx, Δy, Δz, and Δt are the differences in the corresponding spacetime coordinates associated with the inertial frame I. Also consider a second inertial frame I' in which the differences $\Delta x'$, $\Delta y'$, $\Delta z'$ are all zero for the same two events. In this case we have

$$\Delta\tau = \Delta t' \tag{7-54}$$

that is, the invariant timelike interval is equal to the actual time interval registered by a single clock fixed in I'. We might consider this clock to be associated with a particle fixed in I', but moving uniformly relative to I. In this case, the time τ is known as the *proper time* of the particle.

If we consider infinitesimal temporal and spatial intervals, we can write Eq. (7-53) in the form

$$d\tau = \sqrt{1 - \frac{v^2}{c^2}}\,dt \tag{7-55}$$

where v is the particle velocity relative to the inertial frame I. Frequently this result is generalized to include the proper time τ associated with a particle undergoing a more general motion involving accelerations.† Then Eq. (7-55)

---

†The hypothesis that accelerations do not directly affect the rate of a clock is commonly made, but may not be strictly true. We shall, however, neglect any further refinements of the theory in this discussion.

is integrated to yield

$$\tau = \int_0^t \sqrt{1 - \frac{v^2}{c^2}}\, dt \tag{7-56}$$

assuming that τ and t are both zero at the initial instant.

Now suppose that two events are separated by a *spacelike interval*, implying that $s^2 < 0$. One can define an interval of *proper distance*

$$\Delta\sigma = \sqrt{\Delta x^2 + \Delta y^2 + \Delta z^2 - c^2\,\Delta t^2} \tag{7-57}$$

which, of course, has the same value for the two given events in all inertial frames. In particular, if an inertial frame is chosen such that the two events are simultaneous, then the proper distance interval is the same as the ordinary spatial separation.

An interesting approach to the concept of an invariant interval was given by Minkowski in 1908 when he suggested using the coordinates

$$x_1 = x, \qquad x_2 = y, \qquad x_3 = z, \qquad x_4 = ict \tag{7-58}$$

in which case we obtain from Eq. (7-57) that

$$\Delta\sigma^2 = \Delta x_1^2 + \Delta x_2^2 + \Delta x_3^2 + \Delta x_4^2 \tag{7-59}$$

This has the symmetric form of the distance between two points in a four-dimensional Euclidean space, and gave rise to the idea of a four-dimensional spacetime in which space and time are merged. Although the coordinates corresponding to space and time differ in the various inertial frames, the *interval* is the same in all frames.

It should be noted, however, that the negative sign associated with Δt^2 in Eq. (7-57) allows the interval to be zero in certain cases even though none of the components vanishes. This is in contrast with the situation in Euclidean geometry which requires that all components be zero if the length of the corresponding vector is zero.

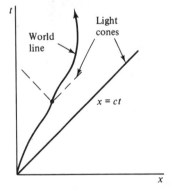

Figure 7-6. The world line of a particle.

The World Line. Let us consider the general motion of a particle. We can think of its motion as a sequence of events. Since an event is defined as a point in spacetime, this sequence of events traces a curve in the four-dimensional space (x, y, z, t). The curve is known as the *world line* for the particle.

As a simple example, consider a particle which leaves the origin at time $t = 0$ and moves in a straight line at a varying speed. Assuming that it moves in the direction of the x axis, a typical world line might be as shown in Fig. 7-6.

Now the invariant interval between any two points on this world line must be timelike in nature since the speed of the particle must be less than the speed of light. This implies that every point lies within the light cones for all previous points which, in turn, sets an upper limit on the angle between the t axis and the tangent to the world line. This limiting slope corresponds to a velocity c.

Now if we consider any infinitesimal invariant interval on the world line of a particle, we see that it also is timelike and is, in fact, an increment $d\tau$ of *proper time* for the particle. Equation (7-56) can then be used to compute the elapsed proper time τ for the particle. Because of the invariant nature of the increments, all inertial observers will calculate exactly the same proper time for the particle. Thus the proper time calculation is not associated with any particular inertial frame. We note from Eq. (7-56), however, that the proper time interval is always less than (or, for the case $v = 0$, equal to) the corresponding interval in the time t of the given inertial frame.

An illustration of the apparent slowness of proper time is given by the observation of the average lifetimes of certain unstable subatomic particles having large velocities. One finds that these lifetimes, as measured in the laboratory, can be many times longer than the corresponding proper lifetimes measured by the innate timekeeping mechanisms of the particles themselves. This effect permits the observation of the trajectories of certain types of particles which would otherwise be unobservable.

Example 7-2. Einstein's clock paradox, sometimes called the twin paradox, has been the focus of much discussion in the literature of relativity. Let us state the paradox in the form of a story concerning twins A and B.

Suppose the twins A and B are both initially on the earth, which is assumed to be an inertial frame. Twin A remains at the earth, but twin B travels with a relativistic speed v to a distant point P and then returns to the earth with the same constant speed (Fig. 7-7). Assuming that the measurements of

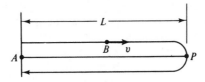

Figure 7-7. The path of the travelling twin.

distance and time are taken relative to the inertial frame of the earth, the time required for the round trip is

$$T_A = \frac{2L}{v} \qquad (7\text{-}60)$$

according to A.

On the other hand, a clock carried by twin B registers his own proper time. If we can assume that the reading on his clock does not change as he turns around instantaneously at P, then his total elapsed time for the trip, using Eq. (7-56), is

$$T_B = \sqrt{1 - \beta^2}\, T_A = \frac{2L}{v}\sqrt{1 - \beta^2} \tag{7-61}$$

where $\beta = v/c$.

Hence, upon B's return to the earth, a comparison of readings on the two clocks reveals that the time required for the journey according to B is shorter than that according to A by the factor $(1 - \beta^2)^{1/2}$. Also, since the biological time for each twin is the same as his own proper time, we find that the travelling twin B ages less than twin A who remained on the earth.

If we now consider the trip from B's viewpoint, we find that A moves away with a velocity v and then returns at the same speed. Hence the paradox: Using the previous reasoning, why should not B find that A's clock is running slow?

The first point to notice is that the dynamical effects of the motions of A and B with respect to each other are *not symmetrical*. If each twin carries an accelerometer, for example, A's accelerometer will give a continuous reading of zero. But B's accelerometer will show a nonzero reading when accelerating at the beginning of the trip, upon turning around at the destination P, and also upon slowing to a stop as it returns to A.

As a further check, we note that from B's viewpoint the line AP goes by twice, first with A receding until P is opposite B, and then with A approaching. In each case the relative speed is v, so the length of AP appears to be $(1 - \beta^2)^{1/2}L$ because of the relativistic contraction. Hence the time required for the round trip according to B is twice the time for one passage of AP, or

$$T_B = \frac{2L}{v}\sqrt{1 - \beta^2}$$

as before.

As an illustration of the difference in the aging processes of the twins, suppose that twin B travels at 80 percent of the speed of light, and the distance L is 4 light years, the approximate distance of the nearest star. Then Eqs. (7-60) and (7-61) yield $T_A = 10$ years and $T_B = 6$ years. Hence twin B ages appreciably less than twin A. In fact, twin B covers the round trip which has a total length of 8 light years in only 6 years. This may seem to imply a speed greater than c, in contradiction to the original assumption. It is actually a valid result, however, since the distance is measured in the earth's frame and the time is measured in a different frame, namely, a frame moving with twin B.

Addition of Velocities. One of the consequences of the Lorentz transformation equations is the restriction that the velocity of a particle relative to *any inertial frame* cannot exceed the speed of light c. Furthermore, the velocity of a photon relative to any inertial frame must be the same value c. It is apparent, then, that the usual laws of vector addition do not apply to the addition of velocities in relativistic kinematics. (Fig. 7-8.)

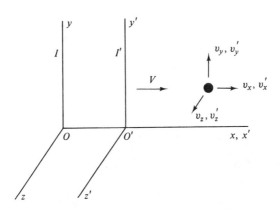

Figure 7-8. Velocity components of a particle.

To obtain the relativistic law of addition of velocities, we start with the Lorentz transformation of Eq. (7-35) written in differential form. We have

$$dx = \frac{dx' + V\,dt'}{\sqrt{1 - \beta^2}}$$

$$dy = dy'$$

$$dz = dz' \tag{7-62}$$

$$dt = \frac{dt' + \dfrac{V}{c^2}\,dx'}{\sqrt{1 - \beta^2}}$$

where $\beta = V/c$.

Now take ratios of these differentials to find the equations for the unprimed velocity components in terms of the primed components. We obtain

$$v_x = \frac{dx}{dt} = \frac{v_x' + V}{1 + \dfrac{Vv_x'}{c^2}}$$

$$v_y = \frac{dy}{dt} = \frac{v_y'\sqrt{1 - \beta^2}}{1 + \dfrac{Vv_x'}{c^2}} \tag{7-63}$$

$$v_z = \frac{dz}{dt} = \frac{v_z'\sqrt{1 - \beta^2}}{1 + \dfrac{Vv_x'}{c^2}}$$

Conversely, the primed velocity components in terms of the unprimed components are the following:

$$v'_x = \frac{v_x - V}{1 - \dfrac{Vv_x}{c^2}}$$

$$v'_y = \frac{v_y\sqrt{1 - \beta^2}}{1 - \dfrac{Vv_x}{c^2}}$$

$$v'_z = \frac{v_z\sqrt{1 - \beta^2}}{1 - \dfrac{Vv_x}{c^2}}$$

(7-64)

Let us assume that the velocity V lies in the range $-c < V < c$. We know that light signals (photons) travel with a velocity c relative to all inertial frames, and every material particle must have a velocity less than c relative to any inertial frame. A check of the velocity addition equations shows that they are consistent with these restrictions.

As a simple example, if I' translates relative to I with a velocity $V = 0.9c$, and if a particle translates relative to I' with a velocity $v'_x = 0.9c$, then we find from Eq. (7-63) that the velocity of the particle relative to I is $v_x = 0.994c$. Thus it is clear that the usual laws for the vector addition of velocities do not apply in this case where two inertial frames are involved. Relative to a single inertial frame, however, vector components may be added in the normal manner to obtain the total velocity with respect to that frame.

With the realization that relativistic velocities do not add in the simple manner of Newtonian kinematics, the question arises whether some other parameter might not be more convenient for representing the velocity. For example, can a parameter be found which uniquely represents the velocity, and which can be added to represent a relativistic addition of velocities? The *velocity parameter* θ meets these conditions and is defined by the equation

$$\tanh \theta = \frac{v}{c} = \beta \tag{7-65}$$

We note that the velocity is zero when θ is zero, and v approaches c as θ goes to infinity. Furthermore $\theta \cong \beta$ for $\theta \ll 1$, so θ is proportional to the velocity for small velocities.

From the defining equation for θ and Eq. (7-36) we see that

$$\cosh \theta = \frac{1}{\sqrt{1 - \beta^2}} = \gamma \tag{7-66}$$

$$\sinh \theta = \frac{\beta}{\sqrt{1 - \beta^2}} \tag{7-67}$$

To illustrate the use of the velocity parameter θ in the relativistic addition of velocities, consider the rectilinear motion of a particle having a velocity

v_2 relative to the frame I' which, in turn, has a velocity v_1 relative to the I frame. From Eq. (7-63) we find that the velocity of the particle relative to the inertial frame I is

$$v = \frac{v_1 + v_2}{1 + \frac{v_1 v_2}{c^2}} \tag{7-68}$$

In terms of the velocity parameter, this equation becomes

$$\tanh \theta = \frac{\tanh \theta_1 + \tanh \theta_2}{1 + \tanh \theta_1 \tanh \theta_2} = \tanh (\theta_1 + \theta_2) \tag{7-69}$$

from which we obtain

$$\theta = \theta_1 + \theta_2 \tag{7-70}$$

Thus, one-dimensional relativistic velocity additions are accomplished by adding the corresponding velocity parameters.

The Lorentz transformation equations can be written conveniently in terms of the velocity parameter. Thus, Eq. (7-37) becomes

$$\begin{Bmatrix} x' \\ ct' \end{Bmatrix} = \begin{bmatrix} \cosh \theta & -\sinh \theta \\ -\sinh \theta & \cosh \theta \end{bmatrix} \begin{Bmatrix} x \\ ct \end{Bmatrix} \tag{7-71}$$

or, conversely,

$$\begin{Bmatrix} x \\ ct \end{Bmatrix} = \begin{bmatrix} \cosh \theta & \sinh \theta \\ \sinh \theta & \cosh \theta \end{bmatrix} \begin{Bmatrix} x' \\ ct' \end{Bmatrix} \tag{7-72}$$

Example 7-3. A rigid rod of rest length l_0 makes an angle ϕ' with the x' axis and is fixed in I' as it translates with a constant velocity V relative to I (Fig. 7-9). Find the length of the rod and the angle between the rod and the x axis, as viewed by an observer in the inertial frame I.

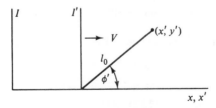

Figure 7-9. A translating rod.

First we notice that the x' and y' components of the length of the rod, as measured in frame I', are

$$\Delta x' = x' = l_0 \cos \phi' \tag{7-73}$$
$$\Delta y' = y' = l_0 \sin \phi'$$

We have observed previously that the y dimensions of a rigid body are unchanged in going from the I' frame to the I frame, but the x dimensions suffer a Lorentz contraction with the factor $(1 - \beta^2)^{1/2}$. Hence we find that

$$\Delta x = \sqrt{1 - \beta^2}\, l_0 \cos \phi'$$
$$\Delta y = l_0 \sin \phi' \tag{7-74}$$

Thus the length in frame I is

$$l = \sqrt{\Delta x^2 + \Delta y^2} = \sqrt{1 - \beta^2 \cos^2 \phi'}\, l_0 \tag{7-75}$$

The angle ϕ between the rod and the x axis, as viewed from frame I, is given by

$$\tan \phi = \frac{\Delta y}{\Delta x} = \frac{\tan \phi'}{\sqrt{1 - \beta^2}} \tag{7-76}$$

In a typical case, then, the rod appears shorter in the frame I, and its angle measured from the x axis is more nearly perpendicular.

Example 7-4. A particle moves relative to the frame I' with a velocity $\mathbf{v'}$ in a direction given by the angle ϕ' measured from the positive x' axis (Fig. 7-10). Find the amplitude and direction of the velocity of this particle relative to the I frame.

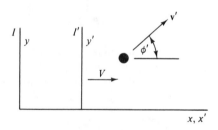

Figure 7-10. The transformation of velocity directions.

The velocity components relative to the I' frame are

$$v'_x = v' \cos \phi'$$
$$v'_y = v' \sin \phi' \tag{7-77}$$

A direct substitution into Eq. (7-63) yields

$$v_x = \frac{v' \cos \phi' + V}{1 + \dfrac{Vv' \cos \phi'}{c^2}}$$

$$v_y = \frac{v' \sin \phi' \sqrt{1 - \beta^2}}{1 + \dfrac{Vv' \cos \phi'}{c^2}} \tag{7-78}$$

where $\beta = V/c$. If we let $\beta' = v'/c$, these results can be written in the form

$$v_x = \left(\frac{\beta + \beta' \cos \phi'}{1 + \beta\beta' \cos \phi'}\right)c$$

$$v_y = \left(\frac{\beta' \sin \phi' \sqrt{1 - \beta^2}}{1 + \beta\beta' \cos \phi'}\right)c$$

(7-79)

The magnitude of the particle velocity relative to I is

$$v = \left(\frac{[\beta^2 + \beta'^2(1 - \beta^2 \sin^2 \phi') + 2\beta\beta' \cos \phi']^{1/2}}{1 + \beta\beta' \cos \phi'}\right)c \qquad (7-80)$$

This velocity vector makes an angle ϕ with the x axis, where

$$\tan \phi = \frac{\beta' \sin \phi' \sqrt{1 - \beta^2}}{\beta + \beta' \cos \phi'} \qquad (7-81)$$

Notice that the angle ϕ is quite different from the result of Example 7-3. Here, for positive β and β', we find that $\phi < \phi'$ for $0 < \phi' < \pi$.

An application of some interest occurs for the case of a photon moving with $\beta' = 1$. Then, as expected, Eq. (7-80) yields $v = c$ for all values of ϕ'. Under these conditions, Eq. (7-81) can be written in the form

$$\cos \phi = \frac{\beta + \cos \phi'}{1 + \beta \cos \phi'} \qquad (7-82)$$

Assuming $\beta > 0$ and $0 < \phi' < \pi$, we find again that $\phi < \phi'$. This implies that if a fast-moving particle emits light equally in all directions relative to I', an observer in frame I will find that most of the light travels in directions defined by the forward hemisphere, the effect being more pronounced as β increases. This is known as the *headlight effect*.

The Relativistic Doppler Effect. The frequency of a received light signal can be different from the transmitted frequency due to the relative motion of the transmitter and receiver. This *Doppler effect* occurs even for nonrelativistic velocities and for other types of wave transmission such as sound, but we shall limit ourselves to a relativistic analysis of electromagnetic waves in a vacuum.

First consider the case of a light source at the origin O' of the inertial frame I' which is receding with a velocity V from an observer at O in the inertial frame I, as shown in Fig. 7-11. Let us assume that the source frequency is f' when measured in I'. A light signal propagates to the left, that is, in the direction of the negative x axis. Suppose the period of this signal is T, as measured in frame I. We notice that the wavefront moves a distance cT to the left and the source moves VT to the right during the time interval T. Hence the wavelength measured in I is

$$\lambda = (c + V)T \qquad (7-83)$$

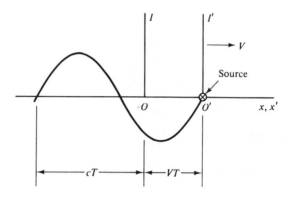

Figure 7-11. Relativistic Doppler effect.

for this case where the wavefront and the source move in opposite directions. The frequency received at O is

$$f = \frac{c}{\lambda} = \frac{1}{(1 + \beta)T} \tag{7-84}$$

where $\beta = V/c$.

Now recall that a clock moving with the source appears to run at a slow rate to an observer in I, that is,

$$T = \frac{T'}{\sqrt{1 - \beta^2}} \tag{7-85}$$

where T' is the oscillation period in I'. Therefore, for an observer translating with I', the source frequency f' is

$$f' = \frac{1}{T'} = \frac{1}{\sqrt{1 - \beta^2}\, T} \tag{7-86}$$

From Eqs. (7-84) and (7-86) we obtain

$$f = \sqrt{\frac{1 - \beta}{1 + \beta}} f' \tag{7-87}$$

This equation applies for a *receding source*.

The analysis of the Doppler effect for an *approaching source* involves a similar procedure except that the sign of β is changed. Hence the result for this case is

$$f = \sqrt{\frac{1 + \beta}{1 - \beta}} f' \tag{7-88}$$

So far we have assumed that the source is moving in a radial direction, that is, either directly away from or directly toward the observer. Now consider a source which has a *transverse velocity V*. In this case there will be no change in the received frequency due to a radial component of velocity. The time dilation effect, however, will remain because of the motion of the source. Hence we obtain from Eq. (7-85) that

$$f = \frac{1}{T} = \sqrt{1 - \beta^2}\, f' \tag{7-89}$$

This change in frequency is known as the *transverse Doppler effect*.

Having separated the effects of radial velocity and time dilation, the Doppler effect can be generalized to the more general case of a source velocity V at an angle α measured from the outward radial direction (Fig. 7-12). We obtain

$$f = \frac{\sqrt{1 - \beta^2}}{1 + \beta \cos \alpha}\, f' \tag{7-90}$$

where the factor $\sqrt{1 - \beta^2}$ represents the time dilation effect, and the term $1 + \beta \cos \alpha$ accounts for the change of wavelength due to the radial motion of the source.

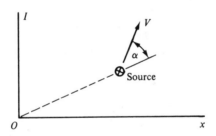

Figure 7-12. A light source in general motion.

Example 7-5. Let us reconsider the twin paradox of Example 7-2. Assume that each twin carries a radio transmitter having a constant frequency f_0 when it is at rest. Twin A compares the number of cycles received from the travelling twin B during the journey with the number of cycles emitted by his own stationary transmitter. Compare the aging rates of the twins, assuming that each twin ages in proportion to the total number of cycles emitted by his own transmitter.

Because of the relativistic Doppler effect, the frequency received at A as twin B recedes with velocity v toward the point P is

$$f_1 = \sqrt{\frac{1 - \beta}{1 + \beta}}\, f_0 \tag{7-91}$$

in accordance with Eq. (7-87), where $\beta = v/c$. Twin A receives this frequency until the signal transmitted at the turnaround point P arrives at A. This time interval is

$$t_1 = \frac{L}{\beta c} + \frac{L}{c} = \frac{(1 + \beta)L}{\beta c} \tag{7-92}$$

On the other hand, all signals transmitted by B on the return trip are received by A at the frequency given by Eq. (7-88), namely,

$$f_2 = \sqrt{\frac{1+\beta}{1-\beta}} f_0 \qquad (7\text{-}93)$$

Now the time interval over which this frequency is received is

$$t_2 = \frac{L}{\beta c} - \frac{L}{c} = \frac{(1-\beta)L}{\beta c} \qquad (7\text{-}94)$$

since we know that the total round trip time according to twin A is

$$t_1 + t_2 = \frac{2L}{\beta c} = \frac{2L}{v} \qquad (7\text{-}95)$$

This is just the total distance travelled divided by the speed.

The total number of cycles received at A from B's transmitter is

$$N_B = f_1 t_1 + f_2 t_2 = \frac{2f_0 L}{\beta c} \sqrt{1-\beta^2} \qquad (7\text{-}96)$$

which, of course, is equal to the number of cycles transmitted by B. The number of cycles emitted by A's transmitter, on the other hand, is equal to its constant frequency f_0 multiplied by the total time interval of Eq. (7-95).

$$N_A = \frac{2f_0 L}{\beta c} \qquad (7\text{-}97)$$

Comparing Eqs. (7-96) and (7-97), we see that

$$N_B = \sqrt{1-\beta^2}\, N_A \qquad (7\text{-}98)$$

This implies once again that the travelling twin B ages less than the earthbound twin A by the factor $(1-\beta^2)^{1/2}$.

7-3. RELATIVISTIC DYNAMICS

In the discussion of relativistic kinematics we found that the transformations between inertial frames for points in spacetime are accomplished by means of the Lorentz transformation equations. In the limit as the relative velocity V approaches zero, these equations become the Galilean transformation equations of Newtonian mechanics.

If we consider now the relativistic definitions of terms such as momentum, energy, and force, it is convenient to require that they approach the Newtonian definitions for small velocities. Furthermore, the definitions must be consistent with the Lorentz transformation equations in spacetime and must contribute to a description of dynamic events which is in accord with experimental observation. So let us proceed with the development of these definitions by considering the motions and interactions of particles.

Momentum. Consider a particle moving with a velocity **v** relative to an inertial frame. If we define the momentum of the particle to be equal to $m\mathbf{v}$, where the mass m is considered to be constant, as in Newtonian mechanics, then we find experimentally that the total momentum is not conserved in a collision of two high-speed particles. It is possible to *define* the relativistic momentum, however, so that the total momentum is conserved in all collisions. This will apply in all inertial frames, and can be shown to be in agreement with experimental results.

In attempting to find the form of the relativistic momentum expression which results in the conservation of momentum, let us mention at the outset certain essential characteristics. We note first that the momentum **p** must have the same direction as the velocity **v**, from symmetry considerations. Otherwise there would be a preferred plane, namely, the plane defined by **p** and **v**, which has no physical basis under the assumption that space is isotropic. Secondly we assume that, for a certain velocity **v**, the momentum of a particle is directly proportional to its *rest mass* m_0, where m_0 (also called the *proper mass*) is the inertial mass measured in a frame in which the particle velocity is negligibly small compared to c. Thus the rest mass of a particle has the same value as its Newtonian mass. Finally, we assume that the momentum varies in some nonlinear fashion with velocity, and the *form* of the defining equation is the same in all inertial frames and is consistent with the relativistic law of addition of velocities. So let us assume that the defining equation for the momentum of a particle has the form

$$\mathbf{p} = m_0 f(v)\mathbf{v} \qquad (7\text{-}99)$$

The explicit form of $f(v)$ can be found by analyzing the particular case of the elastic impact† of two identical frictionless spheres (Fig. 7-13). First consider the collision relative to an inertial frame I which translates with the center of mass. Let us choose the x and y axes so that the motion takes place in the xy plane, and the interaction impulse between the spheres at the moment of impact is parallel to the y axis. For this case of perfectly elastic impact, the y components of velocity are reversed and the x components are unchanged. Hence the speed v of each sphere is unchanged by the collision, and the angles of approach and rebound relative to the x axis are θ, as shown. From the symmetry of the situation, it is apparent that the total momentum is conserved and is equal to zero, regardless of the form of $f(v)$.

Now consider an inertial frame I' [Fig. 7-13(b)] which translates at a constant speed V relative to I, where

$$V = v \cos \theta \qquad (7\text{-}100)$$

---

†The term *elastic impact* implies a conservation of kinetic energy as well as momentum. In the center of mass system, particle velocities are changed in direction only, not in magnitude.

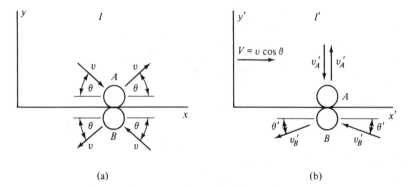

(a)   (b)

Figure 7-13. Elastic collision of identical spheres.

We have chosen V such that sphere A has no motion in the x' direction. As before, however, the y' components of velocity suffer a reversal upon impact while the x' components are unchanged, resulting in an unchanged speed for each sphere. It is clear, then, that the total x' component of momentum is conserved without regard to the form of $f(v)$.

When we turn to the y' components of momentum, however, we are able to obtain the form of $f(v)$. First we use the velocity addition law of Eq. (7-64) to obtain the speed of sphere A relative to frame I', namely,

$$v_A' = \frac{v \sin \theta}{1 - \dfrac{v^2}{c^2} \cos^2 \theta} \sqrt{1 - \frac{v^2}{c^2} \cos^2 \theta}$$

$$= \frac{v \sin \theta}{\sqrt{1 - \dfrac{v^2}{c^2} \cos^2 \theta}} \tag{7-101}$$

In a similar fashion, we find that the velocity components of sphere B before impact are

$$\dot{x}_B' = \frac{-2v \cos \theta}{1 + \dfrac{v^2}{c^2} \cos^2 \theta}$$

$$\dot{y}_B' = \frac{v \sin \theta}{1 + \dfrac{v^2}{c^2} \cos^2 \theta} \sqrt{1 - \frac{v^2}{c^2} \cos^2 \theta} \tag{7-102}$$

In order to simplify the analysis, let us consider the particular case in which $\theta \ll 1$. Then, neglecting terms of order θ^2 or higher, we obtain from Eqs. (7-101) and (7-102) that

$$v_A' = \frac{v\theta}{\sqrt{1 - \dfrac{v^2}{c^2}}} \tag{7-103}$$

$$v'_B = -\dot{x}'_B = \frac{2v}{1 + \dfrac{v^2}{c^2}} \tag{7-104}$$

$$\dot{y}'_B = \frac{\sqrt{1 - \dfrac{v^2}{c^2}}}{1 + \dfrac{v^2}{c^2}} \, v\theta \tag{7-105}$$

Since $v'_A \ll c$, the momentum of sphere A is simply the rest mass multiplied by the velocity, as in Newtonian mechanics. Hence we find that the change in the y' component of the momentum of A, as a result of the collision, is

$$\Delta p'_A = 2m_0 v'_A = \frac{2m_0 v\theta}{\sqrt{1 - \dfrac{v^2}{c^2}}} \tag{7-106}$$

Here we note that $f(v'_A) = 1$.

The corresponding change in the momentum of B is

$$\Delta p'_B = -2m_0 f(v'_B)\dot{y}'_B$$

$$= -2m_0 f(v'_B)v\theta \frac{\sqrt{1 - \dfrac{v^2}{c^2}}}{1 + \dfrac{v^2}{c^2}} \tag{7-107}$$

In this case we see that v'_B is not necessarily small compared with c, and therefore $f(v'_B)$ may be considerably larger than one. The conservation of the total momentum implies that

$$\Delta p'_A + \Delta p'_B = 0 \tag{7-108}$$

Then, from Eqs. (7-106)—(7-108), we can solve for $f(v'_B)$, obtaining

$$f(v'_B) = \frac{1 + \dfrac{v^2}{c^2}}{1 - \dfrac{v^2}{c^2}} \tag{7-109}$$

The right side of Eq. (7-109) is expressed as a function of v, but we are looking for a function of v'_B which has the same value. We find, however, by a direct substitution using Eq. (7-104), that

$$\sqrt{1 - \frac{v'^2_B}{c^2}} = \frac{1 - \dfrac{v^2}{c^2}}{1 + \dfrac{v^2}{c^2}} \tag{7-110}$$

Hence

$$f(v'_B) = \frac{1}{\sqrt{1 - \dfrac{v'^2_B}{c^2}}} \tag{7-111}$$

More generally, for a particle having a velocity v relative to an inertial frame, we find that

$$f(v) = \frac{1}{\sqrt{1 - \dfrac{v^2}{c^2}}} \tag{7-112}$$

Now let us return to Eq. (7-99) which gives the assumed form for $\mathbf{p}$. By substitution from Eq. (7-112) we obtain the following equation for the relativistic momentum:

$$\mathbf{p} = \frac{m_0 \mathbf{v}}{\sqrt{1 - \dfrac{v^2}{c^2}}} = \frac{m_0 \mathbf{v}}{\sqrt{1 - \beta^2}} \tag{7-113}$$

where $\beta = v/c$. It is sometimes convenient to write the momentum in the form

$$\mathbf{p} = m\mathbf{v} \tag{7-114}$$

where the *relativistic mass* m is given by

$$m = \frac{m_0}{\sqrt{1 - \beta^2}} \tag{7-115}$$

We observe that m approaches infinity as the particle velocity v approaches the speed of light, provided that the rest mass m_0 is not zero.

The Cartesian components of the relativistic momentum are

$$p_x = m\dot{x} = \frac{m_0 \dot{x}}{\sqrt{1 - \beta^2}}$$

$$p_y = m\dot{y} = \frac{m_0 \dot{y}}{\sqrt{1 - \beta^2}} \tag{7-116}$$

$$p_z = m\dot{z} = \frac{m_0 \dot{z}}{\sqrt{1 - \beta^2}}$$

In terms of the velocity parameter θ given in Eq. (7-67), we have

$$p = m_0 c \sinh \theta \tag{7-117}$$

Another useful approach is to recall from Eq. (7-55) that

$$d\tau = \sqrt{1 - \beta^2}\, dt$$

Then the momentum components of Eq. (7-116) assume the form

$$p_x = m_0 \frac{dx}{d\tau}$$

$$p_y = m_0 \frac{dy}{d\tau} \tag{7-118}$$

$$p_z = m_0 \frac{dz}{d\tau}$$

where τ is the proper time of the particle. In other words, if $\mathbf{r}$ is the position vector of a particle relative to a given frame, its momentum is

$$\mathbf{p} = m_0 \frac{d\mathbf{r}}{d\tau} \tag{7-119}$$

Energy. Let us approach the definition of the total relativistic energy E of a particle by specifying that its rate of increase is

$$\dot{E} = \dot{\mathbf{p}} \cdot \mathbf{v} \tag{7-120}$$

where $\mathbf{v}$ is the particle velocity. For the case of small velocities, this agrees with the concept from Newtonian mechanics that the energy increases at a rate equal to the rate with which the total external force ($\mathbf{F}$ or $\dot{\mathbf{p}}$) does work on the system. Using the definition of $\mathbf{p}$ from Eq. (7-113), one can obtain the alternate form

$$\dot{E} = \frac{m_0 v \dot{v}}{\left(1 - \dfrac{v^2}{c^2}\right)^{3/2}} \tag{7-121}$$

which can be integrated to yield

$$E = \frac{m_0 c^2}{\sqrt{1 - \dfrac{v^2}{c^2}}} + E_0 \tag{7-122}$$

Now let us arbitrarily set the constant of integration E_0 equal to zero and obtain

$$E = \frac{m_0 c^2}{\sqrt{1 - \dfrac{v^2}{c^2}}} = mc^2 \tag{7-123}$$

Note that the mass of a particle is directly proportional to its total energy which, in turn, depends upon the choice of the reference frame. Furthermore, even for the case of zero velocity, the energy E has a nonzero value $m_0 c^2$ known as the *rest energy*.

Another form of the energy expression is obtained by using the velocity parameter θ given by Eq. (7-66). Then Eq. (7-123) becomes

$$E = m_0 c^2 \cosh \theta \tag{7-124}$$

The *kinetic energy* T of a particle is defined as the difference between the total energy E and the rest energy $m_0 c^2$.

$$T = m_0 c^2 \left(\frac{1}{\sqrt{1 - \dfrac{v^2}{c^2}}} - 1 \right) = m_0 c^2 (\cosh \theta - 1) \tag{7-125}$$

Note that for $v \ll c$ we have the approximation

$$T = m_0 c^2 \left[\left(1 + \frac{1}{2} \frac{v^2}{c^2} + \dots \right) - 1 \right] \cong \frac{1}{2} m_0 v^2 \tag{7-126}$$

in agreement with Newtonian mechanics.

The Momentum-Energy Four-Vector. Let us recall from Eq. (7-118) that

$$p_x = m_0 \frac{dx}{d\tau}$$

$$p_y = m_0 \frac{dy}{d\tau} \qquad (7\text{-}127)$$

$$p_z = m_0 \frac{dz}{d\tau}$$

and then use Eq. (7-55) to write the energy expression in the form

$$\frac{E}{c^2} = m_0 \frac{dt}{d\tau} \qquad (7\text{-}128)$$

Equations (7-127) and (7-128) have a similar form. Furthermore, the rest mass m_0 and the proper time τ are *invariant* with respect to a Lorentz transformation; that is, they have the same values in all inertial frames. It follows, then, that p_x, p_y, p_z, and E/c^2 must transform in the same way as x, y, z, and t, respectively. Hence, if we take the inertial frames I and I' in the usual manner and write expressions corresponding to the Lorentz transformation equations, we obtain

$$p'_x = \frac{p_x - \dfrac{VE}{c^2}}{\sqrt{1 - \beta^2}}$$

$$p'_y = p_y \qquad (7\text{-}129)$$

$$p'_z = p_z$$

$$E' = \frac{E - Vp_x}{\sqrt{1 - \beta^2}}$$

where $\beta = V/c$. The three components of $\mathbf{p}$ and the quantity E/c^2 comprise the *momentum-energy four-vector* analogous to (x, y, z, t).

Eq. (7-129) can be solved for the unprimed quantities, resulting in

$$p_x = \frac{p'_x + \dfrac{VE'}{c^2}}{\sqrt{1 - \beta^2}}$$

$$p_y = p'_y \qquad (7\text{-}130)$$

$$p_z = p'_z$$

$$E = \frac{E' + Vp'_x}{\sqrt{1 - \beta^2}}$$

Since the transformation properties of the momentum-energy four-vector are identical with those of the spacetime four-vector, an invariant expression corresponding to s^2 must exist. Thus we can write

$$E^2 - (p_x^2 + p_y^2 + p_z^2)c^2 = E'^2 - (p_x'^2 + p_y'^2 + p_z'^2)c^2 \qquad (7\text{-}131)$$

For a given particle this equality applies for any two inertial frames, including a *rest frame* in which the particle velocity is zero. This enables us to evaluate the constant, and we obtain

$$E^2 - p^2c^2 = E'^2 - p'^2c^2 = m_0^2c^4 \tag{7-132}$$

Another useful relation between the momentum and energy of a particle is obtained from the defining equations, namely,

$$p = \frac{E}{c}\beta \tag{7-133}$$

where $\beta = v/c$.

Force. One normally concludes that a particle has no resultant force acting on it if it either remains at rest or moves in a straight line with a uniform velocity. In other words, a free particle has a straight world line with respect to any inertial frame. On the other hand, a curved world line indicates the presence of a force. In accordance with this viewpoint, let us define the force **F** acting on a particle to be equal to its rate of change of momentum, that is,

$$\mathbf{F} = \dot{\mathbf{p}} = \frac{d}{dt}\left(\frac{m_0\mathbf{v}}{\sqrt{1-\beta^2}}\right) \tag{7-134}$$

where $\beta = v/c$. For $\beta \ll 1$ we see that this equation reduces to Newton's law of motion.

If we assume that the forces of interaction between two particles are equal and opposite, then Eq. (7-134) implies that the corresponding rates of change of momentum are equal and opposite. It follows, then, that the total momentum is conserved, in agreement with the assumptions made in the development of the defining equation for momentum.

Now let us perform the differentiation indicated in Eq. (7-134). We obtain

$$\mathbf{F} = \frac{m_0\dot{\mathbf{v}}}{\sqrt{1-\beta^2}} + \frac{m_0 v\dot{v}\mathbf{v}}{c^2(1-\beta^2)^{3/2}} \tag{7-135}$$

Because of the presence of the second term on the right, it is clear that the force **F** and the acceleration $\dot{\mathbf{v}}$ do not have the same direction, in general. For two rather common cases, however, the acceleration is parallel to the force.

Consider first the case in which **F** and **v** have the same direction. By symmetry, any acceleration must be longitudinal, that is, in the direction of **v**. Then, using F_l and a_l to designate the *longitudinal* force and acceleration, respectively, we can write Eq. (7-135) in the scalar form

$$F_l = \frac{m_0 a_l}{(1-\beta^2)^{3/2}} \tag{7-136}$$

A second case occurs when **F** and **v** are perpendicular. Here the velocity changes in direction but not in magnitude. Hence the last term of Eq. (7-135) vanishes and we have

$$F_t = \frac{m_0 a_t}{\sqrt{1 - \beta^2}} \tag{7-137}$$

where F_t and a_t refer to the *transverse* force and acceleration, that is, in a direction perpendicular to **v**.

Equations (7-136) and (7-137) can be written in the form

$$F_l = m_l a_l \tag{7-138}$$

$$F_t = m_t a_t \tag{7-139}$$

where m_l and m_t are called the *longitudinal mass* and the *transverse mass*, respectively, and are given by

$$m_l = \frac{m_0}{(1 - \beta^2)^{3/2}} \tag{7-140}$$

$$m_t = \frac{m_0}{\sqrt{1 - \beta^2}} \tag{7-141}$$

From the fact that m_l and m_t are unequal, in general, we note that the ratio of acceleration to force has different values for the longitudinal and transverse components. This confirms once again that the force and acceleration do not have the same direction, in general. Notice that the transverse mass is identical with what we have called the relativistic mass m. This is also the inertial and gravitational mass, and is used in calculations such as finding the location of the center of mass.

An important shortcoming of the concept of a force in relativity is the fact that it is neither an absolute quantity, as in Newtonian mechanics, nor does it have the Lorentz transformation properties of the position or momentum vectors. For example, suppose we consider a force **F** acting at a certain moment on a particle having a rest mass m_0 and a velocity $v = V$ in the direction of the x axis, where V is also the velocity of the inertial frame I'. In other words, let I' be a *momentary rest frame* defined such that the particle velocity relative to I' is zero at the given instant. Under these conditions, it can be shown that the longitudinal and transverse force components transform as follows:

$$\begin{aligned} F_l &= F_l' \\ F_t &= \sqrt{1 - \beta^2}\, F_t' \end{aligned} \tag{7-142}$$

Note that **F'** is the ordinary Newtonian force acting on the particle since its mass is m_0 in the I' frame. **F** is the force in frame I.

The force transformation properties can be improved by introducing the *Minkowski force* or *world force* $\mathbf{F}_M$ which is given by

$$\mathbf{F}_M = \frac{\mathbf{F}}{\sqrt{1 - \beta^2}} \tag{7-143}$$

where $\beta = v/c$ and v is the velocity of the particle. Then the equation of motion corresponding to Eq. (7-134) takes the form

$$\frac{d\mathbf{p}}{d\tau} = \mathbf{F}_M \tag{7-144}$$

and Eq. (7-120) becomes

$$\frac{dE}{d\tau} = \mathbf{F}_M \cdot \mathbf{v} \tag{7-145}$$

Once again we note that the proper time τ is the same for all inertial observers. Therefore $\mathbf{p}$ and E must transform in the same manner as $\mathbf{F}_M$ and $\mathbf{F}_M \cdot \mathbf{v}$. Stated differently, the four-vector $\{\mathbf{F}_M, \mathbf{F}_M \cdot \mathbf{v}/c^2\}$ obeys the same Lorentz transformation equations as $\{\mathbf{p}, E/c^2\}$ or $\{\mathbf{x}, t\}$.

Conservation of Energy. We have defined the relativistic momentum in such a manner that the total momentum is conserved in any collision of two particles, as viewed by an inertial observer. Assuming that a sequence of similar collisions can occur between pairs of particles, and noting that the total momentum of the system is found by summing over the particles, we obtain

$$\sum_{i=1}^{n} \mathbf{p}_{i1} = \sum_{i=1}^{n} \mathbf{p}_{i2} \tag{7-146}$$

where the summations occur at an arbitrary initial time t_1 and final time t_2 in the inertial frame I. If the same set of particles and interactions is observed from I' over an arbitrary interval t'_1 to t'_2, the total momentum is again conserved, that is,

$$\sum_{i=1}^{n} \mathbf{p}'_{i1} = \sum_{i=1}^{n} \mathbf{p}'_{i2} \tag{7-147}$$

If we consider the total energy of a system of *free* particles, we find that it is equal to the sum of the individual total energies of the particles. Since the Lorentz transformation equations for the momentum-energy four-vector are *linear* in these quantities, it follows that the same transformation equations apply to the momentum and energy of the *system* as apply to individual particles. Furthermore, the invariant expression of Eq. (7-132) applies to the system. Thus, at time t_1 or t'_1, that is, before any interactions,

$$E_1^2 - p_1^2 c^2 = E_1'^2 - p_1'^2 c^2 = m_0^2 c^4 \tag{7-148}$$

Similarly, at time t_2 or t'_2, that is, after all interactions,

$$E_2^2 - p_2^2 c^2 = E_2'^2 - p_2'^2 c^2 = m_0^2 c^4 \tag{7-149}$$

where m_0 is the rest mass of the system. Here we use the notation that p_1 is the magnitude of the system momentum at time t_1 measured relative to the frame I, while p'_2 refers to the magnitude of the system momentum at time t'_2 measured relative to frame I', and so forth.

Now subtract Eq. (7-149) from Eq. (7-148) and, with the aid of the momentum conservation law given by Eqs. (7-146) and (7-147), we obtain

$$E_1^2 - E_2^2 = E_1'^2 - E_2'^2 = 0$$

Since the energies are positive quantities, this is equivalent to

$$E_1 = E_2, \qquad E_1' = E_2' \tag{7-150}$$

indicating the *conservation of energy* in each inertial frame. In contrast to Newtonian theory, the conservation of energy obtained here applies in all cases in which momentum is conserved, including inelastic collisions.

Mass and Energy. The Einstein equation $E = mc^2$ shows the equivalence of mass and energy in the sense that they are directly proportional; a knowledge of one implies the value of the other. Furthermore, since the total energy is conserved in an isolated system and is equal to the kinetic energy plus the rest energy, we find that the rest energy increases at the expense of an equal reduction in the kinetic energy, or vice versa. As an example, suppose two particles undergo an inelastic impact. The sum of the particle rest masses will be larger after the impact than they were initially. This increased rest mass reflects an increase in internal energy which often appears as an increase of temperature.

The equivalence of mass and energy also applies to particles such as photons which have a *zero rest mass*. The total energy of a photon associated with radiation of frequency v (hertz) is

$$E = hv \tag{7-151}$$

where $h = 6.626 \times 10^{-34}$ joule sec is Planck's constant. If we equate this energy to mc^2 we find that a photon has an inertial mass

$$m = \frac{hv}{c^2} \tag{7-152}$$

Its momentum is

$$p = mv = \frac{hv}{c} \tag{7-153}$$

since a photon travels with the speed of light c. The well-known radiation pressure which occurs when radiation strikes a surface is explained by the changes in the momentum of the corresponding photons.

Example 7-6. Consider the spheres A and B, each having a rest mass m_0 and moving parallel to the x axis, as shown in Fig. 7-14. Sphere A has an initial velocity $0.9c$ and overtakes sphere B which has a velocity $0.3c$. There is an inelastic collision, after which the spheres stick together and move as a single particle. Solve for the momentum and energy of the individual spheres and of the system, relative to the inertial frames I and I', both before and after impact.

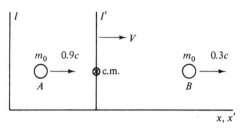

Figure 7-14. Inelastic collision of identical spheres.

First consider the motion relative to I. The relativistic masses of A and B are

$$m_A = \frac{m_0}{\sqrt{1 - (0.9)^2}} = 2.294m_0$$

$$m_B = \frac{m_0}{\sqrt{1 - (0.3)^2}} = 1.048m_0$$

The corresponding momenta are

$$p_A = m_A v_A = 2.065m_0 c$$

$$p_B = m_B v_B = 0.314m_0 c$$

Using capital letters to designate system parameters, we obtain the following total mass and momentum:

$$M = m_A + m_B = 3.342m_0$$

$$P = p_A + p_B = 2.379m_0 c$$

The total energy is obtained by summing the individual energies using Eq. (7-123)

$$E = m_A c^2 + m_B c^2 = 3.342m_0 c^2$$

The kinetic energies of A and B are equal to

$$T_A = m_A c^2 - m_0 c^2 = 1.294m_0 c^2$$

$$T_B = m_B c^2 - m_0 c^2 = 0.048m_0 c^2$$

The total kinetic energy is

$$T = T_A + T_B = 1.342m_0 c^2$$

which, we note, is also equal to $Mc^2 - 2m_0 c^2$. In other words, the total kinetic energy is equal to c^2 multiplied by the difference between the total relativistic mass and the total rest mass of the individual spheres.

Now let us find the center of mass location relative to the frame I. We use relativistic masses in this calculation and obtain

$$Mx_c = m_A x_A + m_B x_B$$

which yields the position x_c of the center of mass at

$$x_c = 0.686x_A + 0.314x_B$$

Notice that the center of mass is not at the midpoint between A and B, even though the spheres have identical rest masses. Furthermore, it will be shown that the relative location of the center of mass is dependent on the choice of reference frame.

The velocity of the center of mass is

$$V = 0.686v_A + 0.314v_B = 0.712c$$

This, of course, is also the velocity of the single body after collision. A check on the total momentum is obtained from

$$P = MV = 2.379m_0c$$

Let us arbitrarily choose the inertial frame I' so that it translates with the center of mass at the constant velocity V. Also, for convenience, we assume that the center of mass is located at $x' = 0$. In order to obtain the particle velocities relative to I' we need to use the relativistic velocity addition law. From Eq. (7-64) we have

$$v'_A = \frac{v_A - V}{1 - Vv_A/c^2} = 0.524c$$

$$v'_B = \frac{v_B - V}{1 - Vv_B/c^2} = -0.524c$$

Because the two spheres have equal rest masses and equal speeds relative to I', their relativistic masses are equal.

$$m'_A = m'_B = \frac{m_0}{\sqrt{1 - (0.524)^2}} = 1.174m_0$$

It is apparent, then, that in the I' frame the center of mass lies midway between the particles, in contrast with its position in I. After the collision, the single resultant body is motionless at $x' = 0$. The mass of this body is equal to the constant total mass, namely,

$$M' = m'_A + m'_B = 2.348m_0$$

Since the velocity is zero, this is also the final rest mass M'_0. The total energy relative to I' is

$$E' = M'c^2 = 2.348m_0c^2$$

Now let us calculate the momentum of each particle in I'. Because of the symmetry in their motions, we find that

$$p'_A = -p'_B = (1.174m_0)(0.524c) = 0.615m_0c$$

The total momentum P' is zero as expected.

The kinetic energies of A and B relative to I' are

$$T'_A = T'_B = m'c^2 - m_0c^2 = 0.174m_0c^2$$

The total kinetic energy is

$$T' = T'_A + T'_B = 0.348m_0c^2$$

In Newtonian mechanics, the total kinetic energy relative to an inertial frame I is equal to the sum of two parts; namely, (1) the translational kinetic energy due to the total mass moving as a particle with the velocity of the center of mass and (2) the kinetic energy due to motion of the system relative to the center of mass. Let us see if this rule also applies in the relativistic case. First consider a particle of relativistic mass M moving with velocity V. Its kinetic energy is

$$T_1 = (M - M_0)c^2 = (1 - \sqrt{1 - \beta^2})Mc^2 = 0.994m_0c^2$$

Note that $M_0 = M'_0 = M' = 2.348m_0$. The second portion of the kinetic energy we have found previously, namely,

$$T_2 = T' = 0.348m_0c^2$$

Adding, we obtain

$$T = T_1 + T_2 = 1.342m_0c^2$$

in agreement with the earlier value. Hence these kinetic energies can be summed as in Newtonian theory.

We have found that the total rest mass of the particles increased from $2m_0$ to $2.348m_0$ as a result of the inelastic impact. This increment of $0.348m_0c^2$ we can now identify with T', the internal kinetic energy due to motion relative to the mass center.

A few additional comments are in order. We have calculated the momentum and energy of the system relative to the inertial frames I and I'. These quantities must tramsform in accordance with Eq. (7-130). A check shows that

$$P = \frac{P' + \dfrac{VE'}{c^2}}{\sqrt{1 - \beta^2}} = 2.379m_0c$$

$$E = \frac{E' + VP'}{\sqrt{1 - \beta^2}} = 3.342m_0c^2$$

in agreement with earlier results.

Finally, let us consider the magnitude of the impulse between the two spheres at the time of the collision. This impulse is equal to the change in momentum of one of the spheres. Considering the motion of sphere B relative to frame I, we obtain

$$\hat{F}_B = \tfrac{1}{2}P - p_B = 0.875m_0c$$

Relative to frame I', however, we have

$$\hat{F}'_B = \tfrac{1}{2}P' - p'_B = -p'_B = 0.615m_0c$$

These impulses are longitudinal in nature, and we have seen that the longitudinal force is invariant with respect to a change in the reference frame. The discrepancy in size of the impulses in I and I' can be explained by the different time scales used in calculating the impulse. Due to the time dilation effect, a clock fixed in I' runs slower than the synchronized clocks of I by the factor $(1 - \beta^2)^{1/2} = 0.702$. This is also the ratio of $\hat{F}'_B$ to $\hat{F}_B$.

The Principle of Equivalence. The principle of equivalence states that the inertial mass and the gravitational mass of a body are equal. The gravitational force acting on a particle in a given gravitational field is proportional to its *gravitational mass* m_g. On the other hand, the inertial force, which measures the resistance of a particle to acceleration, is proportional to its *inertial mass* m_i. For a particle which is freely falling in a vacuum, the gravitational and inertial forces are equal and opposite, that is,

$$m_g\mathbf{f} = -(-m_i\mathbf{a}) \qquad (7\text{-}154)$$

where $\mathbf{f}$ is the gravitational field strength and $\mathbf{a}$ is the acceleration of the particle relative to an inertial frame. Since $m_g = m_i$ according to the principle of equivalence, this implies that all bodies will have the same acceleration in a given gravitational field, regardless of their individual masses and compositions. It has been shown experimentally that the inertial and gravitational masses are, in fact, equal to within one part in 10^{12}.

Another consequence of the principle of equivalence is the fact that a nonrotating frame attached to a body which is falling freely in a uniform gravitational field is the equivalent of a standard inertial frame that is translating uniformly in free space. No local physical experiments can be devised to detect any difference between the frames. This balancing of gravitational and inertial forces is illustrated by the weightlessness experienced by men in an orbiting spacecraft. If the spacecraft is not rotating, it represents a good approximation to an inertial frame in free space. For example, a free particle will move with a uniform speed in a straight line relative to the spacecraft, provided that the path is not too long.

Another illustration of the principle of equivalence is provided by *Einstein's elevator.* Suppose physical experiments are performed within an elevator at rest in a uniform gravitational field of 1 g, that is, the gravitational acceleration matches that at the earth's surface. The results will be indistinguishable from a similar set of experiments performed when the elevator is in free space, but accelerated at a uniform rate of 9.81 meters/sec² relative to an inertial frame. Of course, the idea of representing the effect of an accelerating frame by an equivalent gravitational field applies to Newtonian as well as to relativistic mechanics.

Previously we found that a photon of frequency ν has an inertial mass $h\nu/c^2$. It follows, then, from the principle of equivalence, that it must have an equal gravitational mass, even though the rest mass is zero. The effect of a gravitational field on the motion of photons is shown in the slight bending of light rays which pass very near the sun in travelling from a star to the earth. The apparent positions of these stars, viewed during a total eclipse of the sun, are displaced slightly (approximately 1.75 arc seconds) in a direction away from the center of the sun. Similar effects can be measured at centimeter wavelengths using radio telescopes, without the requirement of an eclipse. Although a qualitative analysis of the curvature of the light path can be obtained by using ordinary Newtonian gravitational theory, an accurate analysis involves general relativity theory.

Lagrangian and Hamiltonian Formulations. So far in our discussion of relativistic dynamics, we have not used the analytical methods of Lagrange and Hamilton. As a start in this direction, let us consider first a single particle of rest mass m_0. Its position is specified by the Cartesian coordinates (x_1, x_2, x_3) measured relative to an inertial frame. In accordance with Eq. (7-116) the momentum components are

$$p_i = \frac{m_0 \dot{x}_i}{\sqrt{1 - \beta^2}} \qquad (i = 1, 2, 3) \tag{7-155}$$

where

$$\beta = \frac{1}{c} \sqrt{\dot{x}_1^2 + \dot{x}_2^2 + \dot{x}_3^2} \tag{7-156}$$

But the basic definition of momentum in terms of the Lagrangian function was given by Eq. (1-137). In terms of Cartesian velocities, it is

$$p_i = \frac{\partial L}{\partial \dot{x}_i} \tag{7-157}$$

If we assume that this equation remains valid in the relativistic case, we obtain from Eqs. (7-155) and (7-157) that

$$\frac{\partial L}{\partial \dot{x}_i} = \frac{m_0 \dot{x}_i}{\sqrt{1 - \beta^2}} \tag{7-158}$$

In general, the Lagrangian function has the form $L(x, \dot{x}, t)$. To be consistent with Eq. (7-158), however, it must take the more explicit form

$$L(x, \dot{x}, t) = -m_0 c^2 \sqrt{1 - \beta^2} - V(x, t)$$
$$= -m_0 c^2 \sqrt{1 - \frac{1}{c^2}(\dot{x}_1^2 + \dot{x}_2^2 + \dot{x}_3^2)} - V(x, t) \tag{7-159}$$

where the negative sign of the last term is taken for later convenience.

The basic equation of motion for a particle, given by Eq. (7-134), can be written in the form

$$\dot{p}_i = \frac{d}{dt}\left(\frac{m_0 \dot{x}_i}{\sqrt{1 - \beta^2}}\right) = F_i \tag{7-160}$$

Let us assume that all the force components F_i are obtained from a potential function $V(x, t)$ in accordance with

$$F_i = -\frac{\partial V}{\partial x_i} \qquad (7\text{-}161)$$

Then it is apparent that the Lagrangian function given in Eq. (7-159) yields the correct relativistic equations of motion when substituted into the standard form of Lagrange's equation, namely,

$$\frac{d}{dt}\left(\frac{\partial L}{\partial \dot{x}_i}\right) - \frac{\partial L}{\partial x_i} = 0 \qquad (7\text{-}162)$$

Notice that the Lagrangian function in this relativistic case is *not equal to* $T - V$.

The Hamiltonian function is related to the Lagrangian function in the usual manner, namely,

$$H = \sum_i p_i \dot{x}_i - L \qquad (7\text{-}163)$$

or

$$H = \frac{m_0 c^2}{\sqrt{1 - \beta^2}} + V(x, t) \qquad (7\text{-}164)$$

After substituting for β in terms of the p's, we obtain

$$H(x, p, t) = c\sqrt{m_0^2 c^2 + p_1^2 + p_2^2 + p_3^2} + V(x, t) \qquad (7\text{-}165)$$

Notice that the Hamiltonian function is equal to the sum of the kinetic energy, potential energy, and the rest energy.

Another problem of interest concerns the motion of a charged particle in an electromagnetic field. The nonrelativistic case was discussed previously in Sec. 3-4. In order to generalize these results to the case of relativistic particle velocities, we note first that the force law is unchanged; that is, it is obtained from the velocity-dependent potential (mks units)

$$U(x, \dot{x}, t) = e(\phi - \mathbf{v} \cdot \mathbf{A}) \qquad (7\text{-}166)$$

where $\mathbf{v}$ is the velocity of the particle, e is the particle charge, ϕ is the scalar potential, and $\mathbf{A}$ is the vector potential. If we assume that any gravitational fields are negligible, and omit the corresponding potential function, the Lagrangian function of Eq. (7-159) is changed to

$$L(x, \dot{x}, t) = -m_0 c^2 \sqrt{1 - \beta^2} - e\phi + e\mathbf{v} \cdot \mathbf{A} \qquad (7\text{-}167)$$

The standard form of Lagrange's equation, as given in Eq. (7-162), then produces the differential equations of motion. Notice that the momentum components are

$$p_i = \frac{\partial L}{\partial \dot{x}_i} = \frac{m_0 \dot{x}_i}{\sqrt{1 - \beta^2}} + eA_i \qquad (7\text{-}168)$$

showing that the electromagnetic field contributes to the momentum.

In a similar fashion, Eqs. (7-163) and (7-167) are used to obtain the Hamiltonian function for a charged particle in an electromagnetic field, with the result

$$H = \frac{m_0 c^2}{\sqrt{1 - \beta^2}} + e\phi \qquad (7\text{-}169)$$

After eliminating the $\dot{x}$'s in favor of p's by using Eq. (7-168), we obtain a Hamiltonian which is perhaps most easily written in the form

$$H = c\sqrt{m_0^2 c^2 + (\mathbf{p} - e\mathbf{A})^2} + e\phi \qquad (7\text{-}170)$$

The first term of this result is identical with that of Eq. (7-165) except that each p_i is replaced by $(p_i - eA_i)$. Note that $\mathbf{A} = \mathbf{A}(x, t)$ and $\phi = \phi(x, t)$.

7-4. ACCELERATED SYSTEMS

In our discussion of special relativity, we have been concerned with the motion of particles relative to inertial frames which have constant velocities relative to each other. It is well-known, however, that the special theory of relativity can be applied to the analysis of particle motions in the various high-energy accelerators with good results. How is it possible that we are able to extend the application of the theory in this manner?

First of all, we use the idea of a *momentary rest frame*, that is, an inertial frame whose velocity at the given instant is equal to the velocity of the particle. Let us designate the momentary rest frame by I' (Fig. 7-15). As viewed from this frame, the mass of the particle is its rest mass m_0, and the particle acceleration is equal to the ratio F'/m_0, where F' is the applied force on the particle.

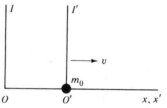

Figure 7-15. The momentary rest frame I'.

Secondly, we make the important assumption that the rate of a standard clock travelling with the particle is not directly affected by the acceleration. In other words, an increment of proper time $d\tau$ for the particle is identical with the increment dt' registered by a clock in the momentary rest frame I'. Thus we can write

$$d\tau = dt' \qquad (7\text{-}171)$$

Finally, instead of a single I' frame, let us assume a *sequence* of these momentary rest frames for successive instants of time. Furthermore, we can assume that the particle is at the origin O' of the applicable rest frame at each instant. Thus, the position x' and the velocity $\dot{x}'$ is zero at each instant, but the acceleration $\ddot{x}'$ is not zero, in general.

With this background, then, let us consider the *rectilinear accelerated motion* of a particle along the common x axes of the frames I and I'. Since frame I and any of the I' frames are inertial, the corresponding spacetime coordinates are connected by a Lorentz transformation. We have assumed that x' equals zero at all times, and therefore we find from Eq. (7-35) that

$$dt = \frac{dt'}{\sqrt{1 - \beta^2}} = \frac{d\tau}{\sqrt{1 - \beta^2}} \qquad (7\text{-}172)$$

where $\beta = v/c$. Also, we obtain

$$dx = v\,dt = \frac{v\,d\tau}{\sqrt{1 - \beta^2}} \qquad (7\text{-}173)$$

In terms of the velocity parameter θ, these equations can be written in the form

$$dx = c \sinh \theta \, d\tau \qquad (7\text{-}174)$$

$$dt = \cosh \theta \, d\tau \qquad (7\text{-}175)$$

Upon integration, the resulting equations are

$$x = c \int_0^\tau \sinh \theta \, d\tau \qquad (7\text{-}176)$$

$$t = \int_0^\tau \cosh \theta \, d\tau \qquad (7\text{-}177)$$

where we assume that the particle leaves the origin O in I at $t = \tau = 0$.

Rocket with Constant Acceleration. Let us consider the rectilinear motion of a rocket which has a constant acceleration g, as recorded by an accelerometer carried by the rocket. In other words, there is a constant acceleration g of the rocket relative to its momentary rest frame I'. The acceleration relative to the inertial frame I is not constant, however, as will be shown by the analysis.

Let us assume that the motion is confined to the x axis. The rocket starts from rest at the origin O when $t = \tau = 0$. Suppose that we consider an infinitesimal change in velocity dv' relative to I' during the interval $d\tau$ of proper time. Then

$$dv' = g\,d\tau \qquad (7\text{-}178)$$

From the definition of the velocity parameter given in Eq. (7-65), we see that

$$d\theta' = \frac{dv'}{c} = \frac{g\,d\tau}{c} \qquad (7\text{-}179)$$

since θ' is zero before the addition of the infinitesimal increment. Furthermore, by the velocity addition law of Eq. (7-70), the velocity parameter θ, representing the rocket velocity relative to I, is equal to the sum of θ', representing the motion of I' relative to I, and the increment $d\theta'$. Therefore, the increment $d\theta$ in I is equal to the increment $d\theta'$ in I'. Hence we can write

$$d\theta = \frac{g\,d\tau}{c} \qquad (7\text{-}180)$$

which, for a constant g and the assumed initial conditions, yields

$$\theta = \frac{g\tau}{c} \tag{7-181}$$

Now we can use the results given in Eqs. (7-176) and (7-177) to obtain

$$x = c \int_0^\tau \sinh \frac{g\tau}{c}\, d\tau = \frac{c^2}{g}\left(\cosh \frac{g\tau}{c} - 1\right) \tag{7-182}$$

$$t = \int_0^\tau \cosh \frac{g\tau}{c}\, d\tau = \frac{c}{g} \sinh \frac{g\tau}{c} \tag{7-183}$$

This gives the motion of the rocket relative to the inertial frame I in terms of proper time τ. The velocity is found from Eqs. (7-65) and (7-181).

$$v = c \tanh \theta = c \tanh \frac{g\tau}{c} \tag{7-184}$$

Note that, as τ approaches infinity, x and t also approach infinity, but the velocity v approaches the speed of light.

In order to find the motion in terms of the time t of frame I, we can eliminate τ from Eqs. (7-182) and (7-183) and obtain

$$x = \frac{c^2}{g}\left(\sqrt{1 + \frac{g^2 t^2}{c^2}} - 1\right) \tag{7-185}$$

The velocity is

$$v = \frac{dx}{dt} = \frac{gt}{\sqrt{1 + \frac{g^2 t^2}{c^2}}} \tag{7-186}$$

A further differentiation yields the acceleration

$$a = \frac{dv}{dt} = \frac{g}{\left(1 + \frac{g^2 t^2}{c^2}\right)^{3/2}} \tag{7-187}$$

This acceleration is equal to g for $t = 0$ and decreases with increasing t. It is clear, then, that an acceleration which appears constant to the occupants of a rocket is not constant as viewed by an external inertial observer.

In order to obtain an appreciation of the physical implications of these equations for rocket flight, let us adopt the year as the unit of time and the light-year as the unit of distance. These are related to the more conventional units by

$$1 \text{ yr} = 3.156 \times 10^7 \text{ sec}$$

$$1 \text{ lt-yr} = 9.46 \times 10^{15} \text{ m}$$

$$1 \text{ lt-yr/yr} = c = 2.998 \times 10^8 \text{ m/sec}$$

$$1 \text{ lt-yr/yr}^2 = 9.50 \text{ m/sec}^2$$

$$= 31.17 \text{ ft/sec}^2$$

$$= 0.969 \text{ earth g's}$$

Now consider the particular case where we take $g = 1$ lt-yr/yr^2 which is approximately equal to the acceleration of gravity on the earth's surface. Suppose the rocket starts from rest and maintains this constant acceleration for one year of proper time; that is, the time interval registered on the pilot's watch is one year. Then g, τ, and c all have a unit magnitude, and Eq. (7-182) yields the final position

$$x = \cosh 1 - 1 = 0.543 \text{ lt-yr}$$

From Eq. (7-184) the final velocity is

$$v = c \tanh 1 = 0.762c$$

measured in the frame I. The corresponding time, measured in I, is

$$t = \frac{c}{g} \sinh 1 = 1.175 \text{ yr}$$

It is interesting to note that for the case of *Galilean kinematics*, the corresponding values would be

$$x = 0.500 \text{ lt-yr}$$
$$v = c$$
$$t = 1 \text{ yr}$$

Hence the effect of relativity theory is to *increase* the distance travelled in a given proper time, even though the velocity relative to the frame I is reduced.

Example 7-7. Suppose a round trip is to be made by rocket from the earth to a nearby star, Alpha Centauri, which is about 4 light-years distant. The rocket is capable of a constant acceleration $g = 9.50$ m/sec^2 (1 lt-yr/yr^2) relative to its momentary rest frame. What is the time required for the trip?

For a round trip which requires the least time, the rocket thrusts continuously. First, it is accelerated at the given constant acceleration until the midpoint $x = 2$ lt-yr is reached. Then the thrust is reversed, and the rocket decelerates for the last half of the outward journey, arriving at Alpha Centauri with zero velocity. The return trip is accomplished in a similar fashion.

If we refer to Eq. (7-182) and solve for the proper time τ of the first thrusting period ($x = 2$ lt-yr), we obtain

$$\tau = \frac{c}{g} \cosh^{-1}\left(\frac{gx}{c^2} + 1\right) = 1.76 \text{ yr}$$

From Eq. (7-183), the corresponding time t registered by clocks fixed in I is

$$t = \frac{c}{g} \sinh \frac{g\tau}{c} = 2.83 \text{ yr}$$

Hence the total proper time required for the round trip is $4 \times 1.76 = 7.04$ years. On the other hand, the total time for the round trip in the frame I is $4 \times 2.83 = 11.32$ years.

Let us consider these results in the context of the twin paradox. Imagine that twin A remains at the earth and twin B goes on a round trip to Alpha Centauri. Under these conditions, the travelling twin B ages by only 7.04 years on his trip, even though he has covered a total distance of 8 light-years. Twin A, on the other hand, ages by 11.32 years during the absence of B.

Rocket with Constant Thrust. Consider a rocket with a constant thrust F_0 and a decreasing rest mass $m_0(1 - \alpha\tau)$, where m_0 is the initial rest mass and the positive constant α is proportional to the rate at which mass is expelled from the rocket. We wish to solve for the rectilinear motion of this rocket with respect to an inertial frame I, for τ in the interval $0 \leq \tau < 1/\alpha$. Assume that the rocket starts from rest at the origin O at $t = \tau = 0$ (Fig. 7-15) and let I' be the momentary rest frame.

First we notice that the acceleration of the rocket relative to I' is

$$g(\tau) = \frac{F_0}{m_0(1 - \alpha\tau)} \tag{7-188}$$

Then, using Eq. (7-180), we obtain

$$\theta = \frac{1}{c} \int_0^\tau g(\tau)\, d\tau = \frac{F_0}{m_0 c\alpha} \ln\left(\frac{1}{1 - \alpha\tau}\right) \tag{7-189}$$

From Eqs. (7-176) and (7-177), we see that

$$\frac{dx}{d\tau} = c \sinh\theta = c \sinh\left[\frac{F_0}{m_0 c\alpha} \ln\left(\frac{1}{1 - \alpha\tau}\right)\right] \tag{7-190}$$

$$\frac{dt}{d\tau} = \cosh\theta = \cosh\left[\frac{F_0}{m_0 c\alpha} \ln\left(\frac{1}{1 - \alpha\tau}\right)\right] \tag{7-191}$$

These equations can be integrated to obtain x and t as functions of τ. In order to simplify matters, however, let us again choose the year as the unit of time and the light-year as the unit of distance. Then the velocity of light will have a unit magnitude. Also, let the initial acceleration F_0/m_0 be 1 lt-yr/yr² , which we have found to be approximately equal to the gravitational acceleration at the earth's surface. Finally, assume $\alpha = 1$, implying that mass is lost at a rate which would result in zero remaining mass after one year. Then, using the exponential form of the hyperbolic functions, Eqs. (7-190) and (7-191) can be written as follows:

$$\frac{dx}{d\tau} = \frac{1}{2}\left[\frac{1}{1 - \tau} - (1 - \tau)\right] \tag{7-192}$$

$$\frac{dt}{d\tau} = \frac{1}{2}\left[\frac{1}{1 - \tau} + (1 - \tau)\right] \tag{7-193}$$

After integrating and applying the initial conditions, we obtain

$$x = -\tfrac{1}{2}\ln(1 - \tau) + \tfrac{1}{4}[(1 - \tau)^2 - 1] \tag{7-194}$$

$$t = -\tfrac{1}{2}\ln(1 - \tau) - \tfrac{1}{4}[(1 - \tau)^2 - 1] \tag{7-195}$$

Suppose we consider the situation when 0.5 yr has elapsed on the rocket since it left the earth (at the origin O of frame I). First, we notice that half the rest mass has been lost, but the thrust is unchanged. Hence, the accelerometer registers 2 g's, as may be verified by substituting $\tau = 0.5$ into Eq. (7-188). Also, from Eqs. (7-194) and (7-195), we find that

$$x = 0.159 \text{ lt-yr}$$

$$t = 0.534 \text{ yr}$$

The velocity of the rocket relative to the earth is found by dividing Eq. (7-192) by Eq. (7-193).

$$v = \frac{dx}{dt} = \frac{1 - (1 - \tau)^2}{1 + (1 - \tau)^2} \tag{7-196}$$

For $\tau = 0.5$ we see that the rocket velocity is $v = 0.600c$.

Let us compare these results with those for a constant-acceleration rocket with the same initial acceleration and proper time τ. From Eqs. (7-182)–(7-184) we obtain

$$x = \cosh \tfrac{1}{2} - 1 = 0.128 \text{ lt-yr}$$

$$t = \sinh \tfrac{1}{2} = 0.521 \text{ yr}$$

$$v = c \tanh \tfrac{1}{2} = 0.462c$$

It can be seen that, in this particular example, the constant-thrust rocket travelled about 24% farther and had a 30% larger final velocity.

For larger values of τ, the ratio of distances travelled will continue to increase in favor of the constant-thrust rocket. The ratio of final velocities, however, will first increase but finally decrease as the velocity of the constant-thrust rocket approaches its limiting value, namely, the speed of light.

REFERENCES

1. TAYLOR, E. F., and J. A. WHEELER, *Spacetime Physics*. San Francisco: W. H. Freeman and Company, 1966. An excellent introduction to relativity, particularly in its explanations of the differences between conventional geometry and the geometry of spacetime.

2. MØLLER, C., *The Theory of Relativity*. London: Oxford University Press, 1952. A good general treatment of the subject. Written in the classical style, it is thorough and usually clear in its presentation.

3. EINSTEIN, A., H. A. LORENTZ, H. MINKOWSKI, and H. WEYL, *The Principle of Relativity*. New York: Dover Publications, Inc., 1952. This is a collection of the English translations of the original papers of unusual importance in the development of the special and general theories of relativity.

4. BERGMANN, P. G., *Introduction to the Theory of Relativity*. Englewood Cliffs, N. J.: Prentice-Hall, Inc., 1942. A widely-used book with good discussions of the

elementary parts of the theory. It moves rapidly, however, to more advanced topics with about half the book being devoted to general relativity and unified field theories.

PROBLEMS

7-1. Particle A is located initially at the origin and moves with a constant velocity $v_x = 0.9c$ along the x axis. Particle B starts at $x = 1$ and moves with a velocity $v_x = -0.9c$ along the same axis. What is the closure rate of the two particles according to an observer in I; that is, at what rate is the distance between the two decreasing? What is the velocity of B, measured in a frame I' translating with A?

7-2. The origins O and O' of the inertial frames I and I' coincide at $t = t' = 0$. I' translates with a constant velocity V relative to I in the positive x direction. At $t = 0$, a photon leaves the common origin and moves with velocity c, hitting a wall at $x = l$ which is fixed in I.
(a) Find the values of x' and t' for the event that the photon hits the wall.
(b) What is the location of the wall relative to O', expressed as a function of t'?

7-3. A relativistic train moves with velocity $V\mathbf{i}$ along a straight track $y = h$ in frame I. When the front end of the train crosses the y axis, a relativistic bullet is fired from the origin O with velocity $\mathbf{v} = v\mathbf{j}$.
(a) How far behind the front end of the train does the bullet strike, according to an observer in frame I?
(b) How far behind its front end does the bullet strike, according to an observer on the train?
(c) What are the x' and y' velocity components of the bullet relative to the train?

7-4. The inertial frame I' moves relative to I with the constant velocity V in the x direction. Find the speed relative to I of a particle moving along the common x axis such that its values of t and t' are equal at all times. Solve for the proper time τ as a function of t.

7-5. Suppose the inertial frame I_1 translates with velocity v_1 relative to I_0 in the direction of the common x axis. Similarly, I_2 translates with velocity v_2 relative to I_1, and the process continues in the same fashion for frames I_3 and I_4. Finally, a particle translates with velocity v_5 relative to I_4. Assuming that $v_1 = v_2 = v_3 = v_4 = v_5 = \frac{1}{2}c$, what is the velocity of the particle relative to I_0?

7-6. An inertial frame I' has a constant velocity V in the common x direction relative to the frame I. Suppose a particle has the velocity components (v'_x, v'_y) relative to I' and the corresponding acceleration components (a'_x, a'_y). Show that the acceleration components of the particle relative to the frame I are

$$a_x = \frac{(1 - \beta^2)^{3/2} a'_x}{(1 + \beta v'_x/c)^3} \qquad a_y = \frac{1 - \beta^2}{(1 + \beta v'_x/c)^2}\left(a'_y - \frac{\beta v'_y/c}{1 + \beta v'_x/c} a'_x \right)$$

where $\beta = V/c$.

7-7. A straight rod of length l_0 makes a constant angle ϕ with the x axis, as measured in the inertial frame I. It translates with velocity components v_x and v_y relative

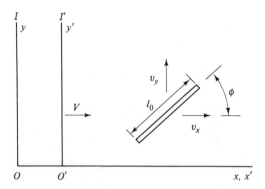

Figure P7-7.

to I. If the frame I' translates with velocity V relative to I, as shown, find the velocity of the rod relative to I'. Show that its inclination ϕ' relative to the x' axis is given by

$$\tan \phi' = \frac{1 - \beta v_x/c}{\sqrt{1 - \beta^2}} \tan \phi + \frac{\beta v_y}{c\sqrt{1 - \beta^2}}$$

where $\beta = V/c$.

7-8. A cylinder of radius r and length l_0 rotates about its axis of symmetry (x axis) with a constant angular velocity ω_0 in I, where $\omega_0 < c/r$.

(a) What is the angular velocity of the cylinder relative to I'?

(b) Suppose lines parallel to the x axis are inscribed on the surface of the cylinder. What is the total angle of twist of the cylinder, as viewed from frame I'?

(c) What is the angle in I' between the tangent to an inscribed line AB and the direction of the x' axis?

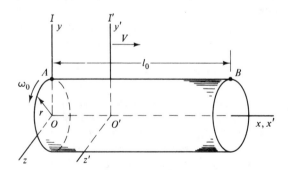

Figure P7-8.

7-9. Consider a straight rod of rest length $2l_0$ which rotates with a constant angular velocity ω in the $x'y'$ plane about its center O' at the origin of the inertial frame I'. The frame I' translates with a constant velocity V relative to the frame I. Show that the rod is curved, according to an observer in I, by solving for y as a function of x and t for points on the rod. For the particular case where $\omega l_0 = V = \frac{1}{2}c$, find the locations of the ends of the rod at $t = 0$.

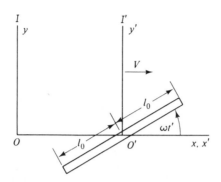

Figure P7-9.

7-10. Suppose a particle has a velocity $v' = \beta'c$ at an angle ϕ', measured from the x' axis in frame I'. Frame I' moves with a velocity $V = \beta c$ relative to I in the direction of the common x axes. Find the angle ϕ that the particle's trajectory makes with the x axis, using *nonrelativistic* theory. How does this angle compare in size with the relativistic result?

7-11. A particle moves along the x axis with a constant velocity $0.8c$. It radiates light with the same intensity in all directions, as viewed by an observer travelling with the particle. According to a fixed observer, what percentage of the light propagates in a direction having a positive x component?

7-12. A rocket moves in a straight line at half the speed of light. When it is at the point of closest approach to a fixed star, it sends a radio signal toward the star. At what angle from the forward direction must the transmitting antenna be aimed, as measured on the rocket? By what angle does this direction differ from the apparent direction of the star, as viewed from the rocket at the time the signal is sent?

7-13. Particles A and B have rest masses $\frac{1}{2}m_0$ and m_0, respectively. They cross the y axis at the same time and have the velocities shown in the figure. Assuming that the particles collide inelastically and stick together, solve for the velocity v and the angle θ of the single particle after impact. What is the final rest mass?

7-14. The rest energy of a proton is 938 Mev (million electron volts). Suppose two protons, each having 2×10^5 Mev total energy, collide head-on. In the fixed center-of-mass frame, the available energy from the collision is the entire 4×10^5 Mev.

Now suppose that, instead of equal and opposite proton velocities, all 4×10^5 Mev is put into a single proton which collides with a second fixed proton. Analyze this collision in a center-of-mass frame. Show that, relative to this frame, only 2.7×10^4 Mev of total energy is available. This illustrates the advantage of using the colliding beam principle in very high energy accelerators.

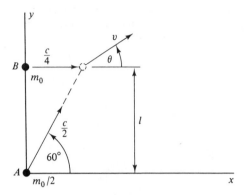

Figure P7-13.

7-15. Consider a relativistic mass-spring system with a rest mass m_0 and spring stiffness k. Let x represent the displacement of the mass. Obtain the Lagrangian function and the differential equation of motion. Integrate this equation and show that the total energy $E = mc^2 + \frac{1}{2}kx^2$ is constant. Let a be the amplitude of the oscillation and show that time and displacement are related by

$$t = \sqrt{\frac{m_0}{k}} \int_{\phi_0}^{\phi} \frac{1 + 2\eta^2 \cos^2 \phi}{\sqrt{1 + \eta^2 \cos^2 \phi}} \, d\phi$$

where $\sin \phi = x/a$ and $\eta^2 = ka^2/4m_0c^2$.

APPENDIX

ANSWERS TO SELECTED PROBLEMS

CHAPTER 1

1-1. $x = A \sin \omega t + l \sin \theta, \quad y = -l \cos \theta$
$(x - A \sin \omega t)^2 + y^2 = l^2$

1-2. $x_1 = z - \frac{1}{2}l\theta, \quad x_2 = z + \frac{1}{2}l\theta$
$Q_z = F_1 + F_2, \quad Q_\theta = \frac{1}{2}l(F_2 - F_1)$

1-3. $x_1 = x - \frac{1}{2}l \cos \theta, \quad y_1 = y - \frac{1}{2}l \sin \theta$
$x_2 = x + \frac{1}{2}l \cos \theta, \quad y_2 = y + \frac{1}{2}l \sin \theta$
$\dfrac{\partial(x_1, y_1, x_2, y_2)}{\partial(x, y, \theta, q_4)} = -l$
$x = \frac{1}{2}(x_1 + x_2), \quad y = \frac{1}{2}(y_1 + y_2)$
$\theta = \tan^{-1} \dfrac{y_2 - y_1}{x_2 - x_1}, \quad q_4 = [(x_2 - x_1)^2 + (y_2 - y_1)^2]^{1/2}$

1-4. $T = mr^2[\omega(\dot\theta + \omega)(1 - \cos \theta) + \frac{1}{2}\dot\theta^2]$
$p_\theta = mr^2[\omega(1 - \cos \theta) + \dot\theta]$

1-5. $T = \frac{1}{2}m\dot{x}^2 + \frac{1}{6}ml^2\dot\theta^2 - \frac{1}{2}ml\dot{x}\dot\theta \sin \theta$
$p_\theta = \frac{1}{3}ml^2\dot\theta - \frac{1}{2}ml\dot{x} \sin \theta$

1-6. $T = \frac{3}{4}m\dot{q}^2 - \frac{3}{2}mr\omega\dot{q} + \frac{1}{2}m\omega^2 q^2 + \frac{3}{4}mr^2\omega^2$

1-7. $\dot\phi = \dfrac{l}{r}\dot\theta + \left(\dfrac{l+r}{r}\right)\dot\psi$
$T = \frac{11}{12}ml^2\dot\theta^2 + \frac{1}{2}ml(l+r)\dot\theta\dot\psi + \frac{1}{2}[I_0 + \frac{1}{2}m(l+r)^2]\dot\psi^2$
$p_\theta = \frac{11}{6}ml^2\dot\theta + \frac{1}{2}ml(l+r)\dot\psi$
$p_\psi = [I_0 + \frac{1}{2}m(l+r)^2]\dot\psi + \frac{1}{2}ml(l+r)\dot\theta$

1-9. Constraint $x^2 + y^2 = r^2$ or $x\delta x + y\delta y = 0$

1-10. Constraint $x^2 + y^2 = l^2$

1-11. $m\ddot{x} - m(x_0 + x)\dot\theta^2 + kx = 0$
$I\ddot\theta + m(x_0 + x)^2\ddot\theta + 2m(x_0 + x)\dot{x}\dot\theta = 0$

1-12. $\theta = \tan^{-1}\frac{1}{2} = 26.57°$

1-13. $\theta = \cos^{-1}\sqrt{\dfrac{27}{32}} = 23.28°$. Stable.

325

CHAPTER 2

2-1. $ml^2\ddot{\theta} + mlA\omega^2 \sin \omega t \sin \theta - mgl \sin \theta = 0$

2-2. $m(r^2 + l^2 - 2rl \cos \theta)\ddot{\theta} + mrl\dot{\theta}^2 \sin \theta - mgr \sin \alpha + mgl \sin (\theta + \alpha) = 0$

2-3. (a) $\sin \theta \, dx - \cos \theta \, dy + \frac{1}{2}l \, d\theta = 0$

   (b) $m\ddot{x} = \lambda \sin \theta, \qquad m\ddot{y} = -\lambda \cos \theta, \qquad \frac{1}{12}ml^2\ddot{\theta} = \frac{1}{2}\lambda l$

2-5. $m\ddot{y} - m\Omega^2 y + ky = 0$

2-6. $m\ddot{q} + \frac{1}{2}\frac{dm}{dq}\dot{q}^2 + \frac{dV}{dq} = 0$

2-8. $t = \frac{1}{\sin \alpha}\sqrt{\frac{2H}{g}}$ both cases

2-9. (a) $m\ddot{x} = 6x\lambda, \qquad m\ddot{y} + mg = -\lambda$

   (b) $\mathbf{C}_{\max} = mg(1 + 12y_0)\mathbf{j}$

2-10. (a) $m\ddot{x} + kx = 2\lambda, \qquad m\ddot{y} + ky = 3\lambda, \qquad m\ddot{z} + kz = 4\lambda$

   (b) $\mathbf{v} = \frac{1}{29}(-10\mathbf{i} - 15\mathbf{j} - 20\mathbf{k})$

2-11. $\frac{1}{3}ml^2\ddot{\theta} - \frac{3\beta_\phi^2 \cos \theta}{ml^2 \sin^3 \theta} - \frac{1}{2}mgl \sin \theta = 0$

2-12. $m\ddot{r} - mr\dot{\phi}^2 \sin^2 \alpha + mg \cos \alpha = 0$

   $mr^2\dot{\phi} \sin^2 \alpha = \beta_\phi$

   $m\ddot{r} - \frac{\beta_\phi^2}{mr^3 \sin^2 \alpha} + mg \cos \alpha = 0$

2-13. (a) $mr^2[2\ddot{\theta}(2 + \cos \phi) + \ddot{\phi}(1 + \cos \phi) - \dot{\phi}(2\dot{\theta} + \dot{\phi}) \sin \phi]$
   $\qquad\qquad + mgr[2 \sin \theta + \sin (\theta + \phi)] = 0$
   $mr^2[\ddot{\phi} + \ddot{\theta}(1 + \cos \phi) + \dot{\theta}^2 \sin \phi] + mgr \sin (\theta + \phi) = 0$

   (b) $\omega_1 = \sqrt{\frac{g}{2r}}, \qquad \omega_2 = \sqrt{\frac{2g}{r}}, \qquad \frac{A_\phi}{A_\theta} = 0, -3$

2-14. $\omega_1 = \sqrt{k/2m}, \qquad \omega_2 = \sqrt{3k/2m}, \qquad A_y/A_x = 0, \infty$

2-15. $\omega_1 = \omega_2 = 0, \qquad \omega_3 = \sqrt{6k/m}$
   $\mathbf{A}^{(1)} = \{1 \quad 1 \quad 1\}, \qquad \mathbf{A}^{(2)} = \{1 \quad 0 \quad -1\}, \qquad \mathbf{A}^{(3)} = \{1 \quad -2 \quad 1\}$

CHAPTER 3

3-1. $\dot{x}_1 = \frac{7\hat{F}_1}{2m}, \qquad \dot{x}_2 = -\frac{\hat{F}_1}{m}, \qquad \dot{x}_3 = \frac{\hat{F}_1}{2m}$

3-2. $\dot{x} = 0.75 \text{ m/sec}, \qquad \dot{y} = -0.75 \text{ m/sec}, \qquad \dot{\theta} = -0.75 \text{ rad/sec}$

3-3. $\dot{x} = \frac{4}{5}v_0, \qquad \dot{y} = \frac{1}{5}v_0, \qquad \dot{\theta} = \frac{6v_0}{5l}, \qquad \hat{C} = \frac{\sqrt{2}}{5}mv_0$

3-4. $\dot{x} = \frac{5}{8}v_0, \qquad \dot{y} = \frac{3}{8}v_0, \qquad \dot{\theta} = \frac{3v_0}{2\sqrt{2}\,l}, \qquad \hat{C} = \frac{5}{8}mv_0$

3-5. $p_\theta = \frac{1}{3}ml^2\dot\theta - \frac{1}{2}mv_0 l \sin\theta, \qquad \dot\theta = \frac{3v_0}{2\sqrt{2}\,l}$

3-6. $v_x = \dfrac{my_0^2 v_0}{I_{zz} + m(x_0^2 + y_0^2)}, \qquad \omega_x = 0$

$v_y = \dfrac{-mx_0 y_0 v_0}{I_{zz} + m(x_0^2 + y_0^2)}, \qquad \omega_y = 0$

$v_z = 0, \qquad \omega_z = \dfrac{my_0 v_0}{I_{zz} + m(x_0^2 + y_0^2)}$

3-7. $\dfrac{\partial a_i}{\partial q_j} = \dfrac{\partial a_j}{\partial q_i}$ for all i, j.

3-8. $T_1 = ml^2\Omega[2\dot q_1(1 + \cos q_1) + (\dot q_1 + \dot q_2)\cos(q_2 - q_1) + \dot q_2(1 + \cos q_2)]$
$\gamma_{12} = -2ml^2\Omega \sin(q_2 - q_1)$

3-9. $(I_a \sin^2\theta + I_t \cos^2\theta)\ddot\psi - I_a\ddot\phi \sin\theta$
$\qquad\qquad\qquad -2(I_t - I_a)(\dot\psi + \omega_0)\dot\theta \sin\theta \cos\theta - I_a\dot\theta\dot\phi \cos\theta = 0$
$I_t\ddot\theta + (I_t - I_a)(\dot\psi + \omega_0)^2 \sin\theta \cos\theta + I_a(\dot\psi + \omega_0)\dot\phi \cos\theta$
$\qquad\qquad\qquad\qquad\qquad\qquad\qquad\qquad\qquad + mgl\cos\theta = 0$
$I_a\ddot\phi - I_a\ddot\psi \sin\theta - I_a(\dot\psi + \omega_0)\dot\theta \cos\theta = 0$

3-10. $\lambda_1^2 = \dfrac{-k}{2m}, \qquad \lambda_2^2 = \dfrac{-3k}{2m}, \qquad \lambda_3^2 = \dfrac{-3k}{m} \qquad$ Stable

3-11. Stable for $\gamma^2 > 4.490\,mk$

3-12. (a) $F = \frac{1}{2}c(\dot x_1 - \dot x_2)^2$
$\qquad m\ddot x_1 + c\dot x_1 + 2kx_1 - c\dot x_2 - kx_2 = 0$
$\qquad m\ddot x_2 + c\dot x_2 + 2kx_2 - c\dot x_1 - kx_1 = 0$
(b) $m^2\lambda^4 + 2mc\lambda^3 + 4mk\lambda^2 + 2ck\lambda + 3k^2 = 0 \qquad$ Neutral stability
(c) $\lambda_{1,2} = -0.07579 \pm i\,0.8668$
$\qquad \lambda_{3,4} = -0.9242 \pm i\,1.763$

3-13. $U = -E_0 e \ln r - \frac{1}{2}B_0 er^2\dot\theta$
$m\ddot r - mr\dot\theta^2 - \dfrac{E_0 e}{r} - B_0 er\dot\theta = 0$
$mr^2\dot\theta + \frac{1}{2}B_0 er^2 = \beta_\theta, \qquad m\dot z = \beta_z$

CHAPTER 4

4-1. $H = \dfrac{p_\theta^2}{2I_t} + \dfrac{p_\phi^2}{2I_a} + \dfrac{(p_\psi + p_\phi \sin\theta)^2}{2I_t \cos^2\theta} + mgl\sin\theta$

$\dot\psi = \dfrac{p_\psi + p_\phi \sin\theta}{I_t \cos^2\theta}, \qquad \dot p_\psi = 0, \qquad \dot\theta = \dfrac{p_\theta}{I_t}$

$\dot p_\theta = -p_\phi\left(\dfrac{p_\psi + p_\phi \sin\theta}{I_t \cos\theta}\right) - \dfrac{(p_\psi + p_\phi \sin\theta)^2}{I_t \cos^3\theta}\sin\theta - mgl\cos\theta$

$\dot\phi = \dfrac{p_\phi}{I_a} + \left(\dfrac{p_\psi + p_\phi \sin\theta}{I_t \cos^2\theta}\right)\sin\theta, \qquad \dot p_\phi = 0$

4-2. $\dot y = p/m, \qquad \dot p = -mg \mp cp^2/m^2 \qquad$ for $p > 0, \qquad p < 0$

4-3. $\dot{\phi} = \dfrac{p_\phi}{I + mr^2 \sin^2 \theta}$, $\qquad p_\phi = (I + mr^2 \sin^2 \theta)\dot{\phi} = \beta_\phi$

$\dot{\theta} = \dfrac{p_\theta}{mr^2}$, $\qquad \dot{p}_\theta = \dfrac{p_\phi^2 mr^2 \sin \theta \cos \theta}{(I + mr^2 \sin^2 \theta)^2} + mgr \sin \theta$

$\dot{\phi}_{\max} = \dfrac{1}{\sqrt{6} - 2} \sqrt{\dfrac{g}{r}}$

4-4. $p_x = mv_0 \cos t^2$, $\qquad p_y = mv_0 \sin t^2$, $\qquad \lambda = -2mv_0 t$

4-7. $I = m[(\tfrac{1}{2}v_0^2 - gy_0)t_1 - v_0 g t_1^2 + \tfrac{1}{3}g^2 t_1^3]$

$I(v_{\text{const}}) = m[(\tfrac{1}{2}v_0^2 - gy_0)t_1 - v_0 g t_1^2 + \tfrac{3}{8}g^2 t_1^3]$

4-8. $m\ddot{x}_1 + \dot{\mu} = 0$, $\qquad m\ddot{x}_2 - \dot{\mu} = 0$, $\qquad m\ddot{x}_3 + \dot{\mu} + mg = 0$

$\ddot{x}_1 = \dfrac{1}{3}g$

4-10. $\mu_1 = \sqrt{\dfrac{\beta_2}{\beta_1}}e^{C/2}$, $\qquad \mu_2 = \sqrt{\dfrac{\beta_1}{\beta_2}}e^{C/2}$, $\qquad f_{\min} = 3.230$

4-11. $x = \dfrac{v_0}{\omega_0}(\cos \omega_0 t - 1)$, $\qquad y = \dfrac{v_0}{\omega_0} \sin \omega_0 t$, $\qquad \theta = \omega_0 t$

$\delta I = -m\omega_0 v_0 \displaystyle\int_0^{t_1} \eta(t) \tan^2 \omega_0 t \, dt$

4-13. No

CHAPTER 5

5-2. $t - t_0 = \displaystyle\int_{\theta_0}^{\theta} \dfrac{I_t \, d\theta}{\sqrt{f(\theta)}}$, $\qquad \psi - \psi_0 = \displaystyle\int_{\theta_0}^{\theta} \dfrac{(\alpha_\psi + \alpha_\phi \sin \theta) \, d\theta}{\cos^2 \theta \sqrt{f(\theta)}}$

where $f(\theta) = 2I_t\left[\alpha_t - \dfrac{\alpha_\phi^2}{2I_a} - \dfrac{(\alpha_\psi + \alpha_\phi \sin \theta)^2}{2I_t \cos^2 \theta} - mgl \sin \theta\right]$

5-3. $t = \dfrac{1}{g \sin \gamma} \sqrt{\dfrac{3}{m}}[\sqrt{\alpha + mgr\theta \sin \gamma} - \sqrt{\alpha}]$

$\theta = \dot{\theta}_0 t + \dfrac{g \sin \gamma}{3r} t^2$

5-4. $x = \dot{x}_0 t$, $\qquad y = \dot{y}_0 t - \tfrac{1}{2}gt^2$

5-5. $t = \displaystyle\int_{r=r_0}^{r} \dfrac{-\tfrac{1}{2}m \, d(r^2)}{\sqrt{-\alpha_\theta^2 + 2m\alpha_t r^2 - mkr^4}}$

$\theta = \displaystyle\int_{r=r_0}^{r} \dfrac{-\tfrac{1}{2}\alpha_\theta \, d(r^2)}{r^2 \sqrt{-\alpha_\theta^2 + 2m\alpha_t r^2 - mkr^4}}$

$r = \left[\dfrac{\alpha_t}{k} + \dfrac{1}{k}\sqrt{\alpha_t^2 - \dfrac{k}{m}\alpha_\theta^2} \cos 2\sqrt{\dfrac{k}{m}}t\right]^{1/2}$

$r = \dfrac{\alpha_\theta}{\sqrt{m}}\left[\alpha_t - \sqrt{\alpha_t^2 - \dfrac{k}{m}\alpha_\theta^2} \cos 2\theta\right]^{-1/2}$

5-6. $\dfrac{q_1 + \sqrt{q_1^2 + a^2}}{q_2 + \sqrt{q_2^2 - b^2}} = A$ (hyperbola)

5-7. $\quad t - t_0 = \displaystyle\int_{\theta_0}^{\theta} \frac{ml^2\,d\theta}{\sqrt{f(\theta)}}, \qquad \phi - \phi_0 = \int_{\theta_0}^{\theta} \frac{\alpha_\phi\,d\theta}{\sin^2\theta\sqrt{f(\theta)}}$

$\quad$ where $f(\theta) = 2\left[ml^2\alpha_t + m^2gl^3\cos\theta - \dfrac{\alpha_\phi^2}{2\sin^2\theta}\right]$

5-8. $\quad W_1(r) = \displaystyle\int \sqrt{2m\alpha_1 - \frac{2\alpha_\theta}{r^2} - 2v_1(r)}\,dr$

$\quad W_2(\theta) = \displaystyle\int \sqrt{2\alpha_\theta - \frac{2\alpha_\phi}{\sin^2\theta} - 2v_2(\theta)}\,d\theta$

$\quad W_3(\phi) = \displaystyle\int \sqrt{2\alpha_\phi - 2v_3(\phi)}\,d\phi$

$\quad \psi_1 = v_1(r), \qquad \psi_2 = v_2(\theta), \qquad \psi_3 = v_3(\phi), \qquad \Phi_{ij} = p_i \dfrac{\partial p_i}{\partial \alpha_j}$

CHAPTER 6

6-1. $\quad P = -\sqrt{q^2 + p^2}\,\tan^{-1}\dfrac{q}{p} \qquad \phi = \dfrac{1}{2}Q^2\sin^{-1}\dfrac{q}{Q} + \dfrac{1}{2}q\sqrt{Q^2 - q^2}$

6-2. $\quad \psi = mv_0 q, \qquad F_1 = mv_0 q, \qquad F_2 = mv_0 q + P\sin q$

$\quad F_3 = (mv_0 - p)\sin^{-1}Q$

$\quad F_4 = (mv_0 - p)\cos^{-1}\left(\dfrac{p - mv_0}{P}\right) + \sqrt{P^2 - (p - mv_0)^2}$

6-3. $\quad P_1 = \dfrac{yp_x + xp_y}{x^2 + y^2}, \qquad P_2 = \dfrac{xp_x - yp_y}{x^2 + y^2}$

$\quad F_2 = xyP_1 + \dfrac{1}{2}(x^2 - y^2)P_2$

$\quad K = \dfrac{1}{m}\sqrt{Q_1^2 + Q_2^2}\,(P_1^2 + P_2^2) + k[\sqrt{Q_1^2 + Q_2^2} - Q_2]^{1/2}$

6-4. $\quad F_2 = (P_1 + v_0)(q_1 - v_0 t) + \dfrac{1}{2}P_2^2 e^{-2t}\tan q_2$

$\quad K = \dfrac{1}{2}P_1^2 - \dfrac{1}{2}v_0^2 - Q_2 P_2 + \dfrac{1}{8}(Q_2^2 e^{2t} + P_2^2 e^{-2t})^2 + \dfrac{1}{2}(Q_1 + v_0 t)^2$

$\quad \dot{Q}_1 = P_1, \qquad \dot{Q}_2 = -Q_2 + \dfrac{1}{2}P_2 e^{-2t}(Q_2^2 e^{2t} + P_2^2 e^{-2t})$

$\quad \dot{P}_1 = -(Q_1 + v_0 t), \qquad \dot{P}_2 = P_2 - \dfrac{1}{2}Q_2 e^{2t}(Q_2^2 e^{2t} + P_2^2 e^{-2t})$

6-5. $\quad K = \dfrac{1}{8}(Q^2 e^{-2t} + P^2 e^{2t})^2 + \dfrac{1}{2}\left(\tan^{-1}\dfrac{Pe^{2t}}{Q}\right)^2 + QP$

6-6. $\quad P = \dfrac{1}{2}\tan 2p, \qquad F_1 = \dfrac{1}{2}q\left(\cos^{-1}\sqrt{\dfrac{Q}{q}} - \dfrac{\sqrt{Q(q - Q)}}{q}\right), \qquad F_4 \equiv 0$

6-8. $\quad q = \sqrt{2\alpha}\,\sin(t + \phi), \qquad p = \sqrt{2\alpha}\,\cos(t + \phi)$

$\quad$ where $\phi = \sin^{-1}(q_0/\sqrt{2\alpha})$

6-9. $\quad Q_1 = \dfrac{1}{2}\alpha - q_1$ where $\alpha = \sin^{-1}\dfrac{-2q_2 p_1}{p_2}$

$\quad Q_2 = \pm\sqrt{\dfrac{1}{2}(1 - \cos\alpha) - q_2^2}$

$\quad P_1 = -p_1, \qquad P_2 = \mp\dfrac{p_2}{q_2}\sqrt{\dfrac{1}{2}(1 - \cos\alpha) - q_2^2}$

6-10. $\quad Q_1 = q_1 - v_0 t, \qquad Q_2 = \sqrt{2q_2}\,e^t\cos p_2$

$$P_1 = p_1 - v_0, \qquad P_2 = \sqrt{2q_2}\, e^{-t} \sin p_2$$
$$K - H = 2q_2 \sin p_2 \cos p_2 - v_0(p_1 - v_0)$$

6-11. $K = \frac{1}{2}\sum_i P_i^2 + \sum_i \sum_j \sum_k a_{kj}\dot{a}_{ij}Q_k P_i$

or $K = \frac{1}{2}\sum_i P_i^2 - \sum_i \sum_j \sum_k a_{ij}\dot{a}_{kj}Q_k P_i$

6-12. $F_2^* = \sum_i \sum_j [a_{ij}(P_i^* - d_i)q_j + c_i P_i^*]$

6-13. $Q = t, \qquad T = q, \qquad P = p_t, \qquad P_T = p$

$\mathcal{K} = \lambda \left[P + \dfrac{1}{2m} P_T^2 - Q P_T + \dfrac{1}{2} k T^2 \right]$

6-20. $G = p_r \cos\theta - \dfrac{1}{r} p_\theta \sin\theta, \qquad \delta r = \epsilon \cos\theta, \qquad \delta\theta = -\dfrac{\epsilon}{r}\sin\theta,$

$\delta p_r = -\dfrac{\epsilon}{r^2} p_\theta \sin\theta, \qquad \delta p_\theta = \epsilon\left(p_r \sin\theta + \dfrac{1}{r} p_\theta \cos\theta \right)$

CHAPTER 7

7-1. $1.8\ c, \qquad v'_B = -0.9945\ c$

7-2. (a) $x' = \sqrt{\dfrac{1-\beta}{1+\beta}}\, l, \qquad t' = \sqrt{\dfrac{1-\beta}{1+\beta}}\, \dfrac{l}{c}$

(b) $x' = \sqrt{1-\beta^2}\, l - Vt'$

7-3. (a) Vh/v \qquad (b) $Vh/v\sqrt{1-\beta^2}$ \qquad (c) $v'_x = -V, \qquad v'_y = v\sqrt{1-\beta^2}$

7-4. $v = c \tanh \frac{1}{2}\theta$ where $\tanh\theta = \beta = \dfrac{V}{c} \qquad \tau = [\cosh \frac{1}{2}\theta]^{-1} t$

7-5. $v = 0.9918\ c$

7-8. (a) $\omega' = \omega_0\sqrt{1-\beta^2}$ \qquad (b) $\dfrac{\omega_0 l_0 \beta}{c}$ \qquad (c) $\tan^{-1}(\omega_0 \beta r/c\sqrt{1-\beta^2})$

7-9. $y = \left(\dfrac{x - Vt}{\sqrt{1-\beta^2}} \right) \tan \omega\!\left(\dfrac{t - Vx/c^2}{\sqrt{1-\beta^2}} \right)$

$x = \pm 0.8406\ l_0, \qquad y = \mp 0.2403\ l_0$

7-10. Nonrelativistic $\phi = \tan^{-1} \dfrac{\beta' \sin\phi'}{\beta + \beta' \cos\phi'}$

Relativistic $\phi = \tan^{-1} \dfrac{\beta'\sqrt{1-\beta^2}\, \sin\phi'}{\beta + \beta' \cos\phi'}$

7-11. 90%

7-12. $120°, \qquad 60°$

7-13. $v = 0.2943\ c, \qquad \theta = 31.84°, \qquad 1.539\ m_0$

7-15. $m_0\left(1 - \dfrac{\dot{x}^2}{c^2}\right)^{-\frac{3}{2}} \ddot{x} + kx = 0$

INDEX

A CATALOG OF SELECTED
DOVER BOOKS
IN SCIENCE AND MATHEMATICS

A CATALOG OF SELECTED

DOVER BOOKS

IN SCIENCE AND MATHEMATICS

QUALITATIVE THEORY OF DIFFERENTIAL EQUATIONS, V.V. Nemytskii and V.V. Stepanov. Classic graduate-level text by two prominent Soviet mathematicians covers classical differential equations as well as topological dynamics and ergodic theory. Bibliographies. 523pp. 5⅜ x 8½. 65954-2 Pa. $14.95

MATRICES AND LINEAR ALGEBRA, Hans Schneider and George Phillip Barker. Basic textbook covers theory of matrices and its applications to systems of linear equations and related topics such as determinants, eigenvalues and differential equations. Numerous exercises. 432pp. 5⅜ x 8½. 66014-1 Pa. $10.95

QUANTUM THEORY, David Bohm. This advanced undergraduate-level text presents the quantum theory in terms of qualitative and imaginative concepts, followed by specific applications worked out in mathematical detail. Preface. Index. 655pp. 5⅜ x 8½. 65969-0 Pa. $14.95

ATOMIC PHYSICS (8th edition), Max Born. Nobel laureate's lucid treatment of kinetic theory of gases, elementary particles, nuclear atom, wave-corpuscles, atomic structure and spectral lines, much more. Over 40 appendices, bibliography. 495pp. 5⅜ x 8½. 65984-4 Pa. $13.95

ELECTRONIC STRUCTURE AND THE PROPERTIES OF SOLIDS: The Physics of the Chemical Bond, Walter A. Harrison. Innovative text offers basic understanding of the electronic structure of covalent and ionic solids, simple metals, transition metals and their compounds. Problems. 1980 edition. 582pp. 6⅛ x 9¼. 66021-4 Pa. $16.95

BOUNDARY VALUE PROBLEMS OF HEAT CONDUCTION, M. Necati Özisik. Systematic, comprehensive treatment of modern mathematical methods of solving problems in heat conduction and diffusion. Numerous examples and problems. Selected references. Appendices. 505pp. 5⅜ x 8½. 65990-9 Pa. $12.95

A SHORT HISTORY OF CHEMISTRY (3rd edition), J.R. Partington. Classic exposition explores origins of chemistry, alchemy, early medical chemistry, nature of atmosphere, theory of valency, laws and structure of atomic theory, much more. 428pp. 5⅜ x 8½. (Available in U.S. only) 65977-1 Pa. $11.95

A HISTORY OF ASTRONOMY, A. Pannekoek. Well-balanced, carefully reasoned study covers such topics as Ptolemaic theory, work of Copernicus, Kepler, Newton, Eddington's work on stars, much more. Illustrated. References. 521pp. 5⅜ x 8½. 65994-1 Pa. $12.95

PRINCIPLES OF METEOROLOGICAL ANALYSIS, Walter J. Saucier. Highly respected, abundantly illustrated classic reviews atmospheric variables, hydrostatics, static stability, various analyses (scalar, cross-section, isobaric, isentropic, more). For intermediate meteorology students. 454pp. 6⅛ x 9¼. 65979-8 Pa. $14.95

CATALOG OF DOVER BOOKS

CHALLENGING MATHEMATICAL PROBLEMS WITH ELEMENTARY SOLUTIONS, A.M. Yaglom and I.M. Yaglom. Over 170 challenging problems on probability theory, combinatorial analysis, points and lines, topology, convex polygons, many other topics. Solutions. Total of 445pp. 5⅜ x 8½. Two-vol. set.
Vol. I: 65536-9 Pa. $7.95
Vol. II: 65537-7 Pa. $7.95

FIFTY CHALLENGING PROBLEMS IN PROBABILITY WITH SOLUTIONS, Frederick Mosteller. Remarkable puzzlers, graded in difficulty, illustrate elementary and advanced aspects of probability. Detailed solutions. 88pp. 5⅜ x 8½.
65355-2 Pa. $4.95

EXPERIMENTS IN TOPOLOGY, Stephen Barr. Classic, lively explanation of one of the byways of mathematics. Klein bottles, Moebius strips, projective planes, map coloring, problem of the Koenigsberg bridges, much more, described with clarity and wit. 43 figures. 210pp. 5⅜ x 8½. 25933-1 Pa. $6.95

RELATIVITY IN ILLUSTRATIONS, Jacob T. Schwartz. Clear nontechnical treatment makes relativity more accessible than ever before. Over 60 drawings illustrate concepts more clearly than text alone. Only high school geometry needed. Bibliography. 128pp. 6⅛ x 9¼. 25965-X Pa. $7.95

AN INTRODUCTION TO ORDINARY DIFFERENTIAL EQUATIONS, Earl A. Coddington. A thorough and systematic first course in elementary differential equations for undergraduates in mathematics and science, with many exercises and problems (with answers). Index. 304pp. 5⅜ x 8½. 65942-9 Pa. $8.95

FOURIER SERIES AND ORTHOGONAL FUNCTIONS, Harry F. Davis. An incisive text combining theory and practical example to introduce Fourier series, orthogonal functions and applications of the Fourier method to boundary-value problems. 570 exercises. Answers and notes. 416pp. 5⅜ x 8½. 65973-9 Pa. $11.95

AN INTRODUCTION TO ALGEBRAIC STRUCTURES, Joseph Landin. Superb self-contained text covers "abstract algebra": sets and numbers, theory of groups, theory of rings, much more. Numerous well-chosen examples, exercises. 247pp. 5⅜ x 8½.
65940-2 Pa. $8.95

STARS AND RELATIVITY, Ya. B. Zel'dovich and I. D. Novikov. Vol. 1 of *Relativistic Astrophysics* by famed Russian scientists. General relativity, properties of matter under astrophysical conditions, stars and stellar systems. Deep physical insights, clear presentation. 1971 edition. References. 544pp. 5⅜ x 8½.
69424-0 Pa. $14.95

Prices subject to change without notice.

Available at your book dealer or write for free Mathematics and Science Catalog to Dept. GI, Dover Publications, Inc., 31 East 2nd St., Mineola, N.Y. 11501. Dover publishes more than 250 books each year on science, elementary and advanced mathematics, biology, music, art, literature, history, social sciences and other areas.